Matthias Kreck
Wolfgang Lück

The Novikov Conjecture
Geometry and Algebra

Birkhäuser Verlag
Basel · Boston · Berlin

Authors:

Matthias Kreck
Mathematisches Institut
Ruprecht-Karls-Universität Heidelberg
Im Neuenheimer Feld 288
69120 Heidelberg
Germany
e-mail: kreck@mathi.uni-heidelberg.de

Wolfgang Lück
Mathematisches Institut
Westfälische Wilhelms-Universität Münster
Einsteinstrasse 62
48149 Münster
Germany
e-mail: lueck@math.uni-muenster.de

2000 Mathematical Subject Classification 19J25, 19K35, 46L80, 55N20, 57R20

A CIP catalogue record for this book is available from the
Library of Congress, Washington D.C., USA

Bibliografische Information Der Deutschen Bibliothek
Die Deutsche Bibliothek verzeichnet diese Publikation in der Deutschen Nationalbibliografie;
detaillierte bibliografische Daten sind im Internet über <http://dnb.ddb.de> abrufbar.

ISBN 3-7643-7141-2 Birkhäuser Verlag, Basel – Boston – Berlin

© 2005 Birkhäuser Verlag, P.O. Box 133, CH-4010 Basel, Switzerland
Part of Springer Science+Business Media
Cover design: Micha Lotrovsky, 4106 Therwil, Switzerland
Printed on acid-free paper produced from chlorine-free pulp. TCF ∞
Printed in Germany
ISBN 3-7643-7141-2

9 8 7 6 5 4 3 2 1   www.birkhauser.ch

Contents

Introduction

Manifolds are the central geometric objects in modern mathematics. An attempt to understand the nature of manifolds leads to many interesting questions. One of the most obvious questions is the following.

Let M and N be manifolds: how can we decide whether M and N are homotopy equivalent or homeomorphic or diffeomorphic (if the manifolds are smooth)?

The prototype of a beautiful answer is given by the Poincaré Conjecture. If N is S^n, the n-dimensional sphere, and M is an arbitrary closed manifold, then it is easy to decide whether M is homotopy equivalent to S^n. This is the case if and only if M is simply connected (assuming $n > 1$, the case $n = 1$ is trivial since every closed connected 1-dimensional manifold is diffeomorphic to S^1) and has the homology of S^n. The *Poincaré Conjecture* states that this is also sufficient for the existence of a homeomorphism from M to S^n. For $n = 2$ this follows from the well-known classification of surfaces. For $n > 4$ this was proved by Smale and Newman in the 1960s, Freedman solved the case in $n = 4$ in 1982 and recently Perelman announced a proof for $n = 3$, but this proof has still to be checked thoroughly by the experts. In the smooth category it is not true that manifolds homotopy equivalent to S^n are diffeomorphic. The first examples were published by Milnor in 1956 and together with Kervaire he analyzed the situation systematically in the 1960s.

For spheres one only needs very little information to determine the homeomorphism type: the vanishing of the fundamental group and control of the homology groups. Another natural class of manifolds is given by aspherical manifolds. A *CW*-complex is called *aspherical* if the homotopy groups vanish in dimension > 1, or, equivalently, if its universal covering is contractible. The *Borel Conjecture*, which is closely related to the Novikov Conjecture, implies that the fundamental group determines the homeomorphism type of an aspherical closed manifold.

For more general manifolds with prescribed fundamental group the classification is in general unknown even if the fundamental group is trivial. In this situation it is natural to construct as many invariants as possible hoping that at least for certain particularly important classes of manifolds one can classify them in terms of theses invariants. The most important invariants after homotopy and (co)homology groups are certainly characteristic classes which were defined and systematically treated in the 1950s. There are two types of characteristic classes for smooth manifolds: the *Stiefel–Whitney classes* $w_k(M)$ in $H^k(M; \mathbb{Z}/2)$ and the *Pontrjagin classes* $p_k(M) \in H^{4k}(M; \mathbb{Z})$. The nature of these classes is rather different. The Stiefel–Whitney classes of a closed manifold can be expressed in terms of cohomology operations and so are homotopy invariants, the Pontrjagin classes are diffeomorphism invariants (for smooth manifolds, and only for those they are a priori defined), but not homeomorphism or even homotopy invariants in general. Only very special linear combinations of the Pontrjagin classes are actually homotopy invariants.

For example, the first Pontrjagin class of a closed oriented 4-manifold $p_1(M)$ is a homotopy invariant. The reason is that $\langle p_1(M), [M] \rangle = 3 \cdot \text{sign}(M)$, where $\text{sign}(M)$ is the *signature* of the intersection form on $H^2(M; \mathbb{Q})$. The signature is by construction a homotopy invariant. More generally, Hirzebruch defined a certain rational polynomial in the Pontrjagin classes (for a definition of Pontrjagin classes see [171]), the *L-class*

$$\mathcal{L}(M) = \mathcal{L}(p_1(M), p_2(M), \ldots) \in \bigoplus_{i \geq 0} H^{4i}(M; \mathbb{Q}).$$

Its i-th component is denoted by

$$\mathcal{L}_i(M) = \mathcal{L}_i(p_1(M), p_2(M), \ldots, p_i(M)) \in H^{4i}(M; \mathbb{Q}).$$

The famous *Signature Theorem* of Hirzebruch says that the evaluation of $\mathcal{L}_k(M)$ on the fundamental class [M] gives the signature of a $4k$-dimensional manifold M:

$$\text{sign}(M) = \langle \mathcal{L}_k(p_1(M), \ldots, p_k(M)), [M] \rangle.$$

One can show that a polynomial in the Pontrjagin classes gives a homotopy invariant if and only if it is a multiple of the k-th L-class.

This sheds light on the homotopy properties of the polynomial $\mathcal{L}_k(M)$ of a $4k$-dimensional manifold M. But what can one say about the other polynomials $\mathcal{L}_1(M), \mathcal{L}_2(M), \mathcal{L}_3(M), \ldots$? Understanding $\mathcal{L}_i(M)$ is — by Poincaré duality — equivalent to understanding the numerical invariants

$$\langle x \cup \mathcal{L}_i(M), [M] \rangle \in \mathbb{Q} \tag{0.1}$$

for all $x \in H^{n-4i}(M)$, where $n = \dim(M)$. One may ask whether these numerical invariants are homotopy invariant in the following sense: If $g \colon N \to M$ is an orientation preserving homotopy equivalence, then

$$\langle x \cup \mathcal{L}_i(M), [M] \rangle = \langle g^*(x) \cup \mathcal{L}_i(N), [N] \rangle. \tag{0.2}$$

In general, these numerical invariants are not homotopy invariants. The Signature Theorem implies that the expression (0.1) is homotopy invariant for all $x \in H^0(M; \mathbb{Q})$. Novikov proved the remarkable result in the 1960s that for $\dim(M) = 4k + 1$ and $x \in H^1(M)$ the expression (0.1) is homotopy invariant. This motivated Novikov to state the following conjecture.

Let G be a group. Denote by BG its *classifying space* which is up to homotopy uniquely determined by the property that it is an aspherical CW-complex with G as fundamental group. Novikov conjectured that the numerical expression

$$\langle f^*(x) \cup \mathcal{L}_i(M), [M] \rangle \; \in \; \mathbb{Q} \qquad\qquad (0.3)$$

is *homotopy invariant* for every map $f \colon M \to BG$ from a closed oriented n-dimensional manifold M to BG and every class $x \in H^{n-4i}(M; \mathbb{Q})$. More precisely, the famous *Novikov Conjecture* says that if $f' \colon M' \to K$ is another map and $g \colon M \to M'$ is an orientation preserving homotopy equivalence such that $f' \circ g$ is homotopic to f, then

$$\langle f^*(x) \cup \mathcal{L}_i(M), [M] \rangle \; = \; \langle ((f')^*(x) \cup \mathcal{L}_i(M'), [M']) \rangle.$$

Notice that Novikov's result that (0.2) holds in the case $\dim(M) = 4k + 1$ and $x \in H^1(M)$ is a special case of the Novikov Conjecture above since S^1 is a model for $B\mathbb{Z}$ and a cohomology class $x \in H^1(M)$ is the same as a homotopy class of maps $f \colon M \to S^1$, the correspondence is given by associating to the homotopy class of $f \colon M \to S^1$ the pullback $f^*(x)$, where x is a generator of $H^1(S^1)$.

Looking at this conjecture in a naive way one does not see a philosophical reason why it should be true. Even in the case of the polynomial $\mathcal{L}_k$, where $4k$ is the dimension of a manifold, the proof cannot be understood without the signature theorem translating the L-class to a cohomological invariant, the signature. In this situation it is natural to ask for other homotopy invariants (instead of the signature) hoping that one can interpret the expressions (0.3) occurring in the Novikov Conjecture in terms of these invariants. These expressions (0.3) are called *higher signatures*. One can actually express them as signature of certain submanifolds. But this point of view does not give homotopy invariants.

It is natural to collect all higher signatures and form from them a single invariant. This can be done, namely, one considers

$$\text{sign}^G(M, f) \; := \; f_*(\mathcal{L}(M) \cap [M]) \; \in \; \bigoplus_{i \in \mathbb{Z}, i \geq 0} H_{m-4i}(BG; \mathbb{Q}),$$

the image of the Poincaré dual of the L-class under the map induced from f. An approach to proving the Novikov Conjecture could be to construct a homomorphism

$$A^G \colon \bigoplus_{i \in \mathbb{Z}, i \geq 0} H_{m-4i}(BG; \mathbb{Q}) \to L(G)$$

where $L(G)$ is some abelian group, such that $A^G(\text{sign}_G(M))$ is a homotopy invariant. Then the Novikov Conjecture would follow if the map A^G is injective. Such maps will be given by so-called *assembly maps*.

The construction of such a map is rather complicated. A large part of these lecture notes treats the background needed to construct such a map. In particular, one needs the full machinery of surgery theory. We will give an introduction to this important theory. Roughly speaking, surgery deals with the following problem. Let W be a compact m-dimensional manifold whose boundary is either empty or consists of two components M_0 and M_1 and $f \colon W \to X$ a map to a finite CW-complex. If the boundary of W is not empty, we assume that f restricted to M_0 and M_1 is a homotopy equivalence. Then X is a so-called *Poincaré complex*, something we also require if the boundary of W is empty. The question is whether we can replace W and f by W' and f' (bordant to (W, f)) such that f' is a homotopy equivalence. If the boundary of W is not empty, then W' is an *h-cobordism* between M_0 and M_1. In general it is not possible to replace (W, f) by (W', f') with f' a homotopy equivalence. Wall has defined abelian groups $L_m^h(\pi_1(X))$ and an obstruction $\theta(W, f) \in L_m^h(\pi_1(X))$ whose vanishing is a necessary and sufficient condition for replacing (W, f) by (W', f') with f' a homotopy equivalence, if $m > 4$. One actually needs some more control, namely a so-called normal structure on W. All this is explained in Chapters 2, 10–14 and Chapter 17.

Why is it so interesting to obtain an h-cobordism? If X is simply-connected, and the dimension of W is greater than five, the celebrated *h-cobordism theorem* of Smale says that an h-cobordism W is diffeomorphic to the cylinder over M_0. In particular, M_0 and M_1 are diffeomorphic. There is a corresponding result for topological manifolds. In the situation which is relevant for the Novikov Conjecture, X is not simply-connected and then the h-cobordism theorem does not hold. There is an obstruction, the *Whitehead torsion*, sitting in the *Whitehead group* which is closely related to the algebraic K_1-group. If the dimension of the h-cobordism W is larger than five, then the vanishing of this obstruction is necessary and sufficient for W to be diffeomorphic to the cylinder. This is called the *s-cobordism theorem*. The Whitehead group, the obstruction and the idea of the proof of the s-cobordism theorem are treated in Chapters 5–8.

In Chapters 15–16 we define the assembly map and apply it to prove the Novikov Conjecture for finitely-generated free abelian groups.

What we have presented so far summarizes and explains information which was known around 1970. To get a feeling for how useful the Novikov Conjecture is, we apply it to some classification problems in low dimensions (see Chapter 0).

In the rest of the lecture notes we present some of the most important concepts and results concerning the Novikov Conjecture and other closely related conjectures dating from after 1970. This starts with an introduction to spectra (see Chapter 18) and continues with classifying spaces of families, a generalization of aspherical spaces (see Chapter 19). With this we have prepared a frame in which not only the Novikov Conjecture but other similar and very important conjectures can be formulated: the *Farrell–Jones* and the *Baum–Connes Conjectures*. After introducing equivariant homology theories in Chapter 20, these conjectures and their relation to the Novikov Conjecture are discussed in Chapters 21–23. Finally, these lecture notes are finished by Chapter 24 called "Miscellaneous" in which the status of the conjectures is summarized and methods and proofs are presented.

It is interesting to speculate whether the Novikov Conjecture holds for all groups. No counterexamples are known to the authors. An interesting article expressing doubts was published by Gromov [102].

We have added a collection of exercises and hints for their solutions.

From the amount of material presented in these lecture notes it is obvious, that we cannot present all of the details. We have tried to explain those things which are realistic for the very young participants of the seminar to master and we have only said a few words (if anything at all) at other places. People who want to understand the details of this fascinating theory will have to consult other books and often the original literature. We hope that they will find our lecture notes useful, since we explain some of the central ideas and give a guide for learning the beautiful mathematics related to the Novikov Conjecture and other closely related conjectures and results.

We would like to thank the participants of this seminar for their interest and many stimulating discussions and Mathematisches Forschungsinstitut Oberwolfach for providing excellent conditions for such a seminar. We also would like to thank Andrew Ranicki for carefully reading a draft of this notes and many useful comments.

Chapter 0

A Motivating Problem

The classification of manifolds is one of the central problems in mathematics. Since a complete answer is (at least for manifolds of dimension ≥ 4) not possible, one firstly has to fix certain invariants in such a way that the classification is in principle possible. The reason why the classification of manifolds is impossible is reduced to the impossibility of classifying their fundamental groups. Thus as a first invariant one has to fix the fundamental group. Then the optimal answer would be to find invariants which determine the diffeomorphism (homeomorphism or homotopy) type. In recent years, low dimensional manifolds (in dimension up to 7) occurred in various mathematical and non-mathematical contexts. We motivate the Novikov Conjecture by considering the following problem:

Problem 0.1 (Classification of manifolds in low dimensions with $\pi_1(M) \cong \mathbb{Z}^2$ and $\pi_2(M) = 0$). *Classify all connected closed orientable manifolds M in dimensions ≤ 6 with fundamental group $\pi_1(M) \cong \mathbb{Z} \oplus \mathbb{Z}$ and second homotopy group $\pi_2(M) = 0$ up to*

(1) *homotopy equivalence;*

(2) *homeomorphism;*

(3) *diffeomorphism.*

Here and in the following we always mean *orientation preserving maps*.

0.1 Dimensions ≤ 4

Since all closed connected 1-manifolds are diffeomorphic to S^1, there is no example in dimension 1.

In dimension 2 there is only one such manifold, the torus $T^2 = S^1 \times S^1$. Here the classification up to the relations i)–iii) agree.

In dimension 3 there is no manifold with fundamental group $\mathbb{Z} \oplus \mathbb{Z}$. The reason is that the classifying map of the universal covering of such a 3-manifold $f \colon M \to T^2$ is 3-connected. Hence it induces isomorphisms $H_p(f) \colon H_p(M) \xrightarrow{\cong} H_p(T^2)$ and $H^p(f) \colon H^p(T^2) \xrightarrow{\cong} H^p(M)$ for $p \leq 2$. Poincaré duality implies $H^1(M) \cong H_2(M)$. This yields a contradiction since $H_2(T^2) = \mathbb{Z}$ and $H_1(T^2) = \mathbb{Z}^2$.

There is also no such manifold M in dimension 4 by the following argument. As above for 3-manifolds we conclude that $H_2(M) \cong \mathbb{Z}$. Poincaré duality implies for the Euler characteristic $\chi(M) = 1$. Now we note that all finite coverings of M are again manifolds of the type we investigate. Namely the fundamental group is a subgroup of finite index in $\mathbb{Z} \oplus \mathbb{Z}$ and so isomorphic to $\mathbb{Z} \oplus \mathbb{Z}$. And the higher homotopy groups of a covering do not change. Now consider a subgroup of index $k > 1$ in $\pi_1(M)$ and let N be the corresponding covering. Then $\chi(N) = k \cdot \chi(M)$. Since N is a manifold under consideration we have $\chi(N) = 1$. This leads to a contradiction.

0.2   Dimension 6

In dimension 6 one has an obvious example, namely $T^2 \times S^4$. But there are many more examples coming from the following construction.

Example 0.2 (Constructing manifolds by surgery). We start with a simply connected smooth 4-manifold M with trivial second Stiefel–Whitney class $w_2(M)$ and consider $T^2 \times M$. Then we choose disjoint embeddings $(S^2 \times D^4)_i$ into $T^2 \times M$ representing a basis of $\pi_2(M) \cong \pi_2(T^2 \times M)$. For this we first choose maps from S^2 to $T^2 \times M$ representing a basis. The Whitney Embedding Theorem implies that we can choose these maps as disjoint smooth embeddings. Finally we note that since $w_2(T^2 \times M) = 0$, the normal bundle of these embeddings is trivial and we use a tubular neighbourhood to construct the desired embeddings. Now we form a new manifold by deleting the interiors of these embeddings and gluing in $D^3 \times S^3$ to each deleted component. We denote the resulting manifold by $N(M)$. This cutting and pasting process is called *surgery*. Using standard considerations in algebraic topology one shows that $N(M)$ is an oriented manifold with $\pi_1 \cong \mathbb{Z} \oplus \mathbb{Z}$, $\pi_2 = 0$ and $w_2 = 0$ (see Exercise 0.1). We will study surgery in later chapters systematically.

Example 0.3 (A higher signature). We introduce the following invariant for the 6-manifolds N under consideration. The second cohomology is isomorphic to $\mathbb{Z}$, and we choose a generator $x \in H^2(N)$. This generator is well defined up to sign. Let $[M]$ be the fundamental class in $H_6(M)$. Taking the cup product with the Pontrjagin class and evaluating on $[M]$ gives our invariant:

$$\pm \langle x \cup p_1(N), [N] \rangle \ \in \mathbb{Z} \tag{0.4}$$

which is unique up to a sign $\pm$. It is easy to see (see Exercise 0.2) that for the manifold $N(M)$ constructed above this invariant agrees with the first Pontrjagin class of M evaluated at $[M]$ up to sign:

$$\pm \langle x \cup p_1(N(M)), [N(M)] \rangle \;\; = \;\; \pm \langle p_1(M), [M] \rangle. \tag{0.5}$$

The values of $\langle p_1(M), [M] \rangle$ for the different simply connected smooth 4-manifolds are known: Every integer divisible by 48 occurs [231].

We want to understand the relevance of this invariant. We firstly note that it is unchanged if we take the connected sum with $S^3 \times S^3$. Thus it is an invariant of the stable diffeomorphism type, where we call two closed manifolds M and N of dimension $2k$ *stably diffeomorphic*, if there exist integers p and q, such that $M \sharp p(S^k \times S^k)$ is diffeomorphic to $N \sharp q(S^k \times S^k)$, i.e., the manifolds M and N are diffeomorphic after taking the connected sum with p resp. q copies of $S^k \times S^k$. The relevance of the invariant $\langle x \cup p_1(N), [N] \rangle$ is demonstrated by the following result

Theorem 0.6 (Stable Classification of Certain Six-Dimensional Manifolds). *Two smooth 6-dimensional closed orientable manifolds M and N with $\pi_1(M) \cong \pi_1(N) \cong \mathbb{Z} \oplus \mathbb{Z}$ and $\pi_2(M) = \pi_2(N) = 0$ are stably diffeomorphic if and only if*

(1) *in both cases w_2 vanishes or does not vanish;*

(2) $\pm \langle x \cup p_1(M), [M] \rangle \;\; = \;\; \pm \langle x \cup p_1(N), [N] \rangle.$

We will give the proof of this result in Chapter 14. In our context this result leads to the following obvious questions: Is the second invariant also a stable homeomorphism or stable homotopy invariant? Here we define *stably homeomorphic* and *stably homotopy equivalent* in analogy to the definition of stably diffeomorphic by replacing in this definition diffeomorphic by homeomorphic or homotopy equivalent.

The answer is in both cases non-trivial. For homeomorphisms we pass from the Pontrjagin class $p_1(M) \in H^4(M)$ to the rational Pontrjagin class $p_1(M; \mathbb{Q}) \in H^4(N; \mathbb{Q})$. Since $H^4(N)$ is torsionfree we do not lose any information. Then we apply a deep result by Novikov (see Theorem 1.5) saying that the rational Pontrjagin classes are homeomorphism invariants and so stable homeomorphism invariants.

The rational Pontrjagin classes are in general not homotopy invariants (see Example 1.6). But Novikov conjectured that certain numerical invariants, the so-called higher signatures, built from the rational Pontrjagin classes and cohomology classes of the fundamental group are homotopy invariants. The invariant occurring in Theorem 0.6 is one of these invariants. (We will give a proof for free abelian groups in Chapter 16).

It should be noted that in contrast to the Pontrjagin classes the Stiefel–Whitney classes of a manifold are homotopy invariants. Thus the condition $w_2 = 0$ or $w_2 \neq 0$ is invariant under (stable) homotopy equivalences. Thus we conclude

Corollary 0.7. *For two smooth 6-dimensional closed orientable manifolds M and N with $\pi_1(M) \cong \pi_1(N) \cong \mathbb{Z} \oplus \mathbb{Z}$ and $\pi_2(M) = \pi_2(N) = 0$ the classifications up to stable diffeomorphism, stable homeomorphism and stable homotopy equivalence agree. In other words, the invariants from Theorem 0.6 determine also the stable homeomorphism and stable homotopy type.*

Remark 0.8 (Role of the Novikov Conjecture). The following formulation explains the surprising role of the Novikov Conjecture. In general the homotopy classification is a simpler question than the homeomorphism or diffeomorphism classification which one can attack by methods of classical homotopy theory. For the 6-manifolds under consideration the Novikov Conjecture implies that the stable homotopy type determines the stable homeomorphism and even stable diffeomorphism type of these smooth manifolds.

0.3   Dimension 5

Now we study the manifolds in dimension 5. In dimension 5 there are at least two such manifolds, namely $T^2 \times S^3$ and the sphere bundle of the non-trivial oriented 4-dimensional vector bundle over T^2. These manifolds are not homotopy equivalent (see Exercise 0.3). Moreover, using standard techniques from homotopy theory one can show that there are precisely two homotopy types of manifolds under consideration, which are given by these two bundles. The next obvious question is the determination of the homeomorphism and diffeomorphism type of these manifolds. One can show that the diffeomorphism type is determined by the first Pontrjagin class, and since this is a homeomorphism invariant (by Novikov's result mentioned above), this also determines the homeomorphism type. But which values can the Pontrjagin class take? Here again the Novikov Conjecture comes into play. It implies in our situation that the first Pontrjagin class is a homotopy invariant. Since we know all homotopy types and in the examples above the Pontrjagin class is trivial, we conclude that the Pontrjagin class is zero for our manifolds. Thus we have again a surprising result: The homotopy type of these 5-manifolds determines the homeomorphism (and actually diffeomorphism) type! For detailed arguments and more results we refer to [136].

Remark 0.9 (Other fundamental groups). The Novikov Conjecture is also valid for all fundamental groups G of closed oriented surfaces. The proof of Theorem 0.6 also holds for these fundamental groups so that Corollary 0.7 can also be generalized to these fundamental groups. We will investigate these 6-manifolds further in [136].

Chapter 1

Introduction to the Novikov and the Borel Conjecture

In this chapter we give a brief introduction to the Novikov and to the Borel Conjecture.

1.1 The Original Formulation of the Novikov Conjecture

Let G be a (discrete) group. We denote by BG its *classifying space*, which is uniquely determined by the property that it is a connected CW-complex BG together with an identification $\pi_1(BG) \xrightarrow{\cong} G$ whose universal covering $\widetilde{BG}$ is contractible. Let $u \colon M \to BG$ be a map from a closed oriented smooth manifold M to BG. Let

$$\mathcal{L}(M) \in \bigoplus_{k \in \mathbb{Z}, k \geq 0} H^{4k}(M; \mathbb{Q})$$

be the *L-class of M*. Its k-th entry $\mathcal{L}(M)_k \in H^{4k}(M; \mathbb{Q})$ is a certain homogeneous polynomial of degree k in the rational Pontrjagin classes $p_i(M; \mathbb{Q}) \in H^{4i}(M; \mathbb{Q})$ for $i = 1, 2, \ldots, k$ such that the coefficient s_k of the monomial $p_k(M; \mathbb{Q})$ is different from zero. We will give its precise definition later and mention at least the first values

$$
\begin{aligned}
\mathcal{L}(M)_1 &= \frac{1}{3} \cdot p_1(M; \mathbb{Q}); \\
\mathcal{L}(M)_2 &= \frac{1}{45} \cdot \left(7 \cdot p_2(M; \mathbb{Q}) - p_1(M; \mathbb{Q})^2 \right); \\
\mathcal{L}(M)_3 &= \frac{1}{945} \cdot \left(62 \cdot p_3(M; \mathbb{Q}) - 13 \cdot p_1(M; \mathbb{Q}) \cup p_2(M; \mathbb{Q}) + 2 \cdot p_1(M; \mathbb{Q})^3 \right).
\end{aligned}
$$

The L-class $\mathcal{L}(M)$ is determined by all the rational Pontrjagin classes and vice versa. Recall that the *k-th rational Pontrjagin class* $p_k(M, \mathbb{Q}) \in H^{4k}(M; \mathbb{Q})$ is defined as the image of k-th Pontrjagin class $p_k(M)$ under the obvious change of coefficients map $H^{4k}(M; \mathbb{Z}) \to H^{4k}(M; \mathbb{Q})$. The L-class depends on the tangent bundle and thus on the differentiable structure of M. For $x \in \prod_{k \geq 0} H^k(BG; \mathbb{Q})$ define the *higher signature of M associated to x and u* to be

$$\operatorname{sign}_x(M, u) := \langle \mathcal{L}(M) \cup u^* x, [M] \rangle \quad \in \mathbb{Q}. \tag{1.1}$$

Here and in the sequel $[M]$ denotes the fundamental class of a closed oriented d-dimensional manifold M in $H_d(M; \mathbb{Z})$ or its image under the change of coefficients map $H_d(M; \mathbb{Z}) \to H_d(M; \mathbb{Q})$ and $\langle u, v \rangle$ denotes the *Kronecker product*. Recall that for $\dim(M) = 4n$ the *signature* $\operatorname{sign}(M)$ of M is the signature of the non-degenerate bilinear symmetric pairing on the middle cohomology $H^{2n}(M; \mathbb{R})$ given by the intersection pairing $(a, b) \mapsto \langle a \cup b, [M] \rangle$. Obviously $\operatorname{sign}(M)$ depends only on the oriented homotopy type of M. We say that sign_x for $x \in H^*(BG; \mathbb{Q})$ is *homotopy invariant* if for two closed oriented smooth manifolds M and N with reference maps $u \colon M \to BG$ and $v \colon N \to BG$ we have

$$\operatorname{sign}_x(M, u) = \operatorname{sign}_x(N, v),$$

whenever there is an orientation preserving homotopy equivalence $f \colon M \to N$ such that $v \circ f$ and u are homotopic.

Conjecture 1.2 (Novikov Conjecture). *Let G be a group. Then sign_x is homotopy invariant for all $x \in \prod_{k \in \mathbb{Z}, k \geq 0} H^k(BG; \mathbb{Q})$.*

This conjecture appears for the first time in the paper by Novikov [183, §11]. A survey about its history can be found in [91].

1.2 Invariance Properties of the L-Class

One motivation for the Novikov Conjecture comes from the Signature Theorem due to Hirzebruch (see [115], [116]).

Theorem 1.3 (Signature Theorem). *Let M be an oriented closed manifold of dimension n. Then the higher signature $\operatorname{sign}_1(M, u) = \langle \mathcal{L}(M), [M] \rangle$ associated to $1 \in H_0(M)$ and some map $u \colon M \to BG$ coincides with the signature $\operatorname{sign}(M)$ of M, if $\dim(M) = 4n$, and is zero, if $\dim(M)$ is not divisible by four.*

The Signature Theorem 1.3 leads to the question, whether the Pontrjagin classes or the L-classes are homotopy invariants. They are obviously invariants of the diffeomorphism type. It is not true that the Pontrjagin classes $p_k(M) \in H^{4k}(M; \mathbb{Z})$ themselves are homeomorphism invariants.

Remark 1.4 (The integral Pontrjagin classes are not homeomorphism invariants). The first Pontrjagin class $p_1(M) \in H^4(M; \mathbb{Z})$ is a homeomorphism invariant, whereas all higher Pontrjagin classes $p_k(M) \in H^{4k}(M; \mathbb{Z})$ for $k \geq 2$ are not homeomorphism invariants. This will be explained in Theorem 4.8.

On the other hand, there is the following deep result due to Novikov [180], [181], [182].

Theorem 1.5 (Topological invariance of rational Pontrjagin classes). *The rational Pontrjagin classes $p_k(M, \mathbb{Q}) \in H^{4k}(M; \mathbb{Q})$ are topological invariants, i.e., for a homeomorphism $f \colon M \to N$ of closed smooth manifolds we have*

$$H_{4k}(f; \mathbb{Q})(p_k(M; \mathbb{Q})) = p_k(N; \mathbb{Q})$$

for all $k \geq 0$ and in particular $H_*(f; \mathbb{Q})(\mathcal{L}(M)) = \mathcal{L}(N)$.

Example 1.6 (The L-class is not a homotopy invariant). The rational Pontrjagin classes and the L-class are not homotopy invariants as the following example shows. There exists for $k \geq 1$ and large enough $j \geq 0$ a $(j+1)$-dimensional vector bundle $\xi \colon E \to S^{4k}$ with Riemannian metric whose k-th Pontrjagin class $p_k(\xi)$ is not zero and which is trivial as a fibration. The total space SE of the associated sphere bundle is a closed $(4k+j)$-dimensional manifold which is homotopy equivalent to $S^{4k} \times S^j$ and satisfies

$$\begin{aligned} p_k(SE) &= -p_k(\xi) \neq 0; \\ \mathcal{L}(SE)_k &= s_k \cdot p_k(SE) \neq 0, \end{aligned}$$

where $s_k \neq 0$ is the coefficient of p_k in the polynomial defining the L-class. But $p_k(S^{4k} \times S^j)$ and $\mathcal{L}(S^{4k} \times S^j)_k$ vanish since the tangent bundle of $S^{4k} \times S^j$ is stably trivial. In particular SE and $S^{4k} \times S^j$ are simply-connected homotopy equivalent closed manifolds, which are not homeomorphic. This example is taken from [202, Proposition 2.9] and attributed to Dold and Milnor there. See also [202, Proposition 2.10] or [171, Section 20].

Remark 1.7 (The signature as a surgery obstruction). Browder (see [33], [34]) and Novikov [179] showed in 1962 independently a kind of converse to the Signature Theorem 1.3 in dimensions ≥ 5. Namely, if X is a simply-connected $4k$-dimensional Poincaré complex for $k \geq 2$ such that the signature of X is $\langle \mathcal{L}(\nu), [X] \rangle$ for $\mathcal{L}(\nu)$ the L-class of a vector bundle over X with spherical Thom class, then there is a closed oriented $4k$-dimensional manifold M and a homotopy equivalence $f \colon M \to X$ of degree one such that the pullback $f^*\nu$ of ν with f and the stable normal bundle νM of M are stably isomorphic.

Remark 1.8 (The homological version of the Novikov Conjecture). One may understand the Novikov Conjecture as an attempt to figure out how much of the L-class is a homotopy invariant of M. If one considers the oriented homotopy type and the simply-connected case, it is just the expression $\langle \mathcal{L}(M), [M] \rangle$ or, equivalently, the top component of $\mathcal{L}(M)$. In the Novikov Conjecture one asks the same question but now taking the fundamental group into account by remembering the classifying map $u_M \colon M \to B\pi_1(M)$, or, more generally, a reference map $u\colon M \to BG$. The Novikov Conjecture can also be rephrased by saying that for a given group G each pair (M, u) consisting of an oriented closed manifold M together with a reference map $u\colon M \to BG$ the term

$$u_*(\mathcal{L}(M) \cap [M]) \in H_*(BG; \mathbb{Q})$$

depends only on the oriented homotopy type of the pair (M, u). This follows from the elementary computation for $x \in H^*(BG; \mathbb{Q})$

$$\langle \mathcal{L}(M) \cup u^*x, [M] \rangle \;=\; \langle u^*x, \mathcal{L}(M) \cap [M] \rangle \;=\; \langle x, u_*(\mathcal{L}(M) \cap [M]) \rangle.$$

and the fact that the Kronecker product $\langle -, - \rangle$ for rational coefficients is non-degenerate. Notice that $- \cap [M]\colon H^{\dim(M)-n}(M; \mathbb{Q}) \to H_n(M; \mathbb{Q})$ is an isomorphism for all $n \geq 0$ by Poincaré duality. Hence $\mathcal{L}(M) \cap [M]$ carries the same information as $\mathcal{L}(M)$.

Remark 1.9 (The converse of the Novikov Conjecture). A kind of converse to the Novikov Conjecture 1.2 is the following result. Let N be a closed connected oriented smooth manifold of dimension $n \geq 5$. Let $u\colon N \to BG$ be a map inducing an isomorphism on the fundamental groups. Consider any element $l \in \prod_{i \geq 0} H^{4i}(N; \mathbb{Q})$ such that $u_*(l \cap [N]) = 0$ holds in $H_*(BG; \mathbb{Q})$. Then there exists a non-negative integer K such that for any multiple k of K there is a homotopy equivalence $f\colon M \to N$ of closed oriented smooth manifolds satisfying

$$f^*(\mathcal{L}(N) + k \cdot l) = \mathcal{L}(M).$$

A proof can be found for instance in [63, Theorem 6.5]. This shows that the top dimension part of the L-class $\mathcal{L}(M)$ is essentially the only homotopy invariant rational characteristic class for simply-connected closed $4k$-dimensional manifolds.

1.3   The Borel Conjecture

In this chapter we will explain the

Conjecture 1.10 (Borel Conjecture). *Let M and N be aspherical closed topological manifolds. Then*

(1) *Each homotopy equivalence $f\colon M \to N$ is homotopic to a homeomorphism;*

(2) *The manifolds M and N are homeomorphic if and only if they have isomorphic fundamental groups.*

   We say that the Borel Conjecture holds for a group G, *if it is true for all aspherical closed topological manifolds M and N, whose fundamental groups are isomorphic to G.*

A manifold M is *closed* if and only if it is compact and has no boundary. A CW-complex is called *aspherical* if X is connected and $\pi_n(X, x) = 0$ for $n \geq 2$ and one (and hence all) basepoints $x \in X$, or, equivalently, if X is connected and its universal covering is contractible. If X and Y are aspherical CW-complexes, then for each homomorphism $\varphi \colon \pi_1(X, x) \to \pi_1(Y, y)$ there is a map $f \colon X \to Y$ with $f(x) = y$ such that $\pi_1(f, x) = \varphi$ holds. If φ is bijective, then f is automatically a homotopy equivalence. This explains why in the Borel Conjecture 1.10 assertion (1) implies assertion (2).

Remark 1.11 (The Borel Conjecture versus Mostow rigidity). The Borel Conjecture 1.10 is a topological version of the *Mostow Rigidity Theorem*. A special version of it says that for two hyperbolic closed manifolds M and N each homotopy equivalence from M to N is homotopic to an isometric diffeomorphism. In particular M and N are isometrically diffeomorphic if and only if they have isomorphic fundamental groups. Recall that a Riemannian manifold is *hyperbolic* if and only if it is complete and its sectional curvature is constant -1. This is equivalent to the condition that the universal covering is isometrically diffeomorphic to the hyperbolic space $\mathbb{H}^n$ for $n = \dim(M)$. Since $\mathbb{H}^n$ is contractible, each hyperbolic manifold is aspherical. So also the Borel Conjecture 1.10 applies to hyperbolic closed manifolds, but the conclusion is weaker, we only get a homeomorphism instead of an isometric diffeomorphism. On the other hand the assumptions appearing in the Borel Conjecture 1.10 are much weaker.

Remark 1.12 (The Borel Conjecture fails in the smooth category). In general the Borel Conjecture 1.10 becomes false if one considers aspherical closed smooth manifolds and replaces homeomorphisms by diffeomorphisms. Counterexamples have been constructed by Farrell–Jones [78, Theorem 1.1]. Given any $\delta > 0$, they consider an appropriate hyperbolic manifold M and an appropriate exotic sphere Σ such that there is a homeomorphism $f \colon M \sharp \Sigma \to M$ and the manifold $M \sharp \Sigma$ does admit a Riemannian metric of negative sectional curvature which is pinched between $1 - \delta$ and $1 + \delta$, but $M \sharp \Sigma$ and M are not diffeomorphic.

Remark 1.13 (The Borel Conjecture holds only for aspherical manifolds). The condition aspherical is crucial in the Borel Conjecture 1.10, otherwise there are counterexamples as we have seen already in Example 1.6. Older counterexamples are given by so-called lens spaces, some of which are homotopy equivalent but not homeomorphic. A detailed discussion of lens spaces can be found for instance in [53, Chapter V] and [155, Section 2.4].

Remark 1.14 (The Borel Conjecture and the Novikov Conjecture). The Borel Conjecture 1.10 implies the Novikov Conjecture 1.2 in the case, where $u \colon M \to BG$ is a homotopy equivalence, by the topological invariance of the rational Pontrjagin classes (see Theorem 1.5). We will explain in Section 21.5 that the L-theoretic version of the Borel Conjecture 1.10 for a given group G implies the Novikov Conjecture 1.2 for this group G.

Remark 1.15 (The Borel Conjecture and the Poincaré Conjecture). The Borel Conjecture 1.10 for T^3 does imply by an indirect argument (see [85, Remark on page 233]) the *Poincaré Conjecture in dimension* 3, which says that a 3-manifold, which is homotopy equivalent to S^3, is already homeomorphic to S^3. Gabai [96] has a program to show that the Borel Conjecture 1.10 in dimension 3 is actually equivalent to the Poincaré Conjecture in dimension 3.

Remark 1.16 (Status of the Borel Conjecture and the Novikov Conjecture). The Borel Conjecture 1.10 has been proved by Farrell and Jones, if N is a closed Riemannian manifold with non-negative sectional curvature [82, Proposition 0.10 and Lemma 0.12], or if $\pi_1(N)$ is a subgroup of a group G which is a discrete cocompact subgroup of a Lie group with finitely many path components [80]. This implies the Novikov Conjecture 1.2 for such groups G. The Novikov Conjecture 1.2 is even known for many other classes of groups such as for all linear groups, all arithmetic groups and all word hyperbolic groups. Both the Borel Conjecture 1.10 and the Novikov Conjecture 1.2 are open in general. A more detailed discussion of the status of these conjectures will be given in Section 24.1.

Chapter 2

Normal Bordism Groups

2.1 Normal Bordism Groups

If one wants to classify manifolds in terms of invariants one looks in particular for invariants which are comparatively easy to compute or have properties which make their analysis easy. Such a property is the bordism invariance of an invariant. The higher signatures are invariants of this type, but there are many more. Another feature of "good" invariants is that they should have a strong output for classification. This is an additional advantage of bordism invariants. In this chapter we want to introduce the relevant bordism groups systematically.

We want to define for each topological space X together with a stable oriented vector bundle E over X its *normal bordism group*

$$\Omega_n(X; E).$$

This will play a central role in the surgery program.

The concept of a stable vector bundle is delicate. If X is a finite CW-complex one can simply take an oriented vector bundle E of dimension greater than the dimension of X. This bundle should be identified with $E \oplus \mathbb{R}^k$ for arbitrary k. We recommend that the reader takes this point of view to begin. The most important example in our context comes from a smooth submanifold M of $\mathbb{R}^N$, where we consider the tangent bundle $T(M)$ or the normal bundle $\nu(M)$. Passing from $\mathbb{R}^N$ to $\mathbb{R}^{N+k}$ corresponds to passing from $\nu(M)$ to $\nu(M) \oplus \mathbb{R}^k$. Thus we speak of the stable normal bundle. To avoid difficulties the reader should consider smooth manifolds always as submanifolds of $\mathbb{R}^N$ for some large N. Since all manifolds can be embedded into an Euclidean space (and this embedding is unique up to isotopy for $N > 2n + 1$) this is no loss of generality.

Now we define the bordism group.

Elements in $\Omega_n(X; E)$ are represented by triples (M, f, α), where M is a closed n-dimensional smooth manifold, $f \colon M \to X$ a continuous map and α an isomorphism between f^*E and the stable normal bundle $\nu(M)$. Here we have embedded M into $\mathbb{R}^N$ for some large number N. If we stabilize E by passing to $E \oplus \mathbb{R}$ we similarly pass from $\mathbb{R}^N$ to $\mathbb{R}^{N+1}$ and replace α by $\alpha \oplus \mathrm{id}$. Such a triple is called a *normal map* in (X, E).

Two such triples (M, f, α) and (M', f', α') are called bordant if there is a compact manifold W with $\partial W = M + M'$, the map $f + f'$ can be extended to W by a map $g \colon W \to X$, and $\alpha + \alpha'$ can be extended to an isomorphism $\beta \colon \nu(W) \to g^*E$. We note that we have to identify $\nu(W)|_{\partial W}$ with $\nu|_{\partial W} \oplus \mathbb{R}$ and we do this with the help of a normal vector field on the boundary pointing to the interior.

We summarize

$$\Omega_n(X; E) := \{(M, f, \alpha)\}/\text{bordism}$$

As usual, we make $\Omega_n(X; E)$ a group, where the addition is given by disjoint union. The inverse is given by considering $\nu(M) \oplus \mathbb{R}$ and taking the composite of $\alpha \oplus \mathrm{id} \colon \nu(M) \oplus \mathbb{R} \to f^*(E) \oplus \mathbb{R}$ with the reflection at $\nu(M) \oplus \{0\}$. We leave the details of the proof that $\Omega_n(X; E)$ a group to the reader (see Exercise 2.1).

2.2 Rational Computation of Normal Bordism Groups

To make effective use of bordism invariants one has to determine them, and a first step is the computation of the bordism groups. This is even for the simplest case, where X is a point (and so E is the trivial bundle) unknown. Namely then the bordism group is the bordism group of framed manifolds

$$\Omega_n^{fr} = \Omega_n(\{\bullet\}; \mathbb{R}),$$

and by the Pontrjagin construction this is isomorphic to the stable homotopy groups of spheres

$$\pi_{n+k}(S^k)$$

for k large. Here and elsewhere $\{\bullet\}$ denotes the space consisting of one point. For an elementary proof see [168]. The stable homotopy groups of spheres are only known for small dimensions. But after taking the tensor product with $\mathbb{Q}$ the situation simplifies completely and using this we can give a computation of $\Omega_n(X; E) \otimes \mathbb{Q}$. We have a map $\Omega_n(X; E) \to H_n(X)$ by mapping $[M, f, \alpha]$ to $f_*([M])$.

Theorem 2.1 (Rational computation of normal bordism groups). *For a CW-complex X the map*

$$\Omega_m(X; E) \otimes \mathbb{Q} \xrightarrow{\cong} H_m(X; \mathbb{Q}), \quad [M, f, \alpha] \mapsto f_*([M])$$

is an isomorphism.

Proof. We will only treat the case, where X is a finite CW-complex, the general case follows by a limit argument.

We will show the result inductively over the cells of X. If X is a point, then $\Omega_m(\{\bullet\}; E) = \Omega_m^{fr}$, the bordism group of framed manifolds.

As mentioned above, Ω_m^{fr} is isomorphic to the stable homotopy group of spheres

$$\Omega_m^{fr} \cong \pi_{m+k}(S^k)$$

for $k \gg m$.

We abbreviate $\pi_{m+k}(S^k) = \pi_m^s$ for $k \gg m$.

Theorem 2.2 (Rational stable homotopy groups of spheres (Serre)).

$$\pi_m^s \otimes \mathbb{Q} \cong \left\{ \begin{array}{ll} \mathbb{Q} & \text{for } m = 0, \\ 0 & \text{else.} \end{array} \right.$$

The original proof of Serre [217] uses spectral sequences. Serre actually proves a stronger statement, namely that in addition all these groups are finitely generated implying that they are finite in dimension > 0. A recent elementary proof of the above statement using Gysin and Wang sequences instead of spectral sequences was given in [131].

Since $\Omega_0^{fr}(\{\bullet\}) \to H_0(\{\bullet\})$ is obviously bijective, Theorem 2.1 holds for $X = \{\bullet\}$.

For the induction step we suppose that $X = Y \cup D^m$ and the result holds for $\Omega_m(Y, E|_Y)$. Standard considerations (see Exercise 2.2) for bordism groups give an exact sequence:

$$\to \ldots \Omega_m(Y; E|_Y) \to \Omega_m(X; E) \to \Omega_m(X, Y; E) \to \Omega_{m-1}(Y; E|_Y) \to \ldots,$$

where the relative group $\Omega_m(X, Y; E)$ is as usual given by bordism classes of triples (W, f, α), where W is a compact manifold with boundary, $f \colon (W, \partial W) \to (X, Y)$ is a continuous map and α is an isomorphism $\alpha \colon \nu(W) \to f^* E$.

Another standard result is excision (see Exercise 2.3) implying an isomorphism

$$\Omega_k(Y \cup D^m, Y; E) \overset{\cong}{\leftarrow} \Omega_k(D^m, S^{m-1}; E|_{D^m}).$$

Since D^m is contractible, $E|_{D^m}$ is the trivial bundle and so

$$\Omega_k(D^m, S^{m-1}; E|_{D^m}) \cong \Omega_k^{fr}(D^m, S^{m-1}) \cong \widetilde{\Omega}_k^{fr}(S^m) \cong \Omega_{k-m}^{fr}(\{\bullet\}).$$

Thus

$$\Omega_k(X, Y; E) \otimes \mathbb{Q} \overset{\cong}{\longrightarrow} H_k(X; Y; \mathbb{Q})$$

is an isomorphism and Theorem 2.1 follows from the 5-Lemma. $\qquad\square$

2.3　Rational Computation of Oriented Bordism Groups

As an application we prove Thom's famous result about the rational computation of the oriented bordism group $\Omega_m(\{\bullet\}) = \Omega_m$. For a topological space X we consider the bordism group

$$\Omega_m(X)$$

of bordism classes of pairs (M, f), where M is a closed m-dimensional oriented manifold and $f\colon M \to X$ a continuous map. This is a generalized homology theory, the proof is elementary [54].

If X is a point, we can give a different interpretation of Ω_m which looks more complicated at the first glance. If M is a closed manifold, we embed it into R^N for some large N. The normal Gauß map associates to each $x \in M$ the oriented normal vector space ν_x. The space of $(N-m)$-dimensional oriented subspaces of $\mathbb{R}^N$ is the Grassmann manifold $\widetilde{G}_{N,N-m}$. Let E be the tautological bundle over $\widetilde{G}_{N,N-m}$. Then by construction

$$\nu^* E = \nu(M),$$

and so we consider the element

$$[M, \nu, \mathrm{id}] \in \Omega_m(\widetilde{G}_{N,N-m}; E)$$

leading to a homomorphism

$$\Omega_m \to \Omega_m(\widetilde{G}_{N,N-m}; E).$$

On the other hand we have the forgetful map

$$\Omega_m(\widetilde{G}_{N,N-m}; E) \to \Omega_m$$

and it is easy to see that this is an inverse to the first map (see Exercise 2.4). Moreover, if we consider a topological space X and replace Ω_m by $\Omega_m(X)$ and $\Omega_m(\widetilde{G}_{N,N-m}; E)$ by $\Omega_m(\widetilde{G}_{N,N-m} \times X; p_1^* E)$ we obtain a corresponding isomorphism:

Proposition 2.3 (Translation of oriented bordism to a normal bordism group). *We obtain for large N an isomorphism*

$$\Omega_m(X) \cong \Omega_m(\widetilde{G}_{N,N-m} \times X; p_1^* E).$$

From Theorem 2.1 we know that

$$\Omega_m(\widetilde{G}_{N,N-m}; E) \otimes \mathbb{Q} \cong H_m(\widetilde{G}_{N,N-m}; \mathbb{Q}).$$

If we stabilize N by passing from N to $N+1$, the limit of the spaces is called BSO and so we have shown

Theorem 2.4 (Relation of rational oriented bordism groups to the homology of BSO). *The map*

$$\Omega_m(X) \otimes \mathbb{Q} \xrightarrow{\cong} H_m(BSO \times X; \mathbb{Q}), \quad [M, f] \mapsto (\nu \times f)_*([M])$$

is an isomorphism.

A standard inductive argument gives a computation of $H^*(BSO; \mathbb{Q})$:

$$H^*(BSO; \mathbb{Q}) \cong \mathbb{Q}[p_1, p_2, p_3, \ldots],$$

where p_i are given by the Pontrjagin classes of the tautological bundle E over $G_{N,k}$ [171].

Since $\nu^*(p_i) = p_i(\nu(M))$, the i-th Pontrjagin class of the normal bundle of M, we conclude that the Pontrjagin numbers of the normal bundle determine the bordism class of a manifold M. Since the Pontrjagin classes of the normal bundle determine those of the tangent bundle (and vice versa) we conclude Thom's theorem:

Theorem 2.5 (Rational computation of oriented bordism groups (Thom)). *We obtain an isomorphism*

$$\Omega_{4k} \otimes \mathbb{Q} \xrightarrow{\cong} \mathbb{Q}^{\pi(k)}, \quad [M] \mapsto (\langle p_I(M), [M]\rangle)_I$$

where $\pi(k)$ is the number of partitions I of k and for such a partition $I = (i_1, i_2, \ldots, i_s)$ we put $p_I(M) := p_{i_1}(TM) \cup \ldots \cup p_{i_s}(TM)$.

For $m \neq 0 \mod 4$,

$$\Omega_m \otimes \mathbb{Q} = 0.$$

The numbers

$$\langle p_I(M), [M]\rangle$$

are called *Pontrjagin numbers*.

One can give an explicit basis of $\Omega_{4k} \otimes \mathbb{Q}$. One "only" has to find for each k and each partition J of k manifolds M_J such that the matrix with entries

$$\langle p_I, M_J\rangle$$

has non-trivial determinant. Experiments with small dimensions suggest that $M_J := \mathbb{CP}^{2j_1} \times \mathbb{CP}^{2j_2} \times \cdots \times \mathbb{CP}^{2j_r}$ will have this property, where $k = j_1 + \cdots + j_r$. This is actually true but the proof needs a clever idea. We refer to [171] for the proof of:

Theorem 2.6 (Basis for rational oriented bordism groups). *The products of complex projective spaces $\mathbb{CP}^{2i_1} \times \mathbb{CP}^{2i_2} \times \cdots \times \mathbb{CP}^{2i_r}$ for $i_1 + \cdots + i_r = k$ yield a basis of $\Omega_{4k} \otimes \mathbb{Q}$.*

The material in this chapter will be discussed from a different point of view again in Chapter 18.

Chapter 3

The Signature

The oldest topological invariant is the Euler characteristic, the alternating sum of Betti numbers. The relevance of the Euler characteristic is obvious for the classification of closed connected oriented surfaces: Two such surfaces are homeomorphic if and only if the Euler characteristics agree.

The Euler characteristic of a closed 3-manifold is zero and, more generally, the (co)homological picture of 3-manifolds is rather simple, in this case the most interesting algebraic topological information is contained in the fundamental group.

Passing to dimension 4, we will encounter a new invariant, the signature. It assigns to a closed oriented manifold M of dimension $4k$ an integer, $\sigma(M) \in \mathbb{Z}$, and its importance is visible from the following very deep result. A closed oriented 4-manifold M is called *even* if for all $x \in H^2(M; \mathbb{Z}/2)$, we have $x \cup x = 0 \mod 2$, otherwise it is called *odd*.

Theorem 3.1 (Homeomorphism classification of closed 1-connected smooth 4-manifolds (Donaldson, Freedman)). *Two closed simply connected smooth oriented 4-manifolds M and N are homeomorphic if and only if*

i) *both are even or odd,*

ii) *they have equal Euler characteristic,*

iii) *the signatures agree.*

The proof of this result uses Donaldson's [69] theorem, that the intersection form of a smooth closed 4-manifold is either up to sign the standard Euclidean form or indefinite, and Freedman's classification of 1-connected topological 4-manifolds [94]. It has only a slight relation to the Novikov Conjecture. Namely, one of the easy steps of Freedman's proof is that if the intersection form on $H^2(M; \mathbb{Q})$ is not definite, then the conditions i)–iii) determine the homotopy type, a result obtained by Milnor [165] using standard arguments in homotopy theory and some information about unimodular integral quadratic forms (see Exercise 3.1). But homotopy invariants derived from signatures are the theme of the Novikov Conjecture.

3.1   The Definition of the Signature

We will now define the signature and prove its basic properties. Let M be a closed
oriented topological manifold. If the dimension of M is not divisible by 4 we define
the signature to be 0. If the dimension is $4k$ we consider the cup product form
on the middle dimension and evaluate the product on the fundamental class to
obtain the *intersection form*:

$$S(M) : H^{2k}(M;\mathbb{Q}) \times H^{2k}(M;\mathbb{Q}) \to \mathbb{Q}, \quad (\alpha, \beta) \mapsto \langle \alpha \cup \beta, [M] \rangle.$$

The signature, sign, of a symmetric bilinear form over a finite dimensional
$\mathbb{Q}$-vector space is the difference of the number of positive and negative eigenvalues
after applying $- \otimes_{\mathbb{Q}} \mathbb{R}$. Since M is compact, the rational homology groups are
finite dimensional vector spaces. Thus we can define the *signature* of M as

$$\text{sign}(M) := \text{sign}(S(M)).$$

If we replace M by $-M$ then we only replace $[M]$ by $-[M]$. Thus S changes its
sign implying

$$\text{sign}(-M) = -\text{sign}(M).$$

E.g. since $H^{2k}(S^{4k}) = 0$, it is zero on spheres. The cohomology ring of $\mathbb{CP}^{2k}$
is the truncated polynomial ring

$$H^*(\mathbb{CP}^{2k}) = \mathbb{Z}[x]/(x^{k+1} = 0),$$

where X is a generator of H^2, and that $\langle x^{2k}, [\mathbb{CP}^{2k}] \rangle = 1$. Thus we have:

$$\text{sign}(\mathbb{CP}^{2k}) = 1.$$

3.2   The Bordism Invariance of the Signature

An important property of the signature is demonstrated by the fact that it is
bordism invariant:

Theorem 3.2 (Bordism invariance of the signature). *If a closed oriented smooth
manifold M is the boundary of a compact oriented smooth manifold, then its signature vanishes*

$$\text{sign}(M) = 0.$$

The main ingredient of the proof is the following:

Lemma 3.3. *Let W be a compact smooth oriented manifold of dimension $2k + 1$.
Let $j \colon \partial W \to W$ be the inclusion. Then*

$$\ker \left(i_* \colon H_k(\partial W; \mathbb{Q}) \to H_k(W; \mathbb{Q}) \right) \cong \text{im} \left(j^* \colon H^k(W; \mathbb{Q}) \to H^k(\partial W; \mathbb{Q}) = H_k(\partial W; \mathbb{Q}) \right).$$

Here the isomorphism is given by Poincaré duality.

This lemma is a consequence of the Lefschetz duality Theorem [68].

Combining this lemma with the Kronecker isomorphism we conclude that for $j_* : H_k(\partial W; \mathbb{Q}) \to H_k(W; \mathbb{Q})$:

$$\dim(\ker(j_*)) = \dim(\operatorname{im}((j_*)^*)).$$

From linear algebra we know that $\dim(\operatorname{im}(j_*)) = \dim(\operatorname{im}((j_*)^*))$ and we obtain:

$$\dim(\ker(j_*)) = \dim(\operatorname{im}(j_*))$$

and by the dimension formula:

$$\dim(\ker(j_*)) = \frac{1}{2} \cdot \dim(H_k(\partial W)).$$

Applying the lemma again we finally note:

$$\dim(\operatorname{im}(j^*)) = \frac{1}{2} \cdot \dim(H^k(\partial W)).$$

As a last preparation for the proof of Theorem 3.2 we need the following observation from linear algebra. Let $b: V \times V \to \mathbb{Q}$ be a symmetric non-degenerate bilinear form on a finite dimensional $\mathbb{Q}$-vector space. Suppose that there is a subspace $U \subseteq V$ with $\dim(U) = \frac{1}{2} \cdot \dim(V)$ such that, for all $x, y \in U$, we have $b(x, y) = 0$. Then $sign(b) = 0$. The reason is the following. Let $e_1, \ldots, e_n$ be a basis of U. Since the form is non-degenerate, there are elements $f_1, \ldots, f_n$ in V such that $b(f_i, e_j) = \delta_{ij}$ and $b(f_i, f_j) = 0$. This implies that $e_1, \ldots, e_n, f_1, \ldots, f_n$ are linear independent and thus form a basis of V. Now consider $e_1 + f_1, \ldots, e_n + f_n, e_1 - f_1, \ldots, e_n - f_n$ and note that, with respect to this basis, b has the form

$$\begin{pmatrix} 2 & & & & & \\ & \ddots & & & & \\ & & 2 & & & \\ & & & -2 & & \\ & & & & \ddots & \\ & & & & & -2 \end{pmatrix}$$

and so

$$sign(b) = 0.$$

Now we are ready to give the proof of Theorem 3.2.

Proof. We first note that for $\alpha \in \operatorname{im}(j^*)$ and $\beta \in \operatorname{im}(j^*)$ the intersection form $S(\partial W)(\alpha, \beta)$ vanishes. For if, $\alpha = j^*(\bar{\alpha})$ and $\beta = j^*(\bar{\beta})$ then

$$S(\partial W)(\alpha, \beta) = \langle j^*(\bar{\alpha}) \cup j^*(\bar{\beta}), [\partial W] \rangle = \langle \bar{\alpha}) \cup j^*(\bar{\beta}, j_*([\partial W]) \rangle = 0,$$

since $j_*([\partial W]) = 0$.

Thus the intersection form vanishes on $\text{im}(j^*)$. By Poincaré duality the intersection form $S(\partial W) \otimes \mathbb{Q}$ is non-degenerate. Since the dimension of $\text{im}(j^*)$ is $1/2 \cdot \dim(H^k(\partial W))$, the proof is finished by the considerations above from linear algebra. $\square$

3.3   Multiplicativity and other Properties of the Signature

The signature of a disjoint union is the sum of the signatures and so we obtain a homomorphism

$$\text{sign}\colon \Omega_m \to \mathbb{Z}.$$

Here we define the signature as zero if the dimension of the manifold is not divisible by 4.

The direct sum of bordism groups

$$\Omega_* := \bigoplus_m \Omega_m$$

is a ring in which the multiplication is given by cartesian product. It is natural to ask whether the signature is a ring homomorphism. The following result says that this is the case:

Theorem 3.4 (Multiplicativity of the signature). *Let M and N be closed oriented manifolds, then*

$$\text{sign}(M \times N) \;=\; \text{sign}(M) \cdot \text{sign}(N).$$

Proof. This is a standard application of the Künneth Theorem and the algebraic fact from above, that the signature of an unimodular symmetric bilinear form over the rational numbers vanishes if there is a submodule of half rank on which the form is identically zero. The Künneth theorem allows one to decompose the middle cohomology of $M \times N$ as the tensor product of the middle cohomology groups of M and N plus the orthogonal sum given by tensor products of the other terms. And on this orthogonal summand one immediately sees a subspace of half dimension on which the intersection form vanishes. Thus the result follows since the signature of a tensor product of symmetric bilinear forms is the product of the signatures of the two forms (see Exercise 3.2). $\square$

The (co)homological definition of the signature shows some interesting aspects, which we summarize:

(i) If there is an orientation preserving homotopy equivalence then the signatures are equal: the signature is a homotopy invariant (meaning in this book always orientation preserving).

(ii) The signature is a multiplicative bordism invariant.

There are some other properties which one cannot see directly from the definition. For example the signature of a closed manifold M with stably trivial tangent bundle is zero, if the dimension is positive. Actually it is enough to require that the rational Pontrjagin numbers of the tangent bundle of M vanish, since this implies by Theorem 2.5 that the bordism class of M in $\Omega_m \otimes Q$ vanishes and a homomorphism from a Q-vector space V to $\mathbb{Z}$ vanishes if the element maps to zero in $V \otimes \mathbb{Q}$.

Another property of the signature is that for a finite covering of closed oriented manifolds $p \colon N \to M$ of degree k the signature is — as the Euler characteristic — multiplicative:

$$\text{sign}(N) \; = \; k \cdot \text{sign}(M).$$

This will be a consequence of the signature theorem which we prove in the next section.

3.4 Geometric Interpretation of Cohomology and the Intersection Form

We finish this section with a more geometric interpretation of the intersection form of smooth manifolds in terms of homology classes represented by manifolds. This information is not really needed for the rest of the lecture notes, but we find it useful to add this information for those readers who prefer a more geometric view. Other readers can skip this part.

For geometric considerations it is often better to apply Poincaré duality (or Lefschetz duality if the manifold has a boundary) and consider homology instead of cohomology. Then the cup-product has a geometric interpretation in terms of transversal intersections, at least if the homology classes are represented by smooth maps $f \colon X \to M$, where X is a compact oriented smooth manifold of dimension r. Then the corresponding homology class is

$$f_*([X]) \in H_r(M; \mathbb{Z}),$$

the image of the fundamental class. If $\dim(M) = m$, then the corresponding cohomology class sits in $H^{m-r}(M, \mathbb{Z})$.

Not all homology classes of M can be obtained this way but an appropriate multiple of a homology class can always be obtained so. This is a consequence of Theorem 2.1. Namely, if X is a CW complex with finite skeleta we consider the trivial bundle 0 over X. Then, if (M, f, α) is a normal map the isomorphism α is just a trivialization of the normal bundle, and so (M, α) is a *framed manifold* and we write instead of $\Omega_m(X; 0)$ the standard notation $\Omega_m^{fr}(X)$. Then Theorem 2.1 implies:

$$\Omega_m^{fr}(X) \otimes \mathbb{Q} \to H_m(X; \mathbb{Q}), \quad [M, f, \alpha] \mapsto f_*([M])$$

is an isomorphism. Thus an appropriate multiple of each homology class can be represented by a map from a closed oriented (in this case even framed) manifold

M to X. Since a compact manifold is homotopy equivalent to a finite CW-complex [166], the statement follows.

As the signature is defined in terms of rational (co)homology groups $H^*(M;\mathbb{Q})$ resp. $H_*(M;\mathbb{Q})$, we can use that all rational homology classes can geometrically be represented by a rational multiple of classes given by maps $f\colon N \to M$, where N is a framed manifold. Following Quillen, who worked this out for the cohomology theory corresponding to singular bordism [189], we interpret the cohomology groups corresponding to framed cobordism $(\Omega^{fr})^k(M)$ as bordism classes of triples (N, f, α), where (N, α) is a framed manifold of dimension $m - k$, which is in general *not* compact, but f has to be a *proper* map, which in this context means that the preimage of each compact set is compact. As for the homology theory framed bordism one has a natural transformation from $(\Omega^{fr})^n(M)$ to $H^n(M)$ which after taking the tensor product with $\mathbb{Q}$ becomes an isomorphism:

$$(\Omega^{fr})^n(M) \otimes \mathbb{Q} \cong H^n(M;\mathbb{Q}).$$

Recently one of the authors (K.) introduced a generalization of manifolds, called stratifolds. These are manifolds with certain singularities. In this context one can define integral homology and cohomology of manifolds as above for rational (co)homology by using certain stratifolds S instead of framed manifolds. This is not relevant for this book, but the reader might be interested to look at this geometric approach to integral (co)homology [134].

We note that the interpretation of rational cohomology classes as rational multiples of bordism classes of framed manifolds N together with a proper map $f\colon N \to M$ makes Poincaré duality a tautology: if M is closed then, if $f\colon N \to M$ is proper, N has to be closed and so

$$(\Omega^{fr})^n(M) \;=\; \Omega^{fr}_{m-n}(M).$$

To obtain interesting information from this one has to combine it with the universal coefficient theorem expressing cohomology in terms of homology groups. In particular, if we pass to rational (co)bordism groups, we have isomorphisms

$$(\Omega^{fr})^n(M) \otimes \mathbb{Q} \cong H^n(M, \mathbb{Q}) \cong \hom(H_n(M), \mathbb{Q}) \cong \hom(\Omega^{fr}_n(M), \mathbb{Q}),$$

where the map in the universal coefficient theorem is given by the Kronecker isomorphism.

On cohomology one has a ring structure given by the cup product. If we only consider rational cohomology groups interpreted as $(\Omega^{fr})^n(M)$ (or pass to stratifolds instead of manifolds if we want to have corresponding results for integral (co)homology) we can interpret the cup product geometrically.

Let M be a smooth oriented manifolds and $g_1 : N_1 \to M$ and $g_2 : N_2 \to M$ be maps, where N_i are compact framed manifolds (here and in the following we omit the framing from the notation). We note here that unless M is closed these maps do not give cohomology classes, they give so-called *cohomology classes with compact support.* But if M is closed the maps g_i are proper and so we obtain cohomology classes. Now we approximate g_i by smooth transversal maps and call the resulting maps again g_i. Then we define the *cup product* of $[N_1, g_1]$ and $[N_2, g_2]$ with the help of transversal intersection of g_1 and g_2 which we denote by $g_1 \sqcap g_2 := \{(x, y) \in N_1 \times N_2 \mid g_1(x) = g_2(y)\}$. This is again a compact manifold and we consider on it the map $g_1 p_1 \colon g_1 \sqcap g_2 \to M$. This is the definition of the cup product of two cohomology classes with compact support:

$$[N_1, g_1] \cup [N_2, g_2] := [g_1 \sqcap g_2, g_1 p_1].$$

If $\dim(N_1) = m - k$ and $\dim(M_2) = m - r$, then $\dim(g_1 \sqcap g_2) = m - (k + r)$.

If M is closed, then $[N_i, g_i]$ give cohomology classes and we obtain the ordinary cup product:

$$H^k(M; \mathbb{Q}) \times H^r(M; \mathbb{Q}) \to H^{k+r}(M; \mathbb{Q}).$$

We note that if M is a closed manifold and the N_i are only oriented instead of framed we can use the same construction to interpret the cup product of the Poincaré duals y_1 of $(g_1)_*([N_1])$ and y_2 of $(g_2)_*([N_2])$ as the Poincaré dual of the transversal intersection:

$$[M] \cap (y_1 \cup y_2) = (g_1 p_1)_*([g_1 \sqcap g_2]).$$

If M is not necessarily compact and $q_i \colon N_i \to M$ are transversal maps from closed oriented (again it is not necessary to assume that N_i are framed) manifolds such that $\dim(N_1) + \dim(N_2) = m = \dim(M)$ then $g_1 \sqcap g_2$ is a compact oriented 0-dimensional manifold and we consider the sum of the local orientations. This number is called the *intersection number in homology* of $(g_1)_*([N_1])$ and $(g_2)_*([N_2])$ and we abbreviate it by

$$(g_1)_*([N_1]) \circ (g_2)_*([N_2]).$$

This is a bilinear form. Via Poincaré duality it corresponds to the intersection number in cohomology defined above.

Chapter 4

The Signature Theorem and the Novikov Conjecture

4.1 The Signature Theorem

We have shown that the signature gives a homomorphism

$$\text{sign} \colon \Omega_n \to \mathbb{Z}.$$

Since the rational bordism groups $\Omega_n \otimes \mathbb{Q}$ are 0 for $n \neq 0 \mod 4$ and $\mathbb{Q}^{\pi(K)}$ for $n = 4k$ and $\pi(k)$ the number of partitions of k and the Pontrjagin numbers give an isomorphism

$$\Omega_{4k} \otimes \mathbb{Q} \overset{\cong}{\to} \mathbb{Q}^{\pi(k)},$$

it is clear that there is a unique rational linear combination of Pontrjagin numbers which we denote by $\mathcal{L}_k(p_1, p_2, \ldots, p_k) = \sum_{|I|=k} a_I \cdot p_I$, where I is a partition of k and a_I is in $\mathbb{Q}$, such that

$$\text{sign}(M) = \langle \mathcal{L}_k(p_1(TM), p_2(TM), \ldots, p_k(TM)), [M] \rangle$$

for each closed oriented $4k$-dimensional manifold M.

We also consider $\mathcal{L}(p) := 1 + \mathcal{L}_1(p_1) + \mathcal{L}_2(p_1, p_2) + \ldots \in \prod_{i \geq 0} H^{4i}(M; \mathbb{Q})$. Here p_i is the i-th Pontrjagin class. But what is this linear combination? In low dimensions we can find the formula by computing both sides on generators. Namely for closed 4-manifolds we know the rational bordism group $\Omega_4 \otimes \mathbb{Q} \cong \mathbb{Q}$ and so $\mathcal{L}_1$ is known once we compute the signature and Pontrjagin number p_1 of $\mathbb{CP}^2$. We have already seen that the signature of $\mathbb{CP}^2$ is 1. The stable tangent bundle of $\mathbb{CP}^2$ is $\bigoplus_3 H$, the Whitney sum of three copies of the Hopf bundle [171]. The first Pontrjagin class of the Hopf bundle is x^2, where x is our generator of $H^2(\mathbb{CP}^2)$. Thus the Pontrjagin number $\langle p_1(T(\mathbb{CP}^2), [\mathbb{CP}^2] \rangle$ is 3 and we conclude:

$$\mathcal{L}_1(p_1) = \frac{1}{3} p_1.$$

Already this special case has interesting applications, for example that $\mathbb{CP}^2$ equipped with the opposite orientation has no complex structure (see Exercise 4.1).

A similar consideration computing the Pontrjagin numbers of $\mathbb{CP}^2 \times \mathbb{CP}^2$ and of $\mathbb{CP}^4$ and comparing them with the signature, which in both cases is 1, leads to (see Exercise 4.2)

$$\mathcal{L}_2(p_1, p_2) = \frac{1}{45}(7p_2 - p_1^2).$$

In principle one can obtain $\mathcal{L}_k$ by computing the Pontrjagin numbers of products of projective spaces, since they generate $\Omega_{4k} \otimes \mathbb{Q}$ by Theorem 2.5, but this does not lead to a closed formula.

The problem was solved by Hirzebruch using his theory of multiplicative sequences. We motivate this concept by noting that

$$\text{sign}(M \times N) = \text{sign}(M) \cdot \text{sign}(N),$$

and similarly the tangent bundle is multiplicative:

$$T(M \times N) = TM \times TN,$$

implying that the total Pontrjagin class is multiplicative,

$$p(M \times N) = p(M) \times p(N).$$

From this it follows that the L-class $\mathcal{L}(p_1, \ldots, p_k) = L(p)$ fulfills

$$\mathcal{L}(p(E \oplus F)) = \mathcal{L}(p(E)) \cup \mathcal{L}(p(F))$$

for vector bundles E and F.

Now we apply the splitting principle for oriented vector bundles [171]. Let $p \colon E \to X$ be an oriented vector bundle of dimension $2k$, then there is a map $f \colon Y \to X$ such that

i) $f^* \colon H^i(X) \to H^i(Y)$ is injective

ii) $f^*E \cong E_1 \oplus \ldots \oplus E_k$, where the E_i are 2-dimensional vector bundles.

Thus $\mathcal{L}$ is completely determined by $\mathcal{L}(E)$, where E is an oriented 2-dimensional vector bundle. Since the higher Pontrjagin classes $p_2(E), p_3(E), \ldots$ vanish, $\mathcal{L}$ is a formal power series in $p_1(E)$:

$$\mathcal{L}(p(E)) = \sum_k a_k \cdot p_1(E)^k.$$

It turns out that the power series which determines the L-polynomials is $q(z) := \frac{\sqrt{z}}{\tanh(\sqrt{z})}$:

Theorem 4.1 (Hirzebruch signature theorem [115]). *There are unique polynomials* $\mathcal{L}_k(p(E))$ *in the Pontrjagin classes of an oriented vector bundle such that for a 2-dimensional bundle,*

$$\mathcal{L}(p(E)) = \frac{\sqrt{p_1(E)}}{\tanh\sqrt{p_1(E)}},$$

and for each closed oriented 4k-dimensional manifold M,

$$\mathrm{sign}(M) \; = \; \langle \mathcal{L}_k(p_1(M),\ldots,p_k(M)), [M]\rangle.$$

Proof. For details of the proof we either refer to Hirzebruch's original proof [115] or to the presentation of the same argument in [171]. The basic idea is the following. Firstly one shows that the L-polynomials, which in the statement of the theorem are already characterized, really exist. This is a nice purely algebraic consideration using elementary symmetric polynomials. The topological background for this is that if $T^k = (S^1)^k$ is the maximal torus of $SO(2k)$, then we consider the induced map $f\colon B(T^k) \to B(SO(2k))$. The pullback of the universal bundle E over $B(SO(2k))$ to $B(T)$ splits as a sum of line bundles and the rational cohomology ring of $B(SO(2k))$ is mapped isomorphically to the subring of rational cohomology classes invariant under the action of all permutations of the factors of T^k [54]. In other words $B(T^k)$ is a space occurring in the splitting principle for E. If we now stabilize by enlarging k and pass to $H^*(BSO) = \mathbb{Q}[p_1, p_2, \ldots]$ we see that we can study the existence of the L-polynomial completely in terms of the polynomial ring $\mathbb{Q}[p_1, p_2, \ldots]$ and the subring of symmetric polynomials.

Once the existence of the L-polynomials is guaranteed, one has to check that the formula holds. Since $\Omega_{4k}\otimes\mathbb{Q}$ is generated by the product of complex projective spaces one only has to check the formula for these manifolds. Their stable tangent bundles are sums of line bundles: $T\mathbb{CP}^n$ is stably isomorphic to $\bigoplus_{n+1} H$, where H is the Hopf bundle. With this information one reduces the computation of the left side of the signature theorem to a computation with the residue formula, which shows that it agrees with the right side, which is 1, since the signature of even dimensional projective spaces is 1 and the signature is multiplicative. $\square$

Remark 4.2. The L-class is a rational polynomial in the Pontrjagin classes. This description can be inverted: The rational Pontrjagin classes of a bundle can be expressed in terms of the L-class. Thus over the rationals the L-classes are equivalent to the Pontrjagin classes [171].

As a consequence of the signature theorem we obtain the announced multiplicativity of the signature theorem for coverings (see Exercise 4.3).

4.2 Higher Signatures

Now we want to define the higher signatures. We motivate the definition of the higher signatures by the following considerations. Let M be a closed smooth oriented manifold and $N \subseteq M$ a closed oriented submanifold of dimension $4k$ with stably trivial normal bundle ν. Actually, it is enough to require, that all rational Pontrjagin classes of the normal bundle or equivalently the L-class vanish. Then one can compute the signature of N in terms of the cohomology and L-class of M. Namely by the signature theorem we have

$$\operatorname{sign}(N) \;=\; \langle \mathcal{L}(N), [N] \rangle.$$

From the Thom isomorphism we conclude

$$\langle \mathcal{L}(N), [N] \rangle \;=\; \langle \mathcal{L}(N) \cup t_\nu, [D\nu, S\nu] \rangle$$

where $D\nu$ is the disk bundle of the normal bundle, $S\nu$ is the sphere bundle, and t_ν is the Thom class. But $\mathcal{L}(N) = \mathcal{L}(TM \oplus \nu) = i^*(\mathcal{L}(M))$, since the L-class of ν vanishes by assumption, where $i\colon N \to M$ is the inclusion. Finally we consider the Poincaré dual $x \in H^{m-4k}(N)$ of $i_*([N])$ and note that $x \cup \mathcal{L}(M) = t_\nu \cup i^*(\mathcal{L}(M))$. Combining these steps we have shown:

Proposition 4.3 (The signature of a submanifold). *Let $N \subseteq M$ be a submanifold such that the L-class of the normal bundle is trivial. Then*

$$\operatorname{sign}(N) \;=\; \langle x \cup \mathcal{L}(M), [M] \rangle,$$

where $x \in H^{m-4k}(N)$ is the Poincaré dual of $i_*([N])$.

The expression on the right side makes sense even if the Poincaré dual of x is not representable by a smooth submanifold with vanishing L-class of the normal bundle. We call it the *higher signature of M associated to x*:

$$\langle x \cup \mathcal{L}(M), [M] \rangle.$$

If $x = 1 \in H^0(M)$ then we obtain the ordinary signature.

In contrast to the ordinary signature it is not true that the higher signatures associated to all cohomology classes are homotopy invariants. By this we mean that if $h\colon M' \to M$ is a homotopy equivalence then the higher signature of M' with respect to $h^*(x)$ is in general different from the higher signature of M associated to x (Example 1.6).

4.3   The Novikov Conjecture

The Novikov Conjecture states that for special cohomology classes the higher signatures are homotopy invariants, namely for those cohomology classes which are induced from classifying spaces BG, where G is some group. More precisely Novikov conjectured the following:

Conjecture 4.4 (Novikov Conjecture). *Let G be a group and $h\colon M' \to M$ be an orientation preserving homotopy equivalence between closed oriented smooth manifolds and $f\colon M \to BG$ be a map. Then for each class $x \in H^k(BG; \mathbb{Q})$ the higher signatures of M associated to $f^*(x)$ and for N associated to $(fh)^*(x)$ agree.*

We want to reformulate this by introducing the higher signatures in a different way which gives equivalent information to the higher signatures of M associated to $f^*(x)$ for all $x \in H^*(BG; \mathbb{Q})$. Since $H^*(BG; \mathbb{Q}) \cong \hom(H_*(BG; \mathbb{Q})$ we can compute $f^*(x) \cup \mathcal{L}(M)$ in terms of $f_*([M] \cap f^*(x))$. Then

$$\langle f^*(x) \cup \mathcal{L}(M), [M] \rangle \;=\; \langle x, f_*([M] \cap \mathcal{L}(M)) \rangle.$$

Thus the collection of the higher signatures of M associated to all cohomology classes $f^*(x)$ for $x \in H^*(BG; \mathbb{Q})$ is determined by

$$\operatorname{sign}^G(M, f) \quad := \; f_*([M] \cap \mathcal{L}(M)) \;\in\; \bigoplus_{i \in \mathbb{Z}, i \geq 0} H_i(BG; \mathbb{Q}). \tag{4.5}$$

We call $\operatorname{sign}^G(M, f)$ the *higher signature* of (M, f). With this we can reformulate the Novikov Conjecture as:

Conjecture 4.6 (Reformulation of the Novikov Conjecture). *Let G be a group. Then for each map $f \colon M \to BG$ the higher signature is a homotopy invariant, i.e., if $h \colon M' \to M$ is an orientation preserving homotopy equivalence, then*

$$\operatorname{sign}^G(M', fh) \;=\; \operatorname{sign}^G(M, f).$$

4.4 The Pontrjagin Classes are not Homeomorphism Invariants

As previously mentioned, Novikov proved that the rational Pontrjagin classes are homeomorphism invariants. It is natural to ask whether the integral Pontrjagin classes are topological invariants. We were surprised that we could not find an answer to this question in the literature. During the seminar Diarmuid Crowley informed us that the answer is negative (as expected) and that this follows from a recent classification of a certain class of 15-dimensional manifolds. More precisely it follows that the second Pontrjagin class p_2 is not a homeomorphism invariant. Besides this classification the key information they use is a result of Brumfiel. A further investigation of the role of Brumfiel's result for Pontrjagin classes implies that one obtains counterexamples by very elementary considerations (and for all Pontrjagin classes p_i for $i > 1$) which we describe here.

We begin with Brumfiel's result. He studies the relation between the homotopy groups $\pi_{4n}(BO)$ and $\pi_{4n}(BTOP)$, where $BTOP$ is the classifying space of topological vector bundles. More precisely, one considers the group $TOP(n)$ of homeomorphism on $\mathbb{R}^n$ fixing 0. Then TOP is the union of $TOP(n)$ which we identify via $f \times \operatorname{id}$ with a subgroup of $TOP(n+1)$, and $B(TOP)$ is its classifying space. Actually Brumfiel considers $B(PL)$ instead of $B(TOP)$, but for $n > 1$ the groups $\pi_{4n}(BPL)$ and $\pi_{4n}(B\,TOP)$ are isomorphic [130]. Brumfiel shows that for $n > 1$ there is no map $f \colon \pi_{4n}(BTOP) \to \pi_{4n}(BO)$ with $fi_* = \operatorname{id}$, where $i \colon BO \to BTOP$ is induced from the inclusion $BO \to BTOP$. This follows from his main result in [37].

Combined with the following elementary result we obtain the desired information.

Proposition 4.7. *If there is no homomorphism*

$$f \colon \pi_{4n}(BTOP) \to \pi_{4n}(BO),$$

such that $i_* f = \mathrm{id}$, where $i \colon BO \to BTOP$ is induced from the inclusion $BO \to BTOP$, then the n-th Pontrjagin class of smooth manifold is in general not a homeomorphism invariant.

Proof. We note that by Bott periodicity $\pi_{4n}(BO) \cong \mathbb{Z}$, and that $i_* \colon \pi_{4n}(BO) \to \pi_{4n}(BTOP)$ is injective (this can for example be seen from the topological invariance of the rational Pontrjagin classes). Thus f as above exists if and only if $1 \in \pi_{4n}(BO)$ is mapped to a primitive element in $\pi_{4n}(BTOP)$.

Assuming that f does not exist, we see that $f_*(1) = k\alpha$ for some $\alpha \in \pi_{4n}(BTOP)$ and $k > 1$. Now we consider the so-called Moore space

$$D^{4n} \cup_g S^{4n-1}$$

where $g \colon S^{4k-1} \to S^{4k-1}$ is a map of degree k. s If we collapse S^{4n-1} to a point, we obtain a map to S^{4n}. Consider a $4n$-dimensional vector bundle over S^{4n} which corresponds to a generator of $\pi_{4n}(BO) \cong \mathbb{Z}$. We pull this back to a vector bundle over $D^{4n} \cup_g S^{4n-1}$ denoted by E. The n-th Pontrjagin class of E is

$$p_n(E) = 1 \in \mathbb{Z}/k\mathbb{Z} \cong H^{4n}(D^{4n} \cup_g S^{4n-1}).$$

If we consider E as a topological vector bundle, it is trivial, topologically isomorphic to $(D^{4n} \cup_g S^{4n-1}) \times \mathbb{R}^{4n}$. This follows from the Puppe sequence

$$[\Sigma S^{4n-1}, BTOP] \xrightarrow{\cdot k} [S^{4n}, BTOP] \to [D^{4n} \cup_g S^{4n-1}, BTOP] \to [S^{4n-1}, BTOP]$$

and the fact that $1 \in \pi_{4n}(BO)$ is mapped to $k\alpha$ in $\pi_{4n}(BTOP)$.

We actually want to replace this Moore space by a non-compact manifold homotopy equivalent to it. For this we replace S^{4n-1} by $S^{4n-1} \times D^{4n}$ and approximate g which we consider as map to $S^{4n-1} \times \{\bullet\}$ for some $\{\bullet\} \in S^{4n-1}$ by an embedding with trivial normal bundle:

$$g' \colon S^{4n-1} \times D^{4n-1} \hookrightarrow S^{4n-1} \times S^{4n-1}.$$

Now we replace the Moore space by

$$M(k) := S^{4n-1} \times D^{4n-1} \cup_{g'} D^{4n} \times D^{4n-1},$$

where we identify $(x, y) \in S^{4n-1} \times D^{4n-1}$ with $g'(x, y) \in S^{4n-1} \times S^{4n-1}$. After smoothing the corners, this is a smooth compact manifold with boundary of dimension $8n - 2$, which is homotopy equivalent to the Moore space. If we choose

g' appropriately it is stably parallelizable, which we now assume. We denote its interior by $N(k)$, which is again homotopy equivalent to the Moore space.

Now we are finished by pulling the bundle E back to $N(k)$ to obtain a vector bundle over $N(k)$ which as topological bundle is trivial. The total space of this bundle is a smooth manifold of dimension $12n-2$ with non-trivial Pontrjagin class $p_n(N(k)) = p_n(E)$ which is homeomorphic to $N(k) \times \mathbb{R}^{4n}$, a smooth manifold with trivial Pontrjagin class p_n.

We can also obtain compact examples by taking the restrictions of the bundle to $\partial(M(k))$ and the sphere bundle of the Whitney sum of this bundle with a trivial line bundle. $\qquad\Box$

As mentioned above Brumfiel has proved that such a map f does not exist and so we conclude:

Theorem 4.8 (Pontrjagin classes are not homeomorphism invariants). *The Pontrjagin classes $p_n(M)$ of closed smooth manifolds M are not homeomorphism invariants for $n > 1$.*

Finally we note that p_1 is a homeomorphism invariant. For a proof of this we refer to [135].

Chapter 5

The Projective Class Group and the Whitehead Group

In this chapter we give a brief introduction to $K_0(R)$ and $K_1(R)$ of a ring and to the Whitehead group $\mathrm{Wh}(G)$ of a group G.

5.1 The Projective Class Group

In the sequel ring means associative ring with unit. Modules are understood to be left modules unless explicitly stated differently.

Definition 5.1 (Projective class group). *The* projective class group $K_0(R)$ *of a ring R is the abelian group, which has isomorphism classes $[P]$ of finitely generated projective R-modules P as generators and for which for each exact sequence $0 \to P_0 \to P_1 \to P_2 \to 0$ of finitely generated projective R-modules the relation $[P_1] = [P_0] + [P_2]$ holds.*

Remark 5.2 (Universal rank function). An *additive invariant* (A, a) for the category of finitely generated projective R-modules consists of an abelian group A and an assignment, which assigns to each finitely generated projective R-module P an element $a(P) \in A$ such that for each exact sequence $0 \to P_0 \to P_1 \to P_2 \to 0$ of finitely generated projective R-modules we have $a(P_1) = a(P_0) + a(P_2)$. We call an additive invariant (U, u) *universal* if for each additive invariant (A, a) there exists precisely one homomorphism $\varphi \colon U \to A$ such that for each finitely generated projective R-module P we have $\varphi(u(P)) = a(P)$. The universal property implies that the universal additive invariant is unique up to unique isomorphism if it exists. One easily checks that $K_0(R)$ together with the assignment $P \mapsto [P]$ is the universal additive invariant.

One may summarize the statement above by saying that $K_0(R)$ together with the assignment $P \mapsto [P]$ is the universal rank function for finitely generated R-modules.

Remark 5.3 (Grothendieck construction). Let M be an abelian monoid, or equivalently, an abelian semi-group. Then the *Grothendieck construction* of M consists of the following abelian group $\mathrm{Gr}(M)$ together with a map $\varphi \colon M \to \mathrm{Gr}(M)$ of abelian monoids. As a set $\mathrm{Gr}(M)$ is the set of equivalence classes of the equivalence relation $\sim$ on the set $M \times M = \{(m_1, m_2) \mid m_1, m_2 \in M\}$ given by

$$(m_1, m_2) \sim (n_1, n_2) \iff m_1 + n_2 + r = n_1 + m_2 + r \text{ for some } r \in M.$$

Let $[m_1, m_2] \in \mathrm{Gr}(M)$ be the class of $(m_1, m_2) \in M \times M$. Addition on $\mathrm{Gr}(M)$ is given on representatives by

$$[m_1, m_2] + [n_1, n_2] = [m_1 + n_1, m_2 + n_2].$$

The zero element is $[0, 0]$. The inverse of $[m_1, m_2]$ is $[m_2, m_1]$. The map $\varphi \colon M \to \mathrm{Gr}(M)$ sends m to $[m, 0]$.

The Grothendieck construction has the following universal property. For each abelian group A and each map $f \colon M \to A$ of abelian monoids there is precisely one homomorphism of abelian groups $\overline{f} \colon \mathrm{Gr}(M) \to A$ satisfying $\overline{f} \circ \varphi = f$.

Let R-FPM be the abelian monoid of isomorphism classes of finitely generated projective R-modules with the addition given by the direct sum of R-modules. Then the homomorphism $f \colon R\text{-}\mathrm{FPM} \to K_0(R)$ sending $[P]$ to $[P]$ induces an isomorphism of abelian groups

$$\overline{f} \colon \mathrm{Gr}(R\text{-}\mathrm{FPM}) \xrightarrow{\cong} K_0(R).$$

One easily checks that it is an isomorphism of abelian groups.

Example 5.4 (The projective class groups of principal ideal domains). Let R be a principal ideal domain, e.g. $R = \mathbb{Z}$ or R a field. Then each finitely generated R-module M is isomorphic to $M \cong R^n \oplus \bigoplus_{i=1}^{r} R/(r_i)$ for non-trivial elements $r_i \in R$. An R-module $R/(r)$ is never projective for $r \in R, r \neq 0$. Hence any finitely generated projective R-module P is isomorphic to R^n for some n. Let $R_{(0)}$ be the quotient field. Denote by $\dim_{R_{(0)}}(V)$ the dimension of a finite dimensional $R_{(0)}$-vector space V. Now one easily checks that the following maps are well defined isomorphisms, which are inverse to one another

$$i \colon \mathbb{Z} \xrightarrow{\cong} K_0(R), \quad n \mapsto n \cdot [R];$$
$$d \colon K_0(R) \to \mathbb{Z}, \quad [P] \mapsto \dim_{R_{(0)}}\left(R_{(0)} \otimes_R P\right).$$

Definition 5.5. *Let G be a group and let R be a ring. Define the* group ring with coefficients in R *to be the following R-algebra. The underlying R-module is the free R-module with the set G as basis. So elements are given by sums $\sum_{g \in G} r_g \cdot g$*

for elements $r_g \in R$ such that only finitely many coefficients r_g are different from zero. The ring structure is given by

$$\left(\sum_{g \in G} r_g \cdot g\right) \cdot \left(\sum_{h \in G} s_h \cdot h\right) := \sum_{g \in G} \left(\sum_{g_1, g_2, g_1 g_2 = g} r_{g_1} s_{g_2}\right) \cdot g.$$

The unit in RG is given by the element $\sum_{g \in G} r_g \cdot g$, for which $r_g = 1_R$ if g is the unit element in G, and $r_g = 0$ otherwise.

Example 5.6 (Representation ring). The complex representation ring $R(G)$ of a finite group G is isomorphic as an abelian group to $K_0(\mathbb{C}G)$. This follows from the facts that $\mathbb{C}G$ is semisimple and hence each $\mathbb{C}G$-module is projective and a $\mathbb{C}G$-module is finitely generated if and only if its underlying complex vector space is finite-dimensional. The analogous statements are true, if on replaces $\mathbb{C}$ by any field of characteristic prime to the order of G.

In most cases R will be commutative, but this is not necessary for the definition of RG. Notice that a (left) RG-module M is the same as a (left) R-module together with a left G-action such that multiplication with g is an R-linear map $l_g \colon M \to M$.

Let $f \colon R \to S$ be a ring homomorphism. Given an R-module M, its *induction with f* is the S-module $f_*M := S \otimes_R M$. If M is a finitely generated projective R-module, then f_*M is a finitely generated projective S-module. Therefore we obtain a homomorphism $f_* \colon K_0(R) \to K_0(S)$ so that K_0 is a functor from the category of rings to abelian groups. Define the *reduced projective class group* $\widetilde{K}_0(R)$ as the cokernel of the map $j_* \colon K_0(\mathbb{Z}) \to K_0(R)$ for the unique ring homomorphism $j \colon \mathbb{Z} \to R$. This agrees with the quotient of $K_0(R)$ by the subgroup generated by $[R] \in K_0(R)$.

Example 5.7 (Integral group rings of finite groups). Let G be a finite group. Then it is known that $\widetilde{K}_0(\mathbb{Z}G)$ is finite [233, Proposition 9.1]. The explicit structure of $\widetilde{K}_0(\mathbb{Z}[\mathbb{Z}/p])$ for a prime p is only known for a few primes p, one does not know the answer in general [169, pages 29,30]).

Conjecture 5.8 (Vanishing of $\widetilde{K}_0(\mathbb{Z}G)$ for torsionfree G). *Let G be a torsionfree group. Then the reduced projective class group $\widetilde{K}_0(\mathbb{Z}G)$ vanishes.*

5.2 The First Algebraic K-Group

Definition 5.9 (K_1-group). The *first algebraic K-group $K_1(R)$* of a ring R is defined as the abelian group whose generators $[f]$ are conjugacy classes of automorphisms $f \colon P \to P$ of finitely generated projective R-modules P and which satisfies the following relations. For each commutative diagram of finitely generated projective

R-modules with exact rows and automorphisms as vertical arrows

$$
\begin{array}{ccccccccc}
0 & \longrightarrow & P_0 & \xrightarrow{\ i\ } & P_1 & \xrightarrow{\ p\ } & P_2 & \longrightarrow & 0 \\
& & f_0 \downarrow \cong & & f_1 \downarrow \cong & & f_2 \downarrow \cong & & \\
0 & \longrightarrow & P_0 & \xrightarrow{\ i\ } & P_1 & \xrightarrow{\ p\ } & P_2 & \longrightarrow & 0
\end{array}
$$

we get the relation $[f_0] - [f_1] + [f_2] = 0$. For every two automorphisms $f, g \colon P \to P$ of the same finitely generated projective R-module we have the relation $[f \circ g] = [f] + [g]$.

Remark 5.10 ($K_1(R)$ in terms of finitely generated free modules). Let $K_0^f(R)$ and $K_1^f(R)$ respectively be defined analogously to the group $K_0(R)$ of Definition 5.1 and $K_1(R)$ of Definition 5.9 respectively except that one everywhere replaces "finitely generated projective" by "finitely generated free". There are obvious maps $\psi_n(R) \colon K_n^f(R) \to K_n(R)$ for $n = 0, 1$. The map $\psi_0(R)$ is in general not an isomorphism but the map $\psi_1(R)$ is always bijective. The inverse of $\psi_1(R)$ is defined as follows. Let the automorphism $f \colon P \to P$ of the finitely generated projective R-module P represent the element $[f] \in K_0(R)$. Choose a finitely generated projective R-module Q and a finitely generated free R-module F together with an isomorphism $u \colon F \to P \oplus Q$. Then ψ^{-1} sends $[f]$ to $\left[u^{-1} \circ (f \oplus \mathrm{id}_Q) \circ u \right]$.

Next we give a matrix description of $K_1(R)$. Denote by $GL_n(R)$ the group of invertible (n, n)-matrices with entries in R. Define the group $GL(R)$ by the colimit of the system indexed by the natural numbers

$$
GL(1, R) \subseteq GL(2, R) \subseteq \ldots \subseteq GL(n, R) \subseteq GL(n + 1, R) \subseteq \ldots,
$$

where the inclusion $GL(n, R)$ to $GL(n + 1, R)$ is given by stabilization

$$
A \mapsto \begin{pmatrix} A & 0 \\ 0 & 1 \end{pmatrix}.
$$

Let $GL(R)/[GL(R), GL(R)]$ be the abelianization of $GL(R)$.

Denote by $E_n(i, j)$ for $n \geq 1$ and $1 \leq i, j \leq n$ the (n, n)-matrix whose entry at (i, j) is one and is zero elsewhere. Denote by I_n the identity matrix of size n. An elementary (n, n)-matrix is a matrix of the form $I_n + r \cdot E_n(i, j)$ for $n \geq 1$, $1 \leq i, j \leq n$, $i \neq j$ and $r \in R$. Let A be an (n, n)-matrix. The matrix $B = A \cdot (I_n + r \cdot E_n(i, j))$ is obtained from A by adding the i-th column multiplied with r from the right to the j-th column. The matrix $C = (I_n + r \cdot E_n(i, j)) \cdot A$ is obtained from A by adding the j-th row multiplied with r from the left to the i-th row. Let $E(R) \subseteq GL(R)$ be the subgroup generated by all elements in $GL(R)$ which are represented by elementary matrices.

Lemma 5.11. (1) *We have $E(R) = [GL(R), GL(R)]$. In particular $E(R) \subseteq GL(R)$ is a normal subgroup and $GL(R)/[GL(R), GL(R)] = GL(R)/E(R)$.*

(2) *Given an invertible (m, m)-matrix A, an invertible (n, n)-matrix B and an (n, m)-matrix C, we get in $GL(R)/[GL(R), GL(R)]$ the relation*

$$\left[\begin{pmatrix} A & C \\ 0 & B \end{pmatrix} \right] = \left[\begin{pmatrix} A & 0 \\ 0 & B \end{pmatrix} \right] = [A] \cdot [B].$$

Proof. For $n \geq 3$, pairwise distinct numbers $1 \leq i, j, k \leq n$ and $r \in R$ we can write $I_n + r \cdot E_n(i, k)$ as a commutator in $GL(n, R)$, namely

$$\begin{aligned} I_n + r \cdot E_n(i, k) &= (I_n + r \cdot E_n(i, j)) \cdot (I_n + E_n(j, k)) \\ &\quad \cdot (I_n + r \cdot E_n(i, j))^{-1} \cdot (I_n + E_n(j, k))^{-1}. \end{aligned}$$

This implies $E(R) \subseteq [GL(R), GL(R)]$.

Let A and B be two elements in $GL(n, R)$. Let $[A]$ and $[B]$ be the elements in $GL(R)$ represented by A and B. Given two elements x and y in $GL(R)$, we write $x \sim y$ if there are elements e_1 and e_2 in $E(R)$ with $x = e_1 y e_2$, in other words, if the classes of x and y in $E(R)\backslash GL(R)/E(R)$ agree. One easily checks

$$[AB] \sim \left[\begin{pmatrix} AB & 0 \\ 0 & I_n \end{pmatrix} \right] \sim \left[\begin{pmatrix} AB & A \\ 0 & I_n \end{pmatrix} \right]$$
$$\sim \left[\begin{pmatrix} 0 & A \\ -B & I_n \end{pmatrix} \right] \sim \left[\begin{pmatrix} 0 & A \\ -B & 0 \end{pmatrix} \right],$$

since each step is given by multiplication from the right or left with a block matrix of the form $\begin{pmatrix} I_n & 0 \\ C & I_n \end{pmatrix}$ or $\begin{pmatrix} I_n & C \\ 0 & I_n \end{pmatrix}$ and such a block matrix is obviously obtained from I_{2n} by a sequence of column and row operations and hence its class in $GL(R)$ belongs to $E(R)$. Analogously we get

$$[BA] \sim \left[\begin{pmatrix} 0 & B \\ -A & 0 \end{pmatrix} \right].$$

Since the element in $GL(R)$ represented by $\begin{pmatrix} 0 & -I_n \\ I_n & 0 \end{pmatrix}$ belongs to $E(R)$, we conclude

$$\left[\begin{pmatrix} 0 & A \\ -B & 0 \end{pmatrix} \right] \sim \left[\begin{pmatrix} A & 0 \\ 0 & B \end{pmatrix} \right] \sim \left[\begin{pmatrix} 0 & B \\ -A & 0 \end{pmatrix} \right].$$

This shows

$$[AB] \sim \begin{pmatrix} A & 0 \\ 0 & B \end{pmatrix} \sim [BA]. \tag{5.12}$$

This implies for any element $x \in GL(R)$ and $e \in E(R)$ that $xex^{-1} \sim ex^{-1}x = e$ and hence $xex^{-1} \in E(R)$. Therefore $E(R)$ is normal. Given a commutator

$xyx^{-1}y^{-1}$ for $x, y \in GL(R)$, we conclude for appropriate elements e_1, e_2, e_3 in $E(R)$

$$xyx^{-1}y^{-1} = e_1 yx e_2 x^{-1} y^{-1} = e_1 yx x^{-1} y^{-1} (yx) e_2 (yx)^{-1} = e_1 e_3 \in E(R).$$

This finishes the proof of Lemma 5.11. ☐

Define a map $\mu_n \colon GL_n(R) \to K_0^f(R)$ by sending a matrix A to the class of the associated automorphism $R^n \to R^n$. It is a group homomorphism. The maps μ_{n+1} and μ_n are compatible with the stabilization maps. Hence the collection of the maps μ_n defines a group homomorphism $\mu \colon GL(R) \to K_1(R)$. Since $K_1(R)$ is abelian, it induces a homomorphism of abelian groups

$$\mu \colon GL(R)/[GL(R), GL(R)] \quad \to \quad K_1(R). \tag{5.13}$$

Theorem 5.14. *The map* $\mu \colon GL(R)/[GL(R), GL(R)] \xrightarrow{\cong} K_1(R)$ *is bijective.*

Proof. Obviously μ yields a map $\mu \colon GL(R)/[GL(R), GL(R)] \to K_1^f(R)$ and it suffices to construct an inverse $\nu \colon K_1^f(R) \to GL(R)/[GL(R), GL(R)]$ because of Remark 5.10. We want to define ν by sending the class of $[f \colon F \to F]$ to the class of a matrix A given by f after a choice of a basis for F. We have to show that ν is well-defined. Obviously the choice of the basis does not matter since a choice of a different basis has the effect of conjugating A with an invertible matrix which does not affect its class in $GL(R)/[GL(R), GL(R)]$. It remains to show for a block matrix $\begin{pmatrix} A & C \\ 0 & B \end{pmatrix}$ for square-matrices A and B such that A and B are invertible that its class in $GL(R)/[GL(R), GL(R)]$ agrees with the one represented by $[A] \cdot [B]$. This has been proved in Lemma 5.11 (2). ☐

Example 5.15 ($K_1(R)$ and determinants). Let R be a commutative ring. Then the determinant induces a map

$$\det \colon K_1(R) \to R^{\mathrm{inv}}$$

into the multiplicative group of units of R. Regarding a unit as an invertible $(1,1)$-matrix defines a homomorphism

$$j \colon R^{\mathrm{inv}} \to K_1(R)$$

such that $\det \circ j = \mathrm{id}$. If the ring R possesses a Euclidian algorithm, each invertible square matrix A can be transformed by a sequence of elementary row and column operations to the identity matrix. Hence Lemma 5.11 implies that both homomorphisms above are bijective. In particular we get $K_1(\mathbb{Z}) \cong \mathbb{Z}^{\mathrm{inv}} = \{\pm 1\}$ and $K_1(F) \cong F^{\mathrm{inv}}$ for each field F.

However, there exist commutative rings such that $\det$ is not injective. An example is $R = \mathbb{R}[x, y]/(x^2 + y^2 - 1)$ (see [223, page 114]). Let $SL_n(R)$ be $\{A \in GL_n(R) \mid \det(A) = 1\}$. The point is that one can define a non-trivial homomorphism $SL_n(R) \to \pi_1(SL_n(\mathbb{R}))$ as follows. Every element in $\mathbb{R}[x,y]$ is a polynomial and hence defines a map $\mathbb{R}^2 \to \mathbb{R}$. If we restrict it to $S^1 \subseteq \mathbb{R}^2$, we get a function $S^1 \to \mathbb{R}$. Obviously each element in the ideal $(x^2 + y^2 - 1)$ defines the zero-function on S^1. Thus every element in $SL_n(R)$ defines a function $S^1 \subseteq \mathbb{R}^2 \to SL_n(\mathbb{R})$. Finally notice that the map induced by the obvious inclusion $\pi_1(Sl_n(\mathbb{R})) \to \pi_1(Sl_{n+1}(\mathbb{R}))$ is a bijection of cyclic groups of order 2 for $n \geq 3$.

Define the *reduced K_1-group* $\widetilde{K}_1(R)$ as the cokernel of the map $j_*\colon K_1(\mathbb{Z}) \to K_1(R)$ for the unique ring homomorphism $j\colon \mathbb{Z} \to R$. This is the same as the quotient of $K_1(R)$ by the cyclic group of order two generated by the $(1,1)$-matrix (-1).

5.3   The Whitehead Group

Definition 5.16 (The Whitehead group). *Let G be a group. Let $\{\pm g \mid g \in G\}$ be the subgroup of $K_1(\mathbb{Z}G)$ given by the classes of $(1,1)$-matrices of the shape $(\pm g)$ for $g \in G$. Define the* Whitehead group of G *as the quotient $K_1(\mathbb{Z}G)/\{\pm g \mid g \in G\}$.*

For us the following result will be important.

Lemma 5.17. *Let A and B be two invertible matrices over $\mathbb{Z}G$. Then they define the same classes in the Whitehead group $\mathrm{Wh}(G)$ if and only if we can pass from A to B by a sequence of the following operations:*

(1) *B is obtained from A by adding the k-th row multiplied with x from the left to the l-th row for $x \in \mathbb{Z}G$ and $k \neq l$;*

(2) *B is obtained by taking the direct sum of A and the $(1,1)$-matrix $I_1 = (1)$, i.e., B looks like the block matrix $\begin{pmatrix} A & 0 \\ 0 & 1 \end{pmatrix}$;*

(3) *A is the direct sum of B and I_1. This is the inverse operation to (2);*

(4) *B is obtained from A by multiplying the i-th row from the left with a trivial unit , i.e., with an element of the shape $\pm\gamma$ for $\gamma \in G$;*

(5) *B is obtained from A by interchanging two rows or two columns.*

Proof. This follows from Lemma 5.11 and the definition of $\mathrm{Wh}(G)$. $\square$

Example 5.18 (Vanishing of the Whitehead group of the trivial group). We conclude from Example 5.15 that the Whitehead group $\mathrm{Wh}(\{1\})$ of the trivial group is trivial.

Remark 5.19 (The Whitehead group of finite groups). Let G be a finite group. In contrast to $\widetilde{K}_0(\mathbb{Z}G)$ (see Remark 5.7) one has a very good understanding of $\mathrm{Wh}(G)$ (see [184]). For instance one knows that $\mathrm{Wh}(G)$ is finitely generated, its rank as an abelian group is the number of conjugacy classes of unordered pairs $\{g, g^{-1}\}$ in G minus the number of conjugacy classes of cyclic subgroups. The torsion subgroup of $\mathrm{Wh}(G)$ is isomorphic to the kernel $SK_1(G)$ of the change of coefficient homomorphism $K_1(\mathbb{Z}G) \to K_1(\mathbb{Q}G)$. For a finite cyclic group G the Whitehead group $\mathrm{Wh}(G)$ is torsionfree. For instance the Whitehead group $\mathrm{Wh}(\mathbb{Z}/p)$ of a cyclic group of order p for an odd prime p is the free abelian group of rank $(p-3)/2$ and $\mathrm{Wh}(\mathbb{Z}/2) = 0$. The Whitehead group of each symmetric group S_n is trivial.

Conjecture 5.20 (Vanishing of the Whitehead group of a torsionfree G). *Let G be a torsionfree group. Then its Whitehead group $\mathrm{Wh}(G)$ vanishes.*

Remark 5.21 (K-groups under free amalgamated products). The functors sending a group G to $K_0(\mathbb{Z}[G])$ and $\mathrm{Wh}(G)$ does not behave well under direct products of groups but they behave nicely under free amalgamated products. Namely, for two groups G and H the inclusions of G and H into $G * H$ induce isomorphisms (see [226]).

$$\widetilde{K}_0(G) \oplus \widetilde{K}_0(H) \xrightarrow{\cong} \widetilde{K}_0(G * H),$$
$$\mathrm{Wh}(G) \oplus \mathrm{Wh}(H) \xrightarrow{\cong} \mathrm{Wh}(G * H).$$

5.4   The Bass–Heller–Swan Decomposition

There is an important relation between K_0 and K_1 coming from the Bass–Heller–Swan decomposition. It consists of an isomorphism

$$K_1(R[\mathbb{Z}]) \cong K_0(R) \oplus K_1(R) \oplus NK_1(R) \oplus NK_1(R). \tag{5.22}$$

Here the group $NK_1(R)$ is defined as the cokernel of the split injection $K_1(R) \to K_1(R[t])$. It can be identified with the cokernel of the split injection $K_0(R) \to K_0(\mathrm{Nil}(R))$, $[P] \mapsto [0: P \to P]$, where $K_0(\mathrm{Nil}(R))$ denotes the K_0-group of nilpotent endomorphisms of finitely generated projective R-modules. The groups $K_0(\mathrm{Nil}(R))$ are known as *Nil-groups* and often also denoted by $\mathrm{Nil}_0(R)$.

For a regular ring R one obtains isomorphisms

$$K_0(i) \oplus j \colon K_0(R) \oplus K_1(R) \cong K_1(R[\mathbb{Z}]);$$
$$K_0(i) \colon K_0(R) \cong K_0(R[\mathbb{Z}]),$$

where $i \colon R \to R[\mathbb{Z}]$ is the obvious inclusion of rings and j sends the class of a finitely generated projective R-module P to the class of the $R[\mathbb{Z}]$-automorphism $l_z \otimes_R \mathrm{id}_P \colon R[\mathbb{Z}] \otimes_R P \xrightarrow{\cong} R[\mathbb{Z}] \otimes_R P$ for $l_z \colon R[\mathbb{Z}] \to R[\mathbb{Z}]$ given by left multiplication with the generator $z \in \mathbb{Z}$. If R is regular, then $R[t]$ and $R[t, t^{-1}] = R[\mathbb{Z}]$ are regular. Hence for a regular ring R we get

$$K_1(R[\mathbb{Z}^n]) \cong K_1(R) \oplus \bigoplus_{i=1}^{n} K_0(R);$$
$$K_0(R[\mathbb{Z}^n]) \cong K_0(R).$$

We conclude in the special case $R = \mathbb{Z}$ that $\widetilde{K}_0(\mathbb{Z}[\mathbb{Z}^n]) = \mathrm{Wh}(\mathbb{Z}^n) = 0$ holds for $n \geq 0$.

More information about K_0 and K_1 can be found for instance in [169, pages 29,30], [208], [223].

Chapter 6

Whitehead Torsion

In this chapter we will assign to a homotopy equivalence $f\colon X \to Y$ of finite CW-complexes its Whitehead torsion $\tau(f)$ in the Whitehead group $\mathrm{Wh}(\pi(Y))$ associated to Y and discuss its main properties.

6.1 Whitehead Torsion of a Chain Map

In this section we give the definition and prove the basic properties of the Whitehead torsion for chain maps.

We begin with some input about chain complexes. We will always assume for a chain complex C_* that $C_p = 0$ for $p < 0$ holds. Let $f_*\colon C_* \to D_*$ be a chain map of R-chain complexes for some ring R. Define the *mapping cylinder* $\mathrm{cyl}_*(f_*)$ to be the chain complex with p-th differential

$$C_{p-1} \oplus C_p \oplus D_p \xrightarrow{\begin{pmatrix} -c_{p-1} & 0 & 0 \\ -\,\mathrm{id} & c_p & 0 \\ f_{p-1} & 0 & d_p \end{pmatrix}} C_{p-2} \oplus C_{p-1} \oplus D_{p-1}.$$

Define the *mapping cone* $\mathrm{cone}_*(f_*)$ to be the quotient of $\mathrm{cyl}_*(f_*)$ by the obvious copy of C_*. Hence the p-th differential of $\mathrm{cone}_*(f_*)$ is

$$C_{p-1} \oplus D_p \xrightarrow{\begin{pmatrix} -c_{p-1} & 0 \\ f_{p-1} & d_p \end{pmatrix}} C_{p-2} \oplus D_{p-1}.$$

Given a chain complex C_*, define its *suspension* ΣC_* to be the quotient of $\mathrm{cone}_*(\mathrm{id}_{C_*})$ by the obvious copy of C_*, i.e., the chain complex with p-th differential

$$C_{p-1} \xrightarrow{\ -c_{p-1}\ } C_{p-2}.$$

Remark 6.1 (Geometric and algebraic mapping cylinders). These algebraic notions of mapping cylinder, mapping cone and suspension are modelled on their geometric counterparts. Namely, the cellular chain complex of a mapping cylinder of a cellular map of CW-complexes $f\colon X \to Y$ is the mapping cylinder of the chain map induced by f. There are obvious exact sequences such as $0 \to C_* \to \mathrm{cyl}(f_*) \to \mathrm{cone}(f_*) \to 0$ and $0 \to D_* \to \mathrm{cone}_*(f_*) \to \Sigma C_* \to 0$. They correspond to the obvious geometric cofibrations for maps of spaces $f\colon X \to Y$ given by $X \to \mathrm{cyl}(f) \to \mathrm{cone}(f)$ and $Y \to \mathrm{cone}(f) \to \Sigma X$. The associated long exact homology sequences of the exact sequences above correspond to the Puppe sequences associated to the cofibration sequences above.

We call an R-chain complex C_* *finite* if there is a number N with $C_p = 0$ for $p > N$ and each R-chain module C_p is a finitely generated R-module. We call an R-chain complex C_* *projective* resp. *free* resp. *based free* if each R-chain module C_p is projective resp. free resp. free with a preferred basis.

A *chain contraction* γ_* for an R-chain complex C_* is a collection of R-homomorphisms $\gamma_p\colon C_p \to C_{p+1}$ for $p \in \mathbb{Z}$ such that $c_{p+1} \circ \gamma_p + \gamma_{p-1} \circ c_p = \mathrm{id}_{C_p}$ holds for all $p \in \mathbb{Z}$. We call an R-chain complex *contractible*, if it possesses a chain contraction.

The proof of the next result is left to the reader.

Lemma 6.2. (1) *A projective chain complex is contractible if and only if $H_p(C_*) = 0$ for $p \geq 0$;*

(2) *Let $f\colon C_* \to D_*$ be a chain map of projective chain complexes. Then the following assertions are equivalent:*

  (a) *The chain map f_* is a chain homotopy equivalence;*

  (b) *$\mathrm{cone}_*(f_*)$ is contractible;*

  (c) *$H_p(f_*)\colon H_p(C_*) \to H_p(D_*)$ is bijective for all $p \geq 0$.*

Suppose that C_* is a finite based free R-chain complex which is *contractible*, i.e., which possesses a chain contraction. Put $C_{\mathrm{odd}} = \bigoplus_{p \in \mathbb{Z}} C_{2p+1}$ and $C_{\mathrm{ev}} = \bigoplus_{p \in \mathbb{Z}} C_{2p}$. Let γ_* and δ_* be two chain contractions. Define R-homomorphisms

$$(c_* + \gamma_*)_{\mathrm{odd}} : C_{\mathrm{odd}} \quad \to \quad C_{\mathrm{ev}};$$
$$(c_* + \delta_*)_{\mathrm{ev}} : C_{\mathrm{ev}} \quad \to \quad C_{\mathrm{odd}}.$$

Let A be the matrix of $(c_* + \gamma_*)_{\mathrm{odd}}$ with respect to the given bases. Let B be the matrix of $(c_* + \delta_*)_{\mathrm{ev}}$ with respect to the given bases. Put $\mu_n := (\gamma_{n+1} - \delta_{n+1}) \circ \delta_n$ and $\nu_n := (\delta_{n+1} - \gamma_{n+1}) \circ \gamma_n$. One easily checks by a direct computation:

Lemma 6.3. *Under the assumptions above the R-maps $(\mathrm{id} + \mu_*)_{\mathrm{odd}}$, $(\mathrm{id} + \nu_*)_{\mathrm{ev}}$ and both compositions $(c_* + \gamma_*)_{\mathrm{odd}} \circ (\mathrm{id} + \mu_*)_{\mathrm{odd}} \circ (c_* + \delta_*)_{\mathrm{ev}}$ and $(c_* + \delta_*)_{\mathrm{ev}} \circ (\mathrm{id} + \nu_*)_{\mathrm{ev}} \circ (c_* + \gamma_*)_{\mathrm{odd}}$ are given by upper triangular matrices, whose diagonal entries are identity maps.*

Hence A and B are invertible and their classes $[A], [B] \in \widetilde{K}_1(R)$ satisfy $[A] = -[B]$. Since $[B]$ is independent of the choice of γ_*, the same is true for $[A]$. Thus we can associate to a finite based free contractible R-chain complex C_* an element

$$\tau(C_*) \;=\; [A] \quad \in \widetilde{K}_1(R), \tag{6.4}$$

which depends only on C_* with the preferred R-basis but not on the choice of the chain contraction.

Example 6.5 (One-dimensional chain complexes). Suppose that C_* is 1-dimensional and the ring R is a field F. Then there is only one non-trivial differential $c_1 \colon C_1 \to C_0$. Let A be the matrix associated to it with respect to the given bases. The determinant induces an isomorphism (see Example 5.15)

$$d \colon \widetilde{K}_1(F) \to F^{\mathrm{inv}}/\{\pm 1\}$$

It sends $\tau(C_*)$ to the element given by $\det(A)$.

Let $f_* \colon C_* \to D_*$ be a homotopy equivalence of finite based free R-chain complexes. Its mapping cone $\mathrm{cone}(f_*)$ is a contractible finite based free R-chain complex. Define the *Whitehead torsion* of f_* by

$$\tau(f_*) \;:=\; \tau(\mathrm{cone}_*(f_*)) \quad \in \widetilde{K}_1(R). \tag{6.6}$$

We call a sequence of finite based free R-chain complexes $0 \to C_* \xrightarrow{i_*} D_* \xrightarrow{q_*} E_* \to 0$ *based exact* if for any $p \in \mathbb{Z}$ the basis B for D_p can be written as a disjoint union $B' \coprod B''$ such that the image of the basis of C_p under i_p is B' and the image of B'' under q_p is the basis for E_p.

Lemma 6.7. (1) *Consider a commutative diagram of finite based free R-chain complexes whose rows are based exact.*

$$
\begin{array}{ccccccccc}
0 & \longrightarrow & C'_* & \longrightarrow & D'_* & \longrightarrow & E'_* & \longrightarrow & 0 \\
 & & {\scriptstyle f_*}\downarrow & & {\scriptstyle g_*}\downarrow & & {\scriptstyle h_*}\downarrow & & \\
0 & \longrightarrow & C_* & \longrightarrow & D_* & \longrightarrow & E_* & \longrightarrow & 0
\end{array}
$$

Suppose that two of the chain maps f_*, g_* and h_* are R-chain homotopy equivalences. Then all three are R-chain homotopy equivalences and

$$\tau(f_*) - \tau(g_*) + \tau(h_*) \;=\; 0.$$

(2) *Let $f_* \simeq g_* \colon C_* \to D_*$ be homotopic R-chain homotopy equivalences of finite based free R-chain complexes. Then*

$$\tau(f_*) \;=\; \tau(g_*).$$

(3) Let $f_* \colon C_* \to D_*$ and $g_* \colon D_* \to E_*$ be R-chain homotopy equivalences of based free R-chain complexes. Then

$$\tau(g_* \circ f_*) = \tau(g_*) + \tau(f_*).$$

Proof. (1) A chain map of projective chain complexes is a homotopy equivalence if and only if it induces an isomorphism on homology (see Lemma 6.2 (1)). The 5-lemma and the long homology sequence of a short exact sequence of chain complexes imply that all three chain maps f_*, h_* and g_* are chain homotopy equivalences if two of them are.

To prove the sum formula, it suffices to show for a based free exact sequence $0 \to C_* \xrightarrow{i_*} D_* \xrightarrow{q_*} E_* \to 0$ of contractible finite based free R-chain complexes that

$$\tau(C_*) - \tau(D_*) + \tau(E_*) \quad = \quad 0. \tag{6.8}$$

Let $u_* \colon F_* \to G_*$ be an isomorphism of contractible finite based free R-chain complexes. Since the choice of a chain contraction does not affect the values of the Whitehead torsion, we can compute $\tau(F_*)$ and $\tau(G_*)$ with respect to chain contractions which are compatible with u_*. Then one easily checks in $\widetilde{K}_1(R)$

$$\tau(G_*) - \tau(F_*) \quad = \quad \sum_{p \in \mathbb{Z}} (-1)^p \cdot [u_p], \tag{6.9}$$

where $[u_p]$ is the element represented by the matrix of u_p with respect to the given bases.

Let ε_* be a chain contraction for E_*. Choose for each $p \in \mathbb{Z}$ an R-homomorphism $\sigma_p \colon E_p \to D_p$ satisfying $p_q \circ \sigma_q = \mathrm{id}$. Define $s_p \colon E_p \to D_p$ by $d_{p+1} \circ \sigma_{p+1} \circ \varepsilon_p + \sigma_p \circ \varepsilon_{p-1} \circ e_p$. One easily checks that the collection of the s_p-s defines a chain map $s_* \colon E_* \to D_*$ with $q_* \circ s_* = \mathrm{id}$. Thus we obtain an isomorphism of contractible based free R-chain complexes

$$i_* \oplus q_* \colon C_* \oplus E_* \to D_*.$$

Since the matrix of $i_p \oplus s_p$ with respect to the given basis is a block matrix of the shape $\begin{pmatrix} I_m & * \\ 0 & I_n \end{pmatrix}$ we get $[i_p \oplus s_p] = 0$ in $\widetilde{K}_1(R)$. Now (6.9) implies $\tau(C_* \oplus D_*) = \tau(E_*)$. Since obviously $\tau(C_* \oplus D_*) = \tau(C_*) + \tau(D_*)$, (6.8) and thus assertion (1) follows.

(2) If $h_* \colon f_* \simeq g_*$ is a chain homotopy, we obtain an isomorphism of based free R-chain complexes

$$\begin{pmatrix} \mathrm{id} & 0 \\ h_{*-1} & \mathrm{id} \end{pmatrix} \colon \mathrm{cone}_*(f_*) = C_{*-1} \oplus D_* \to \mathrm{cone}_*(g_*) = C_{*-1} \oplus D_*.$$

We conclude from (6.9)

$$\tau(g_*) - \tau(f_*) \;=\; \sum_{p \in \mathbb{Z}} (-1)^p \cdot \left[\begin{pmatrix} \mathrm{id} & 0 \\ h_{*-1} & \mathrm{id} \end{pmatrix} \right] \;=\; 0.$$

(3) Define a chain map $h_* \colon \Sigma^{-1} \mathrm{cone}_*(g_*) \to \mathrm{cone}_*(f_*)$ by

$$\begin{pmatrix} 0 & 0 \\ -\mathrm{id} & 0 \end{pmatrix} \;\colon\; D_p \oplus E_{p+1} \to C_{p-1} \oplus D_p.$$

There is an obvious based exact sequence of contractible finite based free R-chain complexes $0 \to \mathrm{cone}_*(f_*) \to \mathrm{cone}(h_*) \to \mathrm{cone}(g_*) \to 0$. There is also a based exact sequence of contractible finite based free R-chain complexes $0 \to \mathrm{cone}_*(g_* \circ f_*) \xrightarrow{i_*} \mathrm{cone}_*(h_*) \to \mathrm{cone}_*(\mathrm{id} \colon D_* \to D_*) \to 0$, where i_p is given by

$$\begin{pmatrix} f_{p-1} & 0 \\ 0 & \mathrm{id} \\ \mathrm{id} & 0 \\ 0 & 0 \end{pmatrix} \;\colon\; C_{p-1} \oplus E_p \to D_{p-1} \oplus E_p \oplus C_{p-1} \oplus D_p.$$

We conclude from assertion (1)

$$\begin{aligned} \tau(h_*) &= \tau(f_*) + \tau(g_*); \\ \tau(h_*) &= \tau(g_* \circ f_*) + \tau(\mathrm{id}_* \colon D_* \to D_*); \\ \tau(\mathrm{id}_* \colon D_* \to D_*) &= 0. \end{aligned}$$

This finishes the proof of Lemma 6.7. $\qquad\qquad\qquad\qquad\qquad\qquad\square$

6.2 The Cellular Chain Complex of the Universal Covering

In order to apply the constructions above for chain complexes to geometry, namely, to cellular maps of CW-complexes, one needs to study the cellular chain complex. Recall that a CW-complex X is a Hausdorff space together with a filtration by skeleta

$$\emptyset = X_{-1} \subseteq X_0 \subseteq X_1 \subseteq \ldots \subseteq X_n \subseteq \ldots \subseteq \bigcup_{n \geq 0} X_n = X$$

such that X carries the colimit topology with respect to this filtration (i.e., a set $C \subseteq X$ is closed if and only if $C \cap X_n$ is closed in X_n for all $n \geq 0$) and X_n is obtained from X_{n-1} for each $n \geq 0$ by attaching n-dimensional cells, i.e., there exists a pushout

$$\begin{array}{ccc} \coprod_{i \in I_n} S^{n-1} & \xrightarrow{\coprod_{i \in I_n} q_i} & X_{n-1} \\ \downarrow & & \downarrow \\ \coprod_{i \in I_n} D^n & \xrightarrow[\coprod_{i \in I_n} Q_i]{} & X_n \end{array}$$

Notice that the pushouts are not part of the structure, only their existence is required. Recall that the cellular chain complex $C_*(X)$ of a CW-complex X is the chain complex

$$\ldots \xrightarrow{c_{n+1}} C_n(X) = H_n(X_n, X_{n-1}) \xrightarrow{c_n} C_{n-1}(X) = H_{n-1}(X_{n-1}, X_{n-2}) \xrightarrow{c_{n-1}} \ldots$$

where H_* denotes singular homology and c_n is the boundary operator in the long exact sequence of the triple (X_n, X_{n-1}, X_{n-2}). After a choice of pushouts above one obtains a composition of isomorphisms

$$\bigoplus_{i \in I_n} H_0(\{\bullet\}) \xrightarrow{\oplus_{i \in I_n} H_n(j_i)} \bigoplus_{i \in I_n} H_0(S^0; \{\bullet\}) \xrightarrow{\oplus_{i \in I_n} \sigma_i} \bigoplus_{i \in I_n} H_n(S^n, \{\bullet\})$$

$$\xleftarrow{\oplus_{i \in I_n} H_n(\mathrm{pr}_i)} \bigoplus_{i \in I_n} H_n(D^n, S^{n-1}) \xrightarrow{\oplus_{i \in I_n} H_n(Q_i, q_i)} H_n(X_n, X_{n-1}) = C_n(X),$$

where j_i denotes the inclusion, σ_i the suspension isomorphism and pr_i the projection. Using the obvious generator in $H_0(\{\bullet\})$, we obtain a $\mathbb{Z}$-basis for the $\mathbb{Z}$-module $C_n(X)$. It depends on the choice of the pushout. We call two such $\mathbb{Z}$-bases equivalent, if they can be obtained from one another by permuting the elements and possibly multiplying some of the elements with -1. One can show that the equivalence class of this basis is independent of the choice of the pushout and hence depends only on the CW-structure on X. Thus $C_*(X)$ inherits a preferred equivalence class of $\mathbb{Z}$-bases, which we will call the *cellular equivalence class of $\mathbb{Z}$-bases*.

Now consider a connected CW-complex X together with the choice of a universal covering $p_X \colon \widetilde{X} \to X$ and a base point $\widetilde{x} \in \widetilde{X}$. Then $\pi = \pi_1(X, p_X(\widetilde{x}))$ acts freely by cellular maps on $\widetilde{X}$. The CW-structure on X given by the filtration by skeleta $\emptyset = X_{-1} \subseteq X_0 \subseteq X_1 \subseteq X_2 \subseteq \ldots$ induces a CW-structure on $\widetilde{X}$ by the filtration $\emptyset = \widetilde{X}_{-1} \subseteq \widetilde{X}_0 \subseteq \widetilde{X}_1 \subseteq \widetilde{X}_2 \subseteq \ldots$ if we put $\widetilde{X}_n = p_X^{-1}(X_n)$. Hence the cellular $\mathbb{Z}$-chain complex $C_*(\widetilde{X})$ inherits the structure of a $\mathbb{Z}\pi$-chain complex. One can choose a π-pushout

$$
\begin{array}{ccc}
\coprod_{i \in I_n} \pi \times S^{n-1} & \xrightarrow{\coprod_{i \in I_n} \widetilde{q}_i} & \widetilde{X}_{n-1} \\
\downarrow & & \downarrow \\
\coprod_{i \in I_n} \pi \times D^n & \xrightarrow{\coprod_{i \in I_n} \widetilde{Q}_i} & \widetilde{X}_n
\end{array}
$$

which induces an isomorphism of $\mathbb{Z}\pi$-modules

$$\bigoplus_{i \in I_n} H_0(\pi) \xrightarrow{\oplus_{i \in I_n} H_n(j_i)} \bigoplus_{i \in I_n} H_0(\pi \times (S^0; \{\bullet\})) \xrightarrow{\oplus_{i \in I_n} \sigma_i} \bigoplus_{i \in I_n} H_n(\pi \times (S^n, \{\bullet\}))$$

$$\xleftarrow{\oplus_{i \in I_n} H_n(\mathrm{pr}_i)} \bigoplus_{i \in I_n} H_n(\pi \times (D^n, S^{n-1})) \xrightarrow{\oplus_{i \in I_n} H_n(\widetilde{Q}_i, \widetilde{q}_i)} H_n(X_n, X_{n-1}) = C_n(X).$$

Using the obvious generator of the $\mathbb{Z}\pi$-module $H_0(\pi) \cong \mathbb{Z}\pi$, we obtain a $\mathbb{Z}\pi$-basis. We call two such $\mathbb{Z}\pi$-bases equivalent, if they can be obtained from one another by permuting the elements and possibly multiplying some of the elements with a unit in $\mathbb{Z}\pi$ of the shape $\pm w$ for some $w \in \pi$. One can show that the equivalence class of this $\mathbb{Z}\pi$-basis is independent of the choice of the π-pushout and hence depends only on the CW-structure on X. Thus $C_*(\widetilde{X})$ inherits a preferred equivalence class of $\mathbb{Z}\pi$-bases, which we will call the *cellular equivalence class of $\mathbb{Z}\pi$-bases*.

Example 6.10 (Cellular chain complex of $\widetilde{S^1}$). Consider S^1 with the CW-complex structure for which the zero-skeleton consists of one point and the one-skeleton is S^1. Then we can identify $\pi = \mathbb{Z}$ and $\widetilde{S^1} = \mathbb{R}$ with the $\mathbb{Z}$-action given by translation. Let $z \in \mathbb{Z}$ be a fixed generator. Then the cellular $\mathbb{Z}[\mathbb{Z}]$-chain complex $C_*(\widetilde{S^1})$ can be identified with the 1-dimensional $\mathbb{Z}[\mathbb{Z}]$-chain complex $\mathbb{Z}[\mathbb{Z}] \xrightarrow{z-1} \mathbb{Z}[\mathbb{Z}]$.

6.3   The Whitehead Torsion of a Cellular Map

Now we can pass to CW-complexes. Let $f: X \to Y$ be a homotopy equivalence of connected finite CW-complexes. Let $p_X: \widetilde{X} \to X$ and $p_Y: \widetilde{Y} \to Y$ be the universal coverings. Fix base points $\widetilde{x} \in \widetilde{X}$ and $\widetilde{y} \in \widetilde{Y}$ such that f maps $x = p_X(\widetilde{x})$ to $y = p_Y(\widetilde{y})$. Let $\widetilde{f}: \widetilde{X} \to \widetilde{Y}$ be the unique lift of f satisfying $\widetilde{f}(\widetilde{x}) = \widetilde{y}$. We abbreviate $\pi = \pi_1(Y, y)$ and identify $\pi_1(X, x)$ in the sequel with π by $\pi_1(f, x)$. After the choice of the base points $\widetilde{x}$ and $\widetilde{y}$ we get unique operations of π on $\widetilde{X}$ and $\widetilde{Y}$. The lift $\widetilde{f}$ is π-equivariant. It induces a $\mathbb{Z}\pi$-chain homotopy equivalence $C_*(\widetilde{f}): C_*(\widetilde{X}) \to C_*(\widetilde{Y})$. Equip $C_*(\widetilde{X})$ and $C_*(\widetilde{Y})$ with a $\mathbb{Z}[\pi]$-basis representing the cellular equivalence class of $\mathbb{Z}[\pi]$-bases. We can apply (6.6) to it and thus obtain an element

$$\tau(f) \quad \in \quad \mathrm{Wh}(\pi_1(Y, y)). \tag{6.11}$$

We have defined the Whitehead group as a quotient of $K_1(\mathbb{Z}[\pi])$ by the subgroup generated by the units of the shape $\pm w$ for $w \in \pi$. This ensures that the choice of $\mathbb{Z}[\pi]$-basis within the cellular equivalence class of $\mathbb{Z}[\pi]$-bases does not matter.

So far this definition depends on the various choices of base points. We can get rid of these choices as follows. If y' is a second base point, we can choose a path w from y to y' in Y. Conjugation with w yields a homomorphism $c_w: \pi_1(Y, y) \to \pi_1(Y, y')$ which induces $(c_w)_*: \mathrm{Wh}(\pi_1(Y, y)) \to \mathrm{Wh}(\pi_1(Y, y'))$. If v is a different path from y to y', then c_w and c_v differ by an inner automorphism of $\pi_1(Y, y)$. Since an inner automorphism of $\pi_1(Y, y)$ induces the identity on $\mathrm{Wh}(\pi_1(Y, y))$, we conclude that $(c_w)_*$ and $(c_v)_*$ agree. Hence we get a unique isomorphism $t(y, y'): \mathrm{Wh}(\pi_1(Y, y)) \to \mathrm{Wh}(\pi_1(Y, y'))$ depending only on y and y'. Moreover $t(y, y) = \mathrm{id}$ and $t(y, y'') = t(y', y'') \circ t(y, y')$. Therefore we can define $\mathrm{Wh}(\pi(Y))$ independently of a choice of a base point by $\coprod_{y \in Y} \mathrm{Wh}(\pi_1(Y, y)) / \sim$, where $\sim$ is the obvious equivalence relation generated by $a \sim b \Leftrightarrow t(y, y')(a) = b$

for $a \in \text{Wh}(\pi_1(Y,y))$ and $b \in \text{Wh}(\pi_1(Y,y'))$. Define $\tau(f) \in \text{Wh}(\pi(Y))$ by the element represented by the element introduced in (6.11). Notice that $\text{Wh}(\pi(Y))$ is isomorphic to $\text{Wh}(\pi_1(Y,y))$ for any base point $y \in Y$. It is not hard to check using Lemma 6.7 that $\tau(f)$ depends only on $f \colon X \to Y$ and not on the choice of the universal coverings and base points. Finally we want to drop the assumption that Y is connected. Notice that f induces a bijection $\pi_0(f) \colon \pi_0(X) \to \pi_0(Y)$.

Definition 6.12. *Let $f \colon X \to Y$ be a (cellular) map of finite CW-complexes which is a homotopy equivalence. Define the Whitehead group $\text{Wh}(\pi(Y))$ of Y and the Whitehead torsion $\tau(f) \in \text{Wh}(\pi(Y))$ by*

$$\text{Wh}(\pi(Y)) = \bigoplus\nolimits_{C \in \pi_0(Y)} \text{Wh}(\pi_1(C));$$
$$\tau(f) = \bigoplus\nolimits_{C \in \pi_0(Y)} \tau\left(f|_{\pi_0(f)^{-1}(C)} \colon \pi_0(f)^{-1}(C) \to C\right).$$

In the notation $\text{Wh}(\pi(Y))$ one should think of $\pi(Y)$ as the fundamental groupoid of Y. A map $f \colon X \to Y$ induces a homomorphism $f_* \colon \text{Wh}(\pi(X)) \to \text{Wh}(\pi(Y))$ such that $\text{id}_* = \text{id}$, $(g \circ f)_* = g_* \circ f_*$ and $f \simeq g \Rightarrow f_* = g_*$. We will later see that two cellular homotopy equivalences of finite CW-complexes which are homotopic as cellular maps have the same Whitehead torsion. Hence in the sequel we can and will drop the assumption cellular by the Cellular Approximation Theorem.

Suppose that the following diagram is a pushout

$$\begin{array}{ccc} A & \xrightarrow{\ f\ } & B \\ {\scriptstyle i}\downarrow & & \downarrow{\scriptstyle j} \\ X & \xrightarrow{\ g\ } & Y \end{array}$$

the map i is an inclusion of CW-complexes and f is a *cellular* map of CW-complexes, i.e., respects the filtration given by the CW-structures. Then Y inherits a CW-structure by defining Y_n as the union of $j(B_n)$ and $g(X_n)$. If we equip Y with this CW-structure, we call the pushout above a *cellular pushout*.

Theorem 6.13. (1) *Sum formula*

Let the following two diagrams be cellular pushouts of finite CW-complexes

$$\begin{array}{ccc} X_0 & \xrightarrow{\ i_1\ } & X_1 \\ {\scriptstyle i_2}\downarrow & & \downarrow{\scriptstyle j_1} \\ X_2 & \xrightarrow{\ j_2\ } & X \end{array} \qquad\qquad \begin{array}{ccc} Y_0 & \xrightarrow{\ k_1\ } & Y_1 \\ {\scriptstyle k_2}\downarrow & & \downarrow{\scriptstyle l_1} \\ Y_2 & \xrightarrow{\ l_2\ } & Y \end{array}$$

Put $l_0 = l_1 \circ k_1 = l_2 \circ k_2 \colon Y_0 \to Y$. Let $f_i \colon X_i \to Y_i$ be homotopy equivalences for $i = 0,1,2$ satisfying $f_1 \circ i_1 = k_1 \circ f_0$ and $f_2 \circ i_2 = k_2 \circ f_0$. Denote by

$f\colon X \to Y$ the map induced by f_0, f_1 and f_2 and the pushout property. Then f is a homotopy equivalence and

$$\tau(f) \;=\; (l_1)_*\tau(f_1) + (l_2)_*\tau(f_2) - (l_0)_*\tau(f_0).$$

(2) *Homotopy invariance*

Let $f \simeq g\colon X \to Y$ be homotopic maps of finite CW-complexes. Then the homomorphisms $f_*, g_*\colon \mathrm{Wh}(\pi(X)) \to \mathrm{Wh}(\pi(Y))$ agree. If additionally f and g are homotopy equivalences, then

$$\tau(g) \;=\; \tau(f).$$

(3) *Composition formula*

Let $f\colon X \to Y$ and $g\colon Y \to Z$ be homotopy equivalences of finite CW-complexes. Then

$$\tau(g \circ f) \;=\; g_*\tau(f) + \tau(g).$$

(4) *Product formula*

Let $f\colon X' \to X$ and $g\colon Y' \to Y$ be homotopy equivalences of connected finite CW-complexes. Then

$$\tau(f \times g) \;=\; \chi(X) \cdot j_*\tau(g) + \chi(Y) \cdot i_*\tau(f),$$

where $\chi(X), \chi(Y) \in \mathbb{Z}$ denote the Euler characteristics, $j_*\colon \mathrm{Wh}(\pi(Y)) \to \mathrm{Wh}(\pi(X \times Y))$ is the homomorphism induced by $j\colon Y \to X \times Y, y \mapsto (y, x_0)$ for some base point $x_0 \in X$ and i_* is defined analogously.

(5) *Topological invariance*

Let $f\colon X \to Y$ be a homeomorphism of finite CW-complexes. Then

$$\tau(f) \;=\; 0.$$

Proof. (1), (2) and (3) follow from Lemma 6.7.

(4) Because of assertion (3) we have

$$\tau(f \times g) \;=\; \tau(f \times \mathrm{id}_Y) + (f \times \mathrm{id}_Y)_*\tau(\mathrm{id}_X \times g).$$

Hence it suffices to treat the case $g = \mathrm{id}_Y$. Now one proceeds by induction over the cells of Y using assertions (1), (2) and (3).

(5) This (in comparison with the other assertions much deeper result) is due to Chapman [50], [51]. This finishes the proof of Theorem 6.13. $\qquad\square$

6.4   Simple Homotopy Equivalences

In this section we introduce the concept of a simple homotopy equivalence $f\colon X \to Y$ of finite CW-complexes geometrically.

We have the inclusion of spaces $S^{n-2} \subseteq S_+^{n-1} \subseteq S^{n-1} \subseteq D^n$, where $S_+^{n-1} \subseteq S^{n-1}$ is the upper hemisphere. The pair (D^n, S_+^{n-1}) carries an obvious relative CW-structure. Namely, attach an $(n-1)$-cell to S_+^{n-1} by the attaching map $\mathrm{id}\colon S^{n-2} \to S^{n-2}$ to obtain S^{n-1}. Then we attach to S^{n-1} an n-cell by the attaching map $\mathrm{id}\colon S^{n-1} \to S^{n-1}$ to obtain D^n. Let X be a CW-complex. Let $q\colon S_+^{n-1} \to X$ be a map satisfying $q(S^{n-2}) \subseteq X_{n-2}$ and $q(S_+^{n-1}) \subseteq X_{n-1}$. Let Y be the space $D^n \cup_q X$, i.e., the pushout

$$
\begin{array}{ccc}
S_+^{n-1} & \xrightarrow{\ q\ } & X \\
{\scriptstyle i}\downarrow & & \downarrow{\scriptstyle j} \\
D^n & \xrightarrow[\ g\ ]{} & Y
\end{array}
$$

where i is the inclusion. Then Y inherits a CW-structure by putting $Y_k = j(X_k)$ for $k \leq n-2$, $Y_{n-1} = j(X_{n-1}) \cup g(S^{n-1})$ and $Y_k = j(X_k) \cup g(D^n)$ for $k \geq n$. Notice that Y is obtained from X by attaching one $(n-1)$-cell and one n-cell. Since the map $i\colon S_+^{n-1} \to D^n$ is a homotopy equivalence and cofibration, the map $j\colon X \to Y$ is a homotopy equivalence and cofibration. We call j an *elementary expansion* and say that Y is obtained from X by an elementary expansion. There is a map $r\colon Y \to X$ with $r \circ j = \mathrm{id}_X$. This map is unique up to homotopy relative $j(X)$. We call any such map an *elementary collapse* and say that X is obtained from Y by an elementary collapse.

Definition 6.14. *Let $f\colon X \to Y$ be a map of finite CW-complexes. We call it a simple homotopy equivalence if there is a sequence of maps*

$$
X = X[0] \xrightarrow{f_0} X[1] \xrightarrow{f_1} X[2] \ldots \xrightarrow{f_{n-1}} X[n] = Y
$$

such that each f_i is an elementary expansion or elementary collapse and f is homotopic to the composition of the maps f_i.

Theorem 6.15.   (1) *Let $f\colon X \to Y$ be a homotopy equivalence of finite CW-complexes. Then f is a simple homotopy equivalence if and only if its Whitehead torsion $\tau(f) \in \mathrm{Wh}(\pi(Y))$ vanishes.*

(2) *Let X be a finite CW-complex. Then for any element $x \in \mathrm{Wh}(\pi(X))$ there is an inclusion $i\colon X \to Y$ of finite CW-complexes such that i is a homotopy equivalence and $i_*^{-1}(\tau(i)) = x$.*

Proof. We only show that for a simple homotopy equivalence $f\colon X \to Y$ of finite CW-complexes we have $\tau(f) = 0$. Because of Theorem 6.13 it suffices to prove for an elementary expansion $j\colon X \to Y$ that its Whitehead torsion $\tau(j) \in \mathrm{Wh}(\pi(Y))$

vanishes. We can assume without loss of generality that Y is connected. In the sequel we write $\pi = \pi_1(Y)$ and identify $\pi = \pi_1(X)$ by $\pi_1(f)$. The diagram of based free finite $\mathbb{Z}\pi$-chain complexes

$$
\begin{array}{ccccccccc}
0 & \longrightarrow & C_*(\widetilde{X}) & \xrightarrow{C_*(\widetilde{j})} & C_*(\widetilde{Y}) & \xrightarrow{\mathrm{pr}_*} & C_*(\widetilde{Y},\widetilde{X}) & \longrightarrow & 0 \\
& & \mathrm{id}_* \uparrow & & C_*(\widetilde{j}) \uparrow & & 0_* \uparrow & & \\
0 & \longrightarrow & C_*(\widetilde{X}) & \xrightarrow{\mathrm{id}_*} & C_*(\widetilde{X}) & \xrightarrow{\mathrm{pr}_*} & 0 & \longrightarrow & 0
\end{array}
$$

has based exact rows and $\mathbb{Z}\pi$-chain homotopy equivalences as vertical arrows. We conclude from Lemma 6.7 (1)

$$
\tau\left(C_*(\widetilde{j})\right) = \tau\left(\mathrm{id}_* \colon C_*(\widetilde{X}) \to C_*(\widetilde{X})\right) + \tau\left(0_* \colon 0 \to C_*(\widetilde{Y},\widetilde{X})\right)
$$
$$
= \tau\left(C_*(\widetilde{Y},\widetilde{X})\right).
$$

The $\mathbb{Z}\pi$-chain complex $C_*(\widetilde{Y},\widetilde{X})$ is concentrated in two consecutive dimensions and its only non-trivial differential is $\mathrm{id}\colon \mathbb{Z}\pi \to \mathbb{Z}\pi$ if we identify the two non-trivial $\mathbb{Z}\pi$-chain modules with $\mathbb{Z}\pi$ using the cellular basis. This implies $\tau(C_*(\widetilde{Y},\widetilde{X})) = 0$ and hence $\tau(j) := \tau(C_*(\widetilde{j})) = 0$. $\qquad \square$

The full proof of this result and more information about Whitehead torsion and the following remark can be found for instance in [53], [153, Chapter 2].

Remark 6.16 (Reidemeister torsion). Another interesting torsion invariant is the so-called *Reidemeister torsion*. It is defined for a space whose fundamental group π is finite and for which the homology with coefficients in a certain π-representation vanishes. The Reidemeister torsion was the first invariant in algebraic topology which can distinguish the homeomorphism type of spaces within a given homotopy type. Namely, there exist two so-called lens spaces L_1 and L_2 which are homotopy equivalent but have different Reidemeister torsion. Given a homotopy equivalence $f\colon L_1 \to L_2$, its Whitehead torsion determines the difference of the Reidemeister torsion of L_1 and L_2 and hence can never be zero. This implies that f cannot be a homeomorphism.

Chapter 7

The Statement and Consequences of the s-Cobordism Theorem

In this chapter we want to discuss the following result

Theorem 7.1 (s-Cobordism Theorem). *Let M_0 be a closed connected oriented manifold of dimension $n \geq 5$ with fundamental group $\pi = \pi_1(M_0)$.*

(1) *Let $(W; M_0, f_0, M_1, f_1)$ be an h-cobordism over M_0. Then W is trivial over M_0 if and only if its Whitehead torsion $\tau(W, M_0) \in \mathrm{Wh}(\pi)$ vanishes.*

(2) *For any $x \in \mathrm{Wh}(\pi)$ there is an h-cobordism $(W; M_0, f_0, M_1, f_1)$ over M_0 with $\tau(W, M_0) = x \in \mathrm{Wh}(\pi)$.*

(3) *The function assigning to an h-cobordism $(W; M_0, f_0, M_1, f_1)$ over M_0 its Whitehead torsion yields a bijection from the diffeomorphism classes relative M_0 of h-cobordisms over M_0 to the Whitehead group $\mathrm{Wh}(\pi)$.*

Here are some explanations. An *n-dimensional cobordism* (sometimes also called just bordism) $(W; M_0, f_0, M_1, f_1)$ consists of a compact oriented n-dimensional manifold W, closed $(n-1)$-dimensional manifolds M_0 and M_1, a disjoint decomposition $\partial W = \partial_0 W \coprod \partial_1 W$ of the boundary ∂W of W and orientation preserving diffeomorphisms $f_0 \colon M_0 \to \partial W_0$ and $f_1 \colon M_1^- \to \partial W_1$. Here and in the sequel we denote by M_1^- the manifold M_1 with the reversed orientation and we use on ∂W the orientation with respect to the decomposition $T_x W = T_x \partial W \oplus \mathbb{R}$ coming from an inward normal field for the boundary. If we equip D^2 with the standard orientation coming from the standard orientation on $\mathbb{R}^2$, the induced orientation on $S^1 = \partial D^2$ corresponds to the anti-clockwise orientation on S^1. If we want to specify M_0, we say that W is a *cobordism over M_0*. If $\partial_0 W = M_0$, $\partial_1 W = M_1^-$ and f_0 and f_1 are given by the identity or if f_0 and f_1 are obvious from the context, we briefly write $(W; \partial_0 W, \partial_1 W)$. Two cobordisms (W, M_0, f_0, M_1, f_1) and $(W', M_0, f_0', M_1', f_1')$ over M_0 are *diffeomorphic relative* M_0 if there is an orientation preserving diffeomorphism $F \colon W \to W'$ with $F \circ f_0 = f_0'$. We call an h-cobordism over M_0 *trivial*, if it is diffeomorphic relative M_0 to the trivial h-cobordism $(M_0 \times [0,1]; M_0 \times \{0\}, (M_0 \times \{1\})^-)$. Notice that the choice of the diffeomorphisms f_i does play a role although they are often suppressed in the notation. We call a cobordism $(W; M_0, f_0, M_1, f_1)$ an h-*cobordism*, if the inclusions $\partial_i W \to W$ for $i = 0, 1$ are homotopy equivalences. The *Whitehead torsion* of an h-cobordism $(W; M_0, f_0, M_1, f_1)$ over M_0,

$$\tau(W, M_0) \quad \in \quad \mathrm{Wh}(\pi_1(M_0)), \tag{7.2}$$

is defined to be the preimage of the Whitehead torsion (see Definition 6.12)

$$\tau\left(M_0 \xrightarrow{f_0} \partial_0 W \xrightarrow{i_0} W\right) \in \mathrm{Wh}(\pi_1(W))$$

under the isomorphism

$$(i_0 \circ f_0)_* \colon \mathrm{Wh}(\pi_1(M_0)) \xrightarrow{\cong} \mathrm{Wh}(\pi_1(W)),$$

where $i_0 \colon \partial_0 W \to W$ is the inclusion. Here we use the fact that each closed manifold has a CW-structure, which comes for instance from a triangulation, and that the choice of CW-structure does not matter by the topological invariance of the Whitehead torsion (see Theorem 6.13 (5)).

The s-Cobordism Theorem 7.1 is due to Barden, Mazur, Stallings. Its topological version was proved by Kirby and Siebenmann [130, Essay II]. More information about the s-cobordism theorem can be found for instance in [128], [153, Chapter 1], [167] [211, pages 87-90]. The s-cobordism theorem is known to be false (smoothly) for $n = \dim(M_0) = 4$ in general, by the work of Donaldson [70], but it is true for $n = \dim(M_0) = 4$ for so-called "good" fundamental groups in the topological category by results of Freedman [94], [95]. The trivial group is an example of a "good" fundamental group. Counterexamples in the case $n = \dim(M_0) = 3$ are constructed by Cappell and Shaneson [45]. The Poincaré Conjecture (see Theorem 7.4) is at the time of writing known in all dimensions except dimension 3.

We already know that the Whitehead group of the trivial group vanishes. Thus the s-Cobordism Theorem 7.1 implies

Theorem 7.3 (h-Cobordism Theorem). *Each h-cobordism $(W; M_0, f_0, M_1, f_1)$ over a simply connected closed n-dimensional manifold M_0 with $\dim(W) \geq 6$ is trivial.*

Theorem 7.4 (Poincaré Conjecture). *The Poincaré Conjecture is true for a closed n-dimensional manifold M with $\dim(M) \geq 5$, namely, if M is simply connected and its homology $H_p(M)$ is isomorphic to $H_p(S^n)$ for all $p \in \mathbb{Z}$, then M is homeomorphic to S^n.*

Proof. We only give the proof for $\dim(M) \geq 6$. Since M is simply connected and $H_*(M) \cong H_*(S^n)$, one can conclude from the Hurewicz Theorem and Whitehead Theorem [255, Theorem IV.7.13 on page 181 and Theorem IV.7.17 on page 182] that there is a homotopy equivalence $f \colon M \to S^n$. Let $D_i^n \subseteq M$ for $i = 0, 1$ be two embedded disjoint disks. Put $W = M - (\text{int}(D_0^n) \coprod \text{int}(D_1^n))$. Then W turns out to be a simply connected h-cobordism. Hence we can find a diffeomorphism $F \colon (\partial D_0^n \times [0, 1], \partial D_0^n \times \{0\}, \partial D_0^n \times \{1\}) \to (W, \partial D_0^n, \partial D_1^n)$ which is the identity on $\partial D_0^n = \partial D_0^n \times \{0\}$ and induces some (unknown) diffeomorphism $f_1 \colon \partial D_0^n \times \{1\} \to \partial D_1^n$. By the *Alexander trick* one can extend $f_1 \colon \partial D_0^n = \partial D_0^n \times \{1\} \to \partial D_1^n$ to a homeomorphism $\overline{f_1} \colon D_0^n \to D_1^n$. Namely, any homeomorphism $f \colon S^{n-1} \to S^{n-1}$ extends to a homeomorphism $\overline{f} \colon D^n \to D^n$ by sending $t \cdot x$ for $t \in [0, 1]$ and $x \in S^{n-1}$ to $t \cdot f(x)$. Now define a homeomorphism $h \colon D_0^n \times \{0\} \cup_{i_0} \partial D_0^n \times [0, 1] \cup_{i_1} D_0^n \times \{1\} \to M$ for the canonical inclusions $i_k \colon \partial D_0^n \times \{k\} \to \partial D_0^n \times [0, 1]$ for $k = 0, 1$ by $h|_{D_0^n \times \{0\}} = \text{id}$, $h|_{\partial D_0^n \times [0,1]} = F$ and $h|_{D_0^n \times \{1\}} = \overline{f_1}$. Since the source of h is obviously homeomorphic to S^n, Theorem 7.4 follows.

In the case $\dim(M) = 5$ one uses the fact that M is the boundary of a contractible 6-dimensional manifold W and applies the s-cobordism theorem to W with an embedded disc removed. $\qquad\square$

Remark 7.5 (Exotic spheres). Notice that the proof of the Poincaré Conjecture in Theorem 7.4 works only in the topological category but not in the smooth category. In other words, we cannot conclude the existence of a diffeomorphism $h \colon S^n \to M$. The proof in the smooth case breaks down when we apply the Alexander trick. The construction of $\overline{f}$ given by coning f yields only a homeomorphism $\overline{f}$ and not a diffeomorphism even if we start with a diffeomorphism f. The map $\overline{f}$ is smooth outside the origin of D^n but not necessarily at the origin. We will see that not every diffeomorphism $f \colon S^{n-1} \to S^{n-1}$ can be extended to a diffeomorphism $D^n \to D^n$ and that there exist so-called *exotic spheres*, i.e., closed manifolds which are homeomorphic to S^n but not diffeomorphic to S^n. The classification of these exotic spheres is one of the early very important achievements of surgery theory and one motivation for its further development. For more information about exotic spheres we refer for instance to [129], [144], [149] and [153, Chapter 6].

Remark 7.6 (The surgery program). In some sense the s-Cobordism Theorem 7.1 is one of the first theorems, where diffeomorphism classes of certain manifolds are determined by an algebraic invariant, namely the Whitehead torsion. Moreover, the Whitehead group $\text{Wh}(\pi)$ depends only on the fundamental group $\pi = \pi_1(M_0)$, whereas the diffeomorphism classes of h-cobordisms over M_0 a priori depend on M_0 itself. The s-Cobordism Theorem 7.1 is one step in a program to decide whether two closed manifolds M and N are diffeomorphic, which is in general a very hard question. The idea is to construct an h-cobordism $(W; M, f, N, g)$ with vanishing Whitehead torsion. Then W is diffeomorphic to the trivial h-cobordism over M which implies that M and N are diffeomorphic. So the *surgery program* would be:

(1) Construct a homotopy equivalence $f \colon M \to N$.

(2) Construct a cobordism $(W; M, N)$ and a map $(F, f, \mathrm{id}) \colon (W; M, N) \to (N \times [0, 1], N \times \{0\}, N \times \{1\})$.

(3) Modify W and F relative boundary by so-called surgery such that F becomes a homotopy equivalence and thus W becomes an h-cobordism. During these processes one should make certain that the Whitehead torsion of the resulting h-cobordism is trivial.

The advantage of this approach will be that it can be reduced to problems in homotopy theory and algebra, which can sometimes be handled by well-known techniques. In particular one will sometimes get computable obstructions for two homotopy equivalent manifolds to be diffeomorphic. Often surgery theory has proved to be very useful, when one wants to distinguish two closed manifolds, which have very similar properties. The classification of homotopy spheres is one example. Moreover, surgery techniques also can be applied to problems which are of different nature than of diffeomorphism or homeomorphism classifications.

Chapter 8

Sketch of the Proof of the s-Cobordism Theorem

In this chapter we want to sketch the proof of the s-Cobordism Theorem 7.1. We will restrict ourselves to assertion (1). If a h-cobordism is trivial, one easily checks using Theorem 6.13 that its Whitehead torsion vanishes. The hard part is to show that the vanishing of the Whitehead torsion already implies that it is trivial. More details of the proof can be found for instance in [128], [153, Chapter 1], [167] [211, pages 87-90].

8.1 Handlebody Decompositions

In this section we explain basic facts about handles and handlebody decompositions.

Definition 8.1. *The n-dimensional handle of index q or briefly q-handle is $D^q \times D^{n-q}$. Its core is $D^q \times \{0\}$. The boundary of the core is $S^{q-1} \times \{0\}$. Its cocore is $\{0\} \times D^{n-q}$ and its transverse sphere is $\{0\} \times S^{n-q-1}$.*

Let $(M, \partial M)$ be an n-dimensional manifold with boundary ∂M. If $\varphi^q \colon S^{q-1} \times D^{n-q} \to \partial M$ is an embedding, then we say that the manifold $M + (\varphi^q)$ defined by $M \cup_{\varphi^q} D^q \times D^{n-q}$ is obtained from M by attaching a handle of index q by φ^q.

Obviously $M + (\varphi^q)$ carries the structure of a topological manifold. To get a smooth structure, one has to use the technique of straightening the angle to get rid of the corners at the place, where the handle is glued to M.

Remark 8.2 (The surgery step). The boundary $\partial(M + (\varphi^q))$ can be described as follows. Delete from ∂M the interior of the image of φ^q. We obtain a manifold with boundary together with a diffeomorphism from $S^{q-1} \times S^{n-q-1}$ to its boundary induced by $\varphi^q|_{S^{q-1} \times S^{n-q-1}}$. If we use this diffeomorphism to glue $D^q \times S^{n-q-1}$

to it, we obtain a closed manifold, namely, $\partial(M + (\varphi^q))$. The step from ∂M to $\partial(M + (\varphi^q))$ described above is called a surgery step on the embedding φ^q.

Let W be a compact manifold whose boundary ∂W is the disjoint sum $\partial_0 W \coprod \partial_1 W$. Then we want to construct W from $\partial_0 W \times [0, 1]$ by attaching handles as follows. Notice that the following construction will not change $\partial_0 W = \partial_0 W \times \{0\}$. If $\varphi^q \colon S^{q-1} \times D^{n-q} \to \partial_1 W$ is an embedding, we get by attaching a handle the compact manifold $W_1 = \partial_0 W \times [0, 1] + (\varphi^q)$ which is given by $W \cup_{\varphi^q} D^q \times D^{n-q}$. Its boundary is a disjoint sum $\partial_0 W_1 \coprod \partial_1 W_1$, where $\partial_0 W_1$ is the same as $\partial_0 W$. Now we can iterate this process, where we attach a handle to $\partial_1 W_1$. Thus we obtain a compact manifold with boundary

$$W = \partial_0 W \times [0, 1] + (\varphi_1^{q_1}) + (\varphi_2^{q_2}) + \ldots + (\varphi_r^{q_r}),$$

whose boundary is the disjoint union $\partial_0 W \coprod \partial_1 W$, where $\partial_0 W$ is just $\partial_0 W \times \{0\}$. We call such a description of W as above a *handlebody decomposition* of W relative $\partial_0 W$. We get from Morse theory [114, Chapter 6], [166, part I]:

Lemma 8.3. *Let W be a compact manifold whose boundary ∂W is the disjoint sum $\partial_0 W \coprod \partial_1 W$. Then W possesses a handlebody decomposition relative $\partial_0 W$, i.e., W is up to diffeomorphism relative $\partial_0 W = \partial_0 W \times \{0\}$ of the form*

$$W \cong \partial_0 W \times [0, 1] + \sum_{i=1}^{p_0} (\varphi_i^0) + \sum_{i=1}^{p_1} (\varphi_i^1) + \ldots + \sum_{i=1}^{p_n} (\varphi_i^n).$$

If we want to show that W is diffeomorphic to $\partial_0 W \times [0, 1]$ relative $\partial_0 W = \partial_0 W \times \{0\}$, we must get rid of the handles. For this purpose we have to find possible modifications of the handlebody decomposition, which reduce the number of handles without changing the diffeomorphism type of W relative $\partial_0 W$.

Example 8.4 (Cancelling Handles). Here is a standard situation, where attaching first a q-handle and then a $(q+1)$-handle does not change the diffeomorphism type of an n-dimensional compact manifold W with the disjoint union $\partial_0 W \coprod \partial_1 W$ as boundary ∂W. Let $0 \le q \le n - 1$. Consider an embedding

$$\mu \colon S^{q-1} \times D^{n-q} \cup_{S^{q-1} \times S_+^{n-1-q}} D^q \times S_+^{n-1-q} \to \partial_1 W,$$

where S_+^{n-1-q} is the upper hemisphere in $S^{n-1-q} = \partial D^{n-q}$. Notice that the source of μ is diffeomorphic to D^{n-1}. Let $\varphi^q \colon S^{q-1} \times D^{n-q} \to \partial_1 W$ be its restriction to $S^{q-1} \times D^{n-q}$. Let $\varphi_+^{q+1} \colon S_+^q \times S_+^{n-q-1} \to \partial_1(W + (\varphi^q))$ be the embedding which is given by

$$S_+^q \times S_+^{n-q-1} = D^q \times S_+^{n-q-1} \subseteq D^q \times S^{n-q-1} = \partial(\varphi^q) \subseteq \partial_1(W + (\varphi^q)).$$

It does not meet the interior of W. Let $\varphi_-^{q+1} \colon S_-^q \times S_+^{n-1-q} \to \partial_1(W \cup (\varphi^q))$ be the embedding obtained from μ by restriction to $S_-^q \times S_+^{n-1-q} = D^q \times S_+^{n-1-q}$.

Then φ_-^{q+1} and φ_+^{q+1} fit together to yield an embedding $\psi^{q+1} \colon S^q \times D^{n-q-1} = S_-^q \times S_+^{n-q-1} \cup_{S^{q-1} \times S_+^{n-q-1}} S_+^q \times S_+^{n-q-1} \to \partial_1(W + (\varphi^q))$. It is not difficult to check that $W + (\varphi^q) + (\psi^{q+1})$ is diffeomorphic relative $\partial_0 W$ to W since up to diffeomorphism $W + (\varphi^q) + (\psi^{q+1})$ is obtained from W by taking the boundary connected sum of W and D^n along the embedding μ of $D^{n-1} = S_+^{n-1} = S^{q-1} \times D^{n-q} \cup_{S^{q-1} \times S_+^{n-1-q}} D^q \times S_+^{n-1-q}$ into $\partial_1 W$.

This cancellation of two handles of consecutive index can be generalized as follows.

Lemma 8.5 (Cancellation Lemma). *Let W be an n-dimensional compact manifold whose boundary ∂W is the disjoint sum $\partial_0 W \coprod \partial_1 W$. Let $\varphi^q \colon S^{q-1} \times D^{n-q} \to \partial_1 W$ be an embedding. Let $\psi^{q+1} \colon S^q \times D^{n-1-q} \to \partial_1(W + (\varphi^q))$ be an embedding. Suppose that $\psi^{q+1}(S^q \times \{0\})$ is transversal to the transverse sphere of the handle (φ^q) and meets the transverse sphere in exactly one point. Then there is a diffeomorphism relative $\partial_0 W$ from W to $W + (\varphi^q) + (\psi^{q+1})$.*

Proof. The idea is to use isotopies and the possibility of embedding isotopies into diffeotopies to arrange that the situation looks exactly like in Example 8.4. □

The Cancellation Lemma 8.5 will be our only tool to reduce the number of handles. Notice that one can never get rid of one handle alone, there must always be involved at least two handles simultaneously. The reason is that the Euler characteristic $\chi(W, \partial_0 W)$ is independent of the handle decomposition and can be computed by $\sum_{q \geq 0} (-1)^q \cdot p_q$, where p_q is the number of q-handles (see Section 8.2).

8.2 Handlebody Decompositions and CW-Structures

Next we explain that from a homotopy theoretic point of view a handlebody decomposition is the same as a CW-structure.

Consider a compact n-dimensional manifold W whose boundary is the disjoint union $\partial_0 W \coprod \partial_1 W$. In view of Lemma 8.3 we can write it as

$$W \cong \partial_0 W \times [0,1] + \sum_{i=1}^{p_0} (\varphi_i^0) + \sum_{i=1}^{p_1} (\varphi_i^1) + \ldots + \sum_{i=1}^{p_n} (\varphi_i^n), \qquad (8.6)$$

where $\cong$ means diffeomorphic relative $\partial_0 W$.

Notation 8.7. *Put for* $-1 \leq q \leq n$

$$W_q \quad := \quad \partial_0 W \times [0,1] + \sum_{i=1}^{p_0}(\varphi_i^0) + \sum_{i=1}^{p_1}(\varphi_i^1) + \ldots + \sum_{i=1}^{p_q}(\varphi_i^q);$$

$$\partial_1 W_q \quad := \quad \partial W_q - \partial_0 W \times \{0\};$$

$$\partial_1^\circ W_q \quad := \quad \partial_1 W_q - \coprod_{i=1}^{p_{q+1}} \varphi_i^{q+1}(S^q \times \text{int}(D^{n-1-q})).$$

Notice for the sequel that $\partial_1^\circ W_q \subseteq \partial_1 W_{q+1}$.

Lemma 8.8. *There is a relative CW-complex $(X, \partial_0 W)$ such that there is a bijection between the q-handles of W and the q-cells of $(X, \partial_0 W)$ and a simple homotopy equivalence $f \colon W \to X$, which induces simple homotopy equivalences $f_q \colon W_q \to X_q$ for $q \geq -1$ with $f_{-1} = \text{id}_{\partial_0 W}$, where X_q is the q-skeleton of X.*

Proof. The point is that the projection $\text{pr} \colon (D^q, S^{q-1}) \times D^{n-q} \to (D^q, S^{q-1})$ is a simple homotopy equivalence and one can successively collapse each handle to its core. $\qquad\square$

So, from the homotopic theoretic point of view, the core and the boundary of the core of a handle are the basic contributions of a handle. In particular we see that the inclusions $W_q \to W$ are q-connected, since the inclusion of the q-skeleton $X_q \to X$ is always q-connected for a CW-complex X.

Denote by $p_W \colon \widetilde{W} \to W$ the universal covering with $\pi = \pi_1(W)$ as group of deck transformations. Let $\widetilde{W}_q$ be the preimage of W_q under p. Notice that this is the universal covering for $q \geq 2$, since each inclusion $W_q \to W$ induces an isomorphism on the fundamental groups, but not necessarily for $q \leq 1$. Let $C_*(\widetilde{W}, \widetilde{\partial_0 W})$ be the $\mathbb{Z}\pi$-chain complex, whose q-th chain group is $H_q(\widetilde{W_q}, \widetilde{W_{q-1}})$ and whose q-th differential is the boundary operator of the triple $(\widetilde{W_q}, \widetilde{W_{q-1}}, \widetilde{W_{q-2}})$. Each handle (φ_i^q) determines an element

$$[\varphi_i^q] \in C_q(\widetilde{W}, \widetilde{\partial_0 W}) \tag{8.9}$$

after choosing a lift $(\widetilde{\varphi_i^q}, \widetilde{\varphi_i^q}) \colon (D^q \times D^{n-q}, S^{q-1} \times D^{n-q}) \to (\widetilde{W_q}, \widetilde{W_{q-1}})$ of its characteristic map $(\varphi_i^q, \varphi_i^q) \colon (D^q \times D^{n-q}, S^{q-1} \times D^{n-q}) \to (W_q, W_{q-1})$, namely, the image of the preferred generator in $H_q(D^q \times D^{n-q}, S^{q-1} \times D^{n-q}) \cong H_0(\{*\}) = \mathbb{Z}$ under the map $H_q(\widetilde{\varphi_i^q}, \widetilde{\varphi_i^q})$. This element is only well-defined up to multiplication with an element $\gamma \in \pi$. The elements $\{[\varphi_i^q] \mid i = 1, 2, \ldots, p_q\}$ form a $\mathbb{Z}\pi$-basis for $C_q(\widetilde{W}, \widetilde{\partial_0 W})$. The next result follows directly from the explicit construction of f.

Lemma 8.10. *The map $f \colon W \to X$ induces an isomorphism of $\mathbb{Z}\pi$-chain complexes*

$$C_*(\widetilde{f}) \colon C_*(\widetilde{W}, \widetilde{\partial_0 W}) \quad \xrightarrow{\cong} \quad C_*(\widetilde{X}, \widetilde{\partial_0 W}),$$

which sends the equivalence class of the $\mathbb{Z}\pi$-bases $\{[\varphi_i^q] \mid i = 1, 2, \ldots, p_q\}$ to the cellular equivalence of $\mathbb{Z}\pi$-basis of $C_*(\widetilde{X}, \widetilde{\partial_0 W})$.

8.3 Reducing the Handlebody Decomposition

In the next step we get rid of the handles of index zero and one in the handlebody decomposition (8.6).

Lemma 8.11. *Let W be an n-dimensional manifold for $n \geq 6$ whose boundary is the disjoint union $\partial W = \partial_0 W \coprod \partial_1 W$. Then the following statements are equivalent*

(1) *The inclusion $\partial_0 W \to W$ is 1-connected.*

(2) *We can find a diffeomorphism relative $\partial_0 W$*

$$W \cong \partial_0 W \times [0,1] + \sum_{i=1}^{p_2} (\varphi_i^2) + \sum_{i=1}^{p_3} (\overline{\varphi}_i^3) + \sum_{i=1}^{p_n} (\overline{\varphi}_i^n).$$

Proof. The easy implication (2) $\Rightarrow$ (1) has already been proved in Section 8.2. $\square$

Now consider an h-cobordism $(W; \partial_0 W, \partial_1 W)$. Because of Lemma 8.11 we can write it as

$$W \cong \partial_0 W \times [0,1] + \sum_{i=1}^{p_2} (\varphi_i^2) + \sum_{i=1}^{p_3} (\overline{\varphi}_i^3) + \dots.$$

Lemma 8.12 (Homology Lemma). *Suppose $n \geq 6$. Fix $2 \leq q \leq n-3$ and $i_0 \in \{1, 2, \dots, p_q\}$. Let $f \colon S^q \to \partial_1 W_q$ be an embedding. Then the following statements are equivalent*

(1) *f is isotopic to an embedding $g \colon S^q \to \partial_1 W_q$ such that g meets the transverse sphere of $(\varphi_{i_0}^q)$ transversally and in exactly one point and is disjoint from transverse spheres of the handles (φ_i^q) for $i \neq i_0$.*

(2) *Let $\widetilde{f} \colon S^q \to \widetilde{W}_q$ be a lift of f under $p|_{\widetilde{W}_q} \colon \widetilde{W}_q \to W_q$. Let $[\widetilde{f}]$ be the image of the class represented by $\widetilde{f}$ under the obvious composition*

$$\pi_q(\widetilde{W}_q) \to \pi_q(\widetilde{W}_q, \widetilde{W_{q-1}}) \to H_q(\widetilde{W}_q, \widetilde{W_{q-1}}) = C_q(\widetilde{W}).$$

Then there is $\gamma \in \pi$ with
$$[\widetilde{f}] = \pm \gamma \cdot [\varphi_{i_0}^q].$$

Proof. (1) $\Rightarrow$ (2) We can isotope f such that $f|_{S_+^q} \colon S_+^q \to \partial_1 W_q$ looks like the canonical embedding $S_+^q = D^q \times \{x\} \subseteq D^q \times S^{n-1-q} = \partial(\varphi_{i_0}^q)$ for some $x \in S^{n-1-q}$ and $f(S_-^q)$ does not meet any of the handles (φ_i^q) for $i = 1, 2, \dots, p_q$. One easily checks that then (2) is true.

(2) $\Rightarrow$ (1) We can isotope f such that it is transversal to the transverse spheres of the handles (φ_i^q) for $i = 1, 2, \dots, p_q$. Since the sum of the dimension of the source of f and of the dimension of the transverse spheres is the dimension of $\partial_1 W_q$,

the intersection of the image of f with the transverse sphere of the handle (φ_i^q) consists of finitely many points $x_{i,1}, x_{i,2}, \ldots, x_{i,r_i}$ for $i = 1, 2, \ldots, p_q$. Fix a base point $y \in S^q$. It yields a base point $z = f(y) \in W$. Fix for each handle (φ_i^q) a path w_i in W from a point in its transverse sphere to z. Let $u_{i,j}$ be a path in S^q with the property that $u_{i,j}(0) = y$ and $f(u_{i,j}(1)) = x_{i,j}$ for $1 \le j \le r_i$ and $1 \le i \le p_q$. Let $v_{i,j}$ be any path in the transverse sphere of (φ_i^q) from $x_{i,j}$ to $w_i(0)$. Then the composition $f(u_{i,j}) * v_{i,j} * w_i$ is a loop in W with base point z and thus represents an element denoted by $\gamma_{i,j}$ in $\pi = \pi_1(W, z)$. It is independent of the choice of $u_{i,j}$ and $v_{i,j}$ since S^q and the transverse sphere of each handle (φ_i^q) are simply connected. The tangent space $T_{x_{i,j}} \partial_1 W_q$ is the direct sum of $T_{f^{-1}(x_{i,j})} S^q$ and the tangent space of the transverse sphere $\{0\} \times S^{n-1-q}$ of the handle (φ_i^q) at $x_{i,j}$. All these three tangent spaces come with preferred orientations. We define elements $\varepsilon_{i,j} \in \{\pm 1\}$ by requiring that it is 1 if these orientations fit together and -1 otherwise. Now one easily checks that

$$[\widetilde{f}] = \sum_{i=1}^{p_q} \sum_{j=1}^{r_i} \varepsilon_{i,j} \cdot \gamma_{i,j} \cdot [\varphi_i^q],$$

where $[\varphi_i^q]$ is the element associated to the handle (φ_i^q) after the choice of the path w_i We have by assumption $[\widetilde{f}] = \pm \cdot \gamma \cdot [\varphi_{i_0}^q]$ for some $\gamma \in \pi$. We want to isotope f such that f does not meet the transverse spheres of the handles (φ_i^q) for $i \ne i_0$ and does meet the transverse sphere of $(\varphi_{i_0}^q)$ transversally and in exactly one point. Therefore it suffices to show in the case that the number $\sum_{i=1}^{p_q} r_i$ of all intersection points of f with the transverse spheres of the handles (φ_i^q) for $i = 1, 2, \ldots, p_i$ is bigger than one that we can change f by an isotopy such that this number becomes smaller. We have

$$\pm \gamma \cdot [\varphi_{i_0}^q] = \sum_{i=1}^{p_q} \sum_{j=1}^{r_i} \varepsilon_{i,j} \cdot \gamma_{i,j} \cdot [\varphi_i^q].$$

Recall that the elements $[\varphi_i^q]$ for $i = 1, 2, \ldots, p_q$ form a $\mathbb{Z}\pi$-basis. Hence we can find an index $i \in \{1, 2, \ldots, p_q\}$ and two different indices $j_1, j_2 \in \{1, 2, \ldots, r_i\}$ such that the composition of the paths $f(u_{i,j_1}) * v_{i,j_1} * v_{i,j_2}^- * f(u_{i,j_2}^-)$ is nullhomotopic in W and hence in $\partial_1 W_q$ and the signs ε_{i,j_1} and ε_{i,j_2} are different. Now by the Whitney trick (see [167, Theorem 6.6 on page 71], [256]) we can change f by an isotopy such that the two intersection points x_{i,j_1} and x_{i,j_2} disappear, the other intersection points of f with transverse spheres of the handles (φ_i^q) for $i \in \{1, 2, \ldots, p_q\}$ remain and no further intersection points are introduced. For the application of the Whitney trick we need the assumption $n - 1 \ge 5$. This finishes the proof of the Homology Lemma 8.12. □

Lemma 8.13 (Modification Lemma). *Let $f \colon S^q \to \partial_1^\circ W_q$ be an embedding and let $x_j \in \mathbb{Z}\pi$ be elements for $j = 1, 2 \ldots, p_{q+1}$. Then there is an embedding $g \colon S^q \to \partial_1^\circ W_q$ with the following properties:*

(1) f and g are isotopic in $\partial_1 W_{q+1}$.

(2) *For a given lifting* $\widetilde{f}\colon S^q \to \widetilde{W}_q$ *of* f *one can find a lifting* $\widetilde{g}\colon S^q \to \widetilde{W}_q$ *of* g *such that we get in* $C_q(\widetilde{W})$

$$[\widetilde{g}] \;=\; [\widetilde{f}] + \sum_{j=1}^{p_{q+1}} x_j \cdot d_{q+1}[\varphi_j^{q+1}],$$

where d_{q+1} *is the* $(q+1)$-*th differential in* $C_*(\widetilde{W}, \widetilde{\partial_0 W})$.

Lemma 8.14 (Normal Form Lemma). *Let* $(W; \partial_0 W, \partial_1 W)$ *be an oriented compact h-cobordism of dimension* $n \geq 6$. *Let* q *be an integer with* $2 \leq q \leq n - 3$. *Then there is a handlebody decomposition which has only handles of index* q *and* $(q+1)$, *i.e., there is a diffeomorphism relative* $\partial_0 W$

$$W \;\cong\; \partial_0 W \times [0,1] + \sum_{i=1}^{p_q} (\varphi_i^q) + \sum_{i=1}^{p_{q+1}} (\varphi_i^{q+1}).$$

8.4 Handlebody Decompositions and Whitehead Torsion

Let $(W, \partial_0 W, \partial_1 W)$ be an n-dimensional compact oriented h-cobordism for $n \geq 6$. By the Normal Form Lemma 8.14 we can fix a handlebody decomposition for some fixed number $2 \leq q \leq n - 3$:

$$W \;\cong\; \partial_0 W \times [0,1] + \sum_{i=1}^{p_q} (\varphi_i^q) + \sum_{i=1}^{p_{q+1}} (\varphi_i^{q+1}).$$

Recall that the $\mathbb{Z}\pi$-chain complex $C_*(\widetilde{W}, \widetilde{\partial_0 W})$ is acyclic. Hence the only non-trivial differential $d_{q+1}\colon H_{q+1}(\widetilde{W_{q+1}}, \widetilde{W_q}) \to H_q(\widetilde{W_q}, \widetilde{W_{q-1}})$ is bijective. Recall that $\{[\varphi_i^{q+1}] \mid i = 1, 2 \ldots, p_{q+1}\}$ is a $\mathbb{Z}\pi$-basis for $H_{q+1}(\widetilde{W_{q+1}}, \widetilde{W_q})$ and $\{[\varphi_i^q] \mid i = 1, 2 \ldots, p_q\}$ is a $\mathbb{Z}\pi$-basis for $H_q(\widetilde{W_q}, \widetilde{W_{q-1}})$. In particular $p_q = p_{q+1}$. The matrix A, which describes the differential d_{q+1} with respect to these basis, is an invertible (p_q, p_q)-matrix over $\mathbb{Z}\pi$. Since we are working with left modules, d_{q+1} sends an element $x \in (\mathbb{Z}G)^n$ to $x \cdot A \in \mathbb{Z}G^n$, or equivalently, $d_{q+1}([\varphi_i^{q+1}]) = \sum_{j=1}^{n} a_{i,j}[\varphi_j^q]$.

Lemma 8.15. *Let* $(W, \partial_0 W, \partial_1 W)$ *be an* n-*dimensional compact oriented h-cobordism for* $n \geq 6$ *and* A *be the matrix defined above. Suppose that* A *can be reduced to a matrix of size* $(0, 0)$ *by a sequence of the following operations:*

(1) B *is obtained from* A *by adding the* k-*th row multiplied with* x *from the left to the* l-*th row for* $x \in \mathbb{Z}\pi$ *and* $k \neq l$;

(2) B is obtained by taking the direct sum of A and the $(1,1)$-matrix $I_1 = (1)$,
   i.e., B looks like the block matrix $\begin{pmatrix} A & 0 \\ 0 & 1 \end{pmatrix}$;

(3) A is the direct sum of B and I_1. This is the inverse operation to (2);

(4) B is obtained from A by multiplying the i-th row from the left with a trivial
   unit , i.e., with an element of the shape $\pm\gamma$ for $\gamma \in \pi$;

(5) B is obtained from A by interchanging two rows or two columns.

Then the h-cobordism W is trivial relative $\partial_0 W$.

Proof. Let B be a matrix which is obtained from A by applying one of the operations (1), (2), (3), (4) and (5). It suffices to show that we can modify the given handlebody decomposition in normal form of W with associated matrix A such that we get a new handlebody decomposition in normal form, whose associated matrix is B.

We begin with (1). Consider $W' = \partial_0 W \times [0,1] + \sum_{i=1}^{p_q}(\varphi_i^q) + \sum_{j=1,j\neq l}^{p_{q+1}}(\varphi_j^{q+1})$. Notice that we get from W' our h-cobordism W if we attach the handle (φ_l^{q+1}). By the Modification Lemma 8.13 we can find an embedding $\overline{\varphi}_l^{q+1}\colon S^q \times D^{n-1-q} \to \partial_1 W'$ such that $\overline{\varphi}_l^{q+1}$ is isotopic to φ_l^{q+1} and we get

$$\left[\widetilde{\overline{\varphi}_l^{q+1}|_{S^q\times\{0\}}}\right] = \left[\widetilde{\varphi_l^{q+1}|_{S^q\times\{0\}}}\right] + x \cdot d_{q+1}([\varphi_k^{q+1}]).$$

If we attach to W' the handle $(\overline{\varphi}_l^{q+1})$, the result is diffeomorphic to W relative $\partial_0 W$ since $\overline{\varphi}_l^{q+1}$ and φ_l^{q+1} are isotopic. One easily checks that the associated invertible matrix B is obtained from A by adding the k-th row multiplied with x from the left to the l-th row.

The claim for the operations (2) and (3) follow from the Cancellation Lemma 8.5 and the Homology Lemma 8.12. The claim for the operation (4) follows from the observation that we can replace the attaching map of a handle $\varphi^q\colon S^q \times D^{n-1-q} \to \partial_1 W_q$ by its composition with $f \times \mathrm{id}$ for some diffeomorphism $f\colon S^q \to S^q$ of degree -1 and that the base element $[\varphi_i^q]$ can also be changed to $\gamma \cdot [\varphi_i^q]$ by choosing a different lift along $\widetilde{W}_q \to W_q$. Operation (5) can be realized by interchanging the numeration of the q-handles and $(q+1)$-handles. $\square$

Finally we can explain how Theorem 7.1 (1) follows. One easily checks from the definitions that $\tau(W, M_0) \in \mathrm{Wh}(\pi)$ is represented by $(-1)^q \cdot [A]$. Hence $\tau(W, M_0) = 0$ is equivalent to $[A] = 0$ in $\mathrm{Wh}(\pi)$. Now apply Lemma 5.17. $\square$

Remark 8.16 (Principle of extracting algebraic obstructions from geometry). Following the history we have defined first the Whitehead group and the Whitehead torsion and then explained the proof of the s-Cobordism Theorem. But one can also motivate the definition of the Whitehead group and Whitehead torsion by the s-Cobordism Theorem using the following general strategy in topology.

Suppose we are given a geometric problem, for instance to check that a given *h*-cobordism is trivial. In the first step one tries to figure out all possible geometric constructions which may lead to a solution of the problem. In the case of the *s*-Cobordism Theorem this comes down to the Cancellation Lemma 8.5. Now try to apply these constructions to come as close to a solution as possible. In the situation of the *s*-Cobordism Theorem this is Normal Form Lemma 8.14 together with Lemma 8.15. If one cannot achieve the solution in general, try to extract an algebraic condition which prevents one from achieving the solution. This gives in the case of the *s*-Cobordism Theorem the following. Define on the disjoint union $\coprod_{n \geq 0} GL_n(\mathbb{Z}\pi)$ an equivalence relation by calling A and B equivalent if one can transform A into B by a sequence of operations appearing in Lemma 8.15. Obviously one can solve the problem if the matrix A given by a handlebody decomposition in normal form is equivalent to the unique element in $GL(0, \mathbb{Z}\pi)$ which is by definition a set consisting of one element. Now one checks that the set of equivalence classes is an abelian group, where the group structure comes from taking the direct sum of two matrices. This is exactly $\mathrm{Wh}(\pi)$ and hence we obtain an element in $\mathrm{Wh}(\pi)$ by the class of A. Finally one must show that the algebraic invariant is independent of the geometric steps one may already have done to get as close as possible to the solution. In the case of the *s*-Cobordism Theorem we must show that the class of A depends only on the given *h*-cobordism but not on the choice of the normal form. This is done in the case of the Whitehead torsion by giving a definition which works for every CW-structure and hence in particular for a every handlebody decomposition and already makes sense before any geometric modifications have been made.

This principle can also be used to explain the definition of the Whitehead torsion having the problem in mind whether a given homotopy equivalence of finite CW-complexes is a simple homotopy equivalence. This principle can also be used to explain the idea of the surgery obstruction which will be discussed later. Here the problem is to modify a map $f : M \to X$ from a closed orientable manifold M to a Poincaré complex X so that the resulting map $g : N \to X$ is a homotopy equivalence with some closed manifold as source.

Chapter 9

From the Novikov Conjecture to Surgery

9.1 The Structure Set

As before all manifolds are oriented and homotopy equivalences or diffeomorphisms are orientation preserving.

Given a smooth closed manifold M together with a map $f\colon M \to BG$, the Novikov Conjecture says that if $g\colon N \to M$ is a homotopy equivalence, then the higher signatures of (M, f) and (N, fg) agree. This suggests introducing the following set consisting of equivalence classes of pairs (N, g), where $g\colon N \to M$ is a homotopy equivalence. Two such pairs (N, g) and (N', g') are called equivalent if there is an h-cobordism W between N and N' and an extension of g and g' to a map $W \to M$. We call such an equivalence class a *homotopy smoothing* of M. The terminology would be more plausible if g were a homeomorphism. Then we would call it a *smoothing* since it can be used to define a new smooth structure on M which is characterized by the property that with respect to this new smooth structure g is a diffeomorphism. If g is a homotopy equivalence one can consider (N, g) as a sort of smooth structure on M up to homotopy equivalence. The set of homotopy smoothings is denoted by

$$\mathcal{S}^h(M),$$

the *set of homotopy smoothings* or *the homotopy structure set*.

Remark 9.1 (Simple homotopy smoothings). If we replace homotopy equivalences by simple homotopy equivalences and h-cobordisms by s-cobordisms we obtain the set

$$\mathcal{S}^s(M)$$

or the *simple homotopy structure set*. If we apply the s-cobordism theorem, then, for dimension of $M > 4$, the equivalence relation is the same as diffeomorphism classes. This means that there is a diffeomorphism $\varphi \colon N \to N'$ such that $g'\varphi$ and g are homotopic. It is obvious that the set $\mathcal{S}^s(M)$ plays a central role in the classification of manifolds of dimension > 4. We will relate this set to homotopy theory and algebra via the surgery exact sequence introduced in Chapter 14.

It is also useful to consider the structure sets for topological manifolds which we call $\mathcal{S}^h_{\mathrm{TOP}}(M)$ and $\mathcal{S}^s_{\mathrm{TOP}}(M)$, the topological structure sets.

We put for a CW-complex X

$$h_m(X) \quad := \quad \bigoplus_{i \in \mathbb{Z}} H_{m-4i}(X; \mathbb{Q}). \tag{9.2}$$

Returning to the Novikov Conjecture, if $f \colon M \to BG$ is a map from a closed oriented m-dimensional manifold M to the classifying space BG of the group G, then we can consider the map

$$\mathrm{sign}^G \colon \mathcal{S}^h(M) \quad \to \quad h_m(BG; \mathbb{Q}) \tag{9.3}$$

mapping $[g \colon N \to M]$ to the difference of higher signatures

$$\mathrm{sign}_G(M, f) - \mathrm{sign}_G(N, fg).$$

If this map sign^G is trivial for all manifolds M, the Novikov Conjecture for the group G follows.

9.2   The Assembly Idea

The key ingredient for all approaches to the Novikov Conjecture is the construction of a so-called *assembly map*. There are many versions of assembly maps, some of which will be discussed in later chapters. A few words about the history. In the late 1960's Sullivan applied the simply-connected surgery product formula and the work of Conner and Floyd to describe the surgery obstructions of normal maps of closed manifolds as the image of an assembly map on bordism (see [248, Theorem 13B.5] and [232]. One can regard the surgery obstruction map in [248, 13B.3], $\theta \colon \Omega^{PL}_m(K(\pi,1) \times G/PL, K(\pi,1) \times *) \to L_m(\pi)$ as the original assembly map, even though the domain is somewhat bigger than that of the assembly map defined below.

One of the assembly maps is a map

$$A^G_m \colon h_m(BG; \mathbb{Q}) \to L_m(G) \otimes \mathbb{Q}, \tag{9.4}$$

where $L_m(G)$ is an abelian group associated to the group G. One of the most fundamental properties of this map is that the composition of it with the map sign^G of (9.3)

$$\mathcal{S}^h(M) \xrightarrow{\ \mathrm{sign}^G\ } h_m(BG; \mathbb{Q}) \xrightarrow{\ A^G_m\ } L_m(G) \otimes \mathbb{Q}$$

is trivial for all closed oriented manifolds M with dimension m. Thus the Novikov Conjecture is a consequence of the following conjecture

Conjecture 9.5 (Assembly map conjecture). *The assembly map A_m^G is injective for all groups G and every $m \in \mathbb{Z}$.*

We mention that the Farrell–Jones Conjecture implies for a torsionfree G that the assembly map $A_m^G \colon h_m(BG; \mathbb{Q}) \to L_m(G) \otimes \mathbb{Q}$ is an isomorphism (compare Remark 21.14 and Remark 23.8).

For finite groups G, this conjecture is trivial since $\widetilde{H}_*(BG; \mathbb{Q})$ is trivial. But already for $G = \mathbb{Z}$ this is a non-trivial problem which we will discuss later.

The construction of the assembly map is non-trivial and will be explained in the following sections. The construction of the assembly map is essentially following the presentation in Wall's book [248]). The main difference is that we use the K_3-surface instead — as in Wall's book — an 8-dimensional manifold. Here, we only indicate the main steps.

We obtain by h_* as defined in (9.2) a homology theory satisfying

$$h_m(\{\bullet\}) \cong \begin{cases} \mathbb{Q}, & m = 0 \mod 4, \\ 0 & \text{else.} \end{cases}$$

By construction $h_m(X)$ is 4-periodic, i.e., $h_{m+4}(X) = h_m(X)$.

We give another geometric description of this group using oriented bordism groups. For this recall that Ω_* is a $\mathbb{Z}$-graded algebra with multiplication given by cartesian product. Since the signature is multiplicative, it defines a homomorphism of $\mathbb{Z}$-graded algebras

$$\text{sign} \colon \Omega_* \to \mathbb{Q}_*,$$

if $\mathbb{Q}_*$ is the graded algebra whose entry in dimension i for $i = 0 \mod 4$ is $\mathbb{Z}$ and zero in all other dimensions and whose multiplicative structure is given as follows: If $q \in \mathbb{Q}_{4n}$ and $r \in \mathbb{Q}_{4m}$, qr is given by $qr \in \mathbb{Q}_{4(n+m)}$, i.e., the multiplication is induced from the multiplication of rational numbers. Thus $\mathbb{Q}_*$ is a $\mathbb{Z}$-graded Ω_*-module. Also $\Omega_*(X) = \bigoplus_{i \geq 0} \Omega_i(X)$ is a $\mathbb{Z}$-graded Ω_*-module via

$$[M] \cdot [N, g] := [M \times N, gp_2].$$

We can consider the graded tensor product

$$\Omega_*(X) \otimes_{\Omega_*} \mathbb{Q}_*.$$

which is a $\mathbb{Z}$-graded rational vector space. It turns out to be a homology theory. To see this we recall using Theorem 2.4 that $\Omega_* \otimes_{\mathbb{Z}} \mathbb{Q} \cong H_*(BSO \times X; \mathbb{Q}) \cong H_*(X; \mathbb{Q}) \otimes H_*(BSO; \mathbb{Q}) \cong H_*(X; \mathbb{Q}) \otimes \Omega_*$ and so $\Omega_*(X) \otimes_{\Omega_*} \mathbb{Q}_* \cong H_*(X; \mathbb{Q}) \otimes \mathbb{Q}_*$, a homology theory.

We want to express this homology theory in a different way which relates it to the higher signatures. Consider the natural transformation of homology theories

$$T_*(X) \colon \Omega_*(X) \otimes_{\Omega_*} \mathbb{Q}_* \to h_*(X), \quad [M, f] \otimes q \mapsto q\dot{f}_*([M] \cap \mathcal{L}(M)).$$

This is well-defined, i.e., compatible with the tensor relation and the gradings by the following calculation for $[M, f] \in \Omega_m(X)$ and $[N] \in \Omega_{4r}$, where $p_M \colon M \times N \to N$ and $\mathrm{pr} \colon N \to \{\bullet\}$ are the projections

$$
\begin{aligned}
(f \circ p_M)_* ([M \times N] \cap \mathcal{L}(M \times N)) &= (f \circ p_M)_* (([M] \times [N]) \cap (\mathcal{L}(M) \times \mathcal{L}(N))) \\
&= (f \circ p_M)_* (([M] \cap \mathcal{L}(M)) \times ([N] \cap \mathcal{L}(N))) \\
&= f_* ([M] \cap \mathcal{L}(M)) \times \mathrm{pr}_* ([N] \cap \mathcal{L}(N)) \\
&= f_* ([M] \cap \mathcal{L}(M)) \cdot \mathrm{sign}(N).
\end{aligned}
$$

Here we have used the Signature Theorem 1.3 which implies

$$
\mathrm{pr}_* ([N] \cap \mathcal{L}(N))) = \mathrm{sign}(N) \cdot 1 \in H_0(\{\bullet\}).
$$

If X is the one-point-space $\{\bullet\}$, then there are obvious identifications

$$
\Omega_*(\{\bullet\}) \otimes_{\Omega_*} \mathbb{Q}_* = \Omega_* \otimes_{\Omega_*} \mathbb{Q}_* \cong \mathbb{Q}_*
$$

and

$$
h_*(\{\bullet\}) = \mathbb{Q}_*
$$

under which $T(\{\bullet\})$ becomes the identity. Thus — by induction over the skeleta, the 5-Lemma and a colimit argument — $T_*(X)$ is an isomorphism for all CW-complexes X. We have shown:

Proposition 9.6 (A bordism description of $h_m(X)$). *The higher signature induce an isomorphism for all CW-complexes X*

$$
T_*(X) \colon \Omega_*(X) \otimes_{\Omega_*} \mathbb{Q}_* \xrightarrow{\cong} h_*(X) = \bigoplus_{i \in \mathbb{Z}} H_{*-4i}(X; \mathbb{Q}).
$$

For the description of the assembly map A_*^G, it suffices in view of Proposition 9.6 to describe a map of $\mathbb{Z}$-graded vector spaces

$$
\overline{A}_*^G \colon \Omega_*(BG) \otimes_{\Omega_*} \mathbb{Q}_* \to L_*(G) \otimes_{\mathbb{Z}} \mathbb{Q}.
$$

The assembly map A_*^G corresponds to this under the isomorphism $T_*(BG)$. In chapter 15 we will construct a map

$$
\overline{\overline{A}}_m^G \colon \Omega_m(BG) \to L_m(G). \tag{9.7}
$$

and explain that the map $\overline{\overline{A}}_m^G$ induces the desired map $\overline{A}_*^G$ by putting

$$
\overline{A}_*^G([M, f] \otimes r) = \overline{\overline{A}}_{\dim(M)}^G([M, f]) \otimes r.
$$

This proof boils down to verify the necessary tensor relation over Ω_*, i.e., that

$$
\overline{A}_{m+n}^G([M_1 \times N, f \circ p_1]) = \mathrm{sign}(N) \cdot \overline{A}_m^G([M, f])
$$

holds for $f\colon M \to BG$ and N if $m = \dim(M)$ and $n = \dim(N)$ and $p_1\colon M \times N \to M$ is the projection, and is based on the fact that the L-groups $L_m(G)$ are 4-periodic, i.e., $L_m(G) = L_{m+4}(G)$ and the product formula for the surgery obstruction.

Now we explain the idea behind the definition of the maps $\overline{A}_m^{=G}$. (More details will be given in Section 15.1 after we have presented more details about surgery theory.) It is rather indirect. Let K be the *Kummer surface* or $K3$-surface. This is the quartic in $\mathbb{CP}^3$ consisting of the points

$$K := \{[x_0, x_1, x_2, x_3] \in \mathbb{CP}^3 \mid x_0^4 + x_1^4 + x_2^4 + x_3^4 = 0\}.$$

This is a simply connected closed oriented 4-manifold with vanishing second Stiefel-Whitney class and so $K - x$, the complement of a point, is stably parallelizable (see Exercise 9.1). The signature of K is -16 (see Exercise 9.2). Now we remove two open disjoint discs from K and call the resulting compact parallelizable manifold Q. The boundary ∂Q is $S^3 + S^3$, the sum of two 3-spheres.

Then, for a closed m-dimensional manifold M we consider

$$M \times Q,$$

a compact manifold of dimension $m + 4$ with boundary $M \times S^3 + M \times S^3$. Thus $M \times Q$ is a bordism between $M \times S^3$ and $M \times S^3$.

We will try to replace this manifold up to cobordism rel. boundary by an h-cobordism between $M \times S^3$ and $M \times S^3$. If this is possible, the assembly map maps $[M, f]$ to zero for all f. In general, this is not possible and there is an obstruction in an abelian group $L_{m+4}(\pi_1(M))$. A map $f\colon M \to BG$ is the same (up to homotopy) as a map from $\pi_1(M)$ to G and such a map induces a map

$$f_*\colon L_{m+4}(\pi_1(M)) \to L_{m+4}(G).$$

The latter group is by construction 4-periodic and so we consider the image of the construction under f_* to obtain the desired element

$$\overline{A}_m^{=G}([M, f]) \in L_m(G).$$

From the basic properties of the obstruction to replacing $M \times Q$ by an h-cobordism, it is plausible that the composition

$$\mathcal{S}^h(M) \xrightarrow{\mathrm{sign}^G} h_m(BG) \xrightarrow{A_m^G} L_m(G)$$

is zero. This explains why the assembly map Conjecture 9.5 implies the Novikov Conjecture for the group G.

To finish this chapter, we return to the bordism $M \times Q$ and the question of when it is bordant to an h-cobordism. The method is surgery, a way to "improve" a manifold, in this case $M \times Q$, by cutting out a certain submanifold and gluing another manifold in, like a doctor does with the body of an ill person by replacing a bad part by a healthy one.

Remark 9.8 (The non-periodic assembly map). We have presented the 4-periodic assembly map $A_m^G \colon h_m(BG; \mathbb{Q}) \to L_m(G) \otimes \mathbb{Q}$ in (9.4). Since the L-groups are 4-periodic and we want to get an assembly map which is at least for torsionfree groups an isomorphism, the source must be 4-periodic as well. But one can also produce a connective version whose source is smaller and actually generated by classes given by manifolds.

Namely, replace the 4-periodic $\mathbb{Z}$-graded Ω_*-module $\mathbb{Q}_*$ by the $\mathbb{Z}$-graded Ω_*-submodule $\widehat{\mathbb{Q}}_*$ which is $\mathbb{Q}$ in dimensions $i \geq 0$, $i = 0 \mod 4$ and zero in all other dimensions. Also replace the 4-periodic homology theory $h_*(X)$ defined in (9.2) by the homology theory $\widehat{h}_*$ given by

$$\widehat{h}_m(X) \; := \; \bigoplus_{i \in \mathbb{Z}, i \geq 0} H_{m-4i}(X; \mathbb{Q}). \tag{9.9}$$

In contrast to $h_*(X)$ the homology theory $\widehat{h}_m(X)$ is not 4-periodic. Then analogously to Proposition 9.6 one obtains for every CW-complex an isomorphism for all CW-complexes X

$$\widehat{T}_*(X) \colon \Omega_*(X) \otimes_{\Omega_*} \widehat{\mathbb{Q}}_* \; \xrightarrow{\cong} \; \widehat{h}_*(X) = \bigoplus_{i \in \mathbb{Z}, i \geq 0} H_{*-4i}(X; \mathbb{Q}).$$

The following diagram commutes where the vertical maps are the obvious injective maps induced by inclusions.

$$
\begin{array}{ccc}
\Omega_*(X) \otimes_{\Omega_*} \widehat{\mathbb{Q}}_* & \xrightarrow{\;\widehat{T}_*(X)\;} & \widehat{h}_m(BG) \\
\downarrow & & \downarrow \\
\Omega_*(X) \otimes_{\Omega_*} \mathbb{Q}_* & \xrightarrow[\;T_*(X)\;]{} & h_m(BG)
\end{array}
$$

The map $\overline{\overline{A}}_m^G$ of (9.7) defines analogously to A_m^G an assembly map

$$\widehat{A}_m^G \colon \widehat{h}_m(BG) \to L_m(G) \otimes \mathbb{Q}, \tag{9.10}$$

which is the composition of the 4-periodic assembly map $A_m^G \colon h_m(BG; \mathbb{Q}) \to L_m(G) \otimes \mathbb{Q}$ of (9.4) with the obvious inclusion $\widehat{h}_m(X) \to h_m(X)$.

The advantage of $\widehat{h}_m(BG)$ is that it is generated by classes given by manifolds. Since $\Omega_* \to \widehat{\mathbb{Q}}_*$ is surjective, the canonical map $\Omega_m(X) \to \left(\Omega_*(X) \otimes_{\Omega_*} \widehat{\mathbb{Q}}_* \right)_m$ is surjective and has the subvectorspace $U_m \subseteq \Omega_m(X)$ as kernel which is generated by the classes of the shape $[M \times N, f p_1]$ for closed manifolds M and N with $\dim(M) + \dim(N) = m$ and $\mathrm{sign}(N) = 0$ and maps $f \colon M \to BG$ and $p_1 \colon M \times N \to M$ the projection. So we can rewrite $\widehat{A}_m^G$ as a map

$$\widehat{A}_m^G \colon \Omega_m(X)/U_m \; \to \; L_m(G). \tag{9.11}$$

We give a different interpretation in terms of symmetric signatures of the assembly maps A_m^G of (9.4) and $\widehat{A}_m^G$ of (9.11) when we deal with the Farrell–Jones Conjecture and its relation to the Novikov Conjecture (see Remark 23.8).

Chapter 10

Surgery Below the Middle Dimension I: An Example

10.1 Surgery and its Trace

To introduce the idea of surgery, we consider the following situation, which will lead to an attractive application. Let M be a closed smooth simply-connected n-dimensional manifold and let k be an integer satisfying $k > 1$ and $k < n - k$. Suppose that the homology of M is trivial except in dimension 0, k, $n - k$ and n, where it is $\mathbb{Z}$. Examples of such manifolds are $S^k \times S^{n-k}$, other examples will be given later. By the Hurewicz theorem $\pi_k(M)$ is isomorphic to $H_k(M)$ and so $\pi_k(M) \cong \mathbb{Z}$. Let $\alpha \colon S^k \to M$ be a generator of $\pi_k(M)$. By the Whitney embedding theorem we can represent $[\alpha]$ by an embedding (by assumption K is smaller than half the dimension of M) which we again denote by $\alpha \colon S^k \hookrightarrow M$. Now we make an additional assumption, namely that the normal bundle of $\alpha(S^k)$ in M is trivial. Then a tubular neighborhood gives an extension of α to an embedding

$$\beta \colon S^k \times D^{n-k} \hookrightarrow M.$$

Now we perform surgery using β to obtain a smooth manifold

$$\Sigma := M - \beta(S^k \times \overset{\circ}{D}{}^{n-k}) \cup_{\beta|_{S^k \times S^{n-k-1}}} D^{k+1} \times S^{n-k-1}.$$

We say that Σ is obtained from M via *surgery* on β.

10.2   The Effect on the Fundamental Group and Homology Groups

We want to compute the fundamental group and the homology group of Σ. For this we consider

$$T := M \times [0,1] \cup_\beta D^{k+1} \times D^{n-k}.$$

Here we consider β as a map from $S^k \times D^{n-k}$ to $M \times \{1\}$ and we note that $S^k \times D^{n-k}$ is part of the boundary of $D^{k+1} \times D^{n-k}$. The boundary of T is the disjoint union of $M = M \times \{0\}$ and Σ and we call T the *trace of surgery*.

By construction T is homotopy equivalent to $M \cup_{\beta|_{S^k \times \{0\}}} D^{k+1}$. Thus the Seifert–van Kampen theorem implies that $\pi_1(T) \cong \pi_1(M) \cong \{0\}$. The long exact sequence of the pair $(M \cup_{\beta|_{S^k \times \{0\}}} D^{k+1}, M)$ is:

$$\ldots \to H_{i+1}(M) \to H_{i+1}(M \cup_{\beta|_{S^k \times \{0\}}} D^{k+1}) \to \widetilde{H}_{i+1}(S^{k+1})$$

$$\xrightarrow{\partial} H_i(M) \to H_i(M \cup_{\beta|_{S^k \times \{0\}}} D^{k+1}) \to \widetilde{H}_i(S^{k+1}) \to \ldots.$$

Here we have used excision to replace the relative homology group by $\widetilde{H}_i(S^k)$. By construction $\partial \colon \widetilde{H}_{k+1}(S^{k+1}) \to H_k(M)$ maps the generator to the generator given by $[\alpha]$. Thus we conclude that $H_i(M) \to H_i(T)$ is an isomorphism for $i \neq k$ and $H_k(T) = 0$. The reader should add details to the computation of the fundamental group and the homology groups (see Exercise 10.1)

Next we make an important geometric observation. Let $\gamma \colon D^{k+1} \times S^{n-k-1} \hookrightarrow \Sigma = M - \beta(S^k \times D^{n-k} \cup D^{k+1} \times S^{n-k-1})$ the obvious embedding. Then obviously M is obtained from Σ by surgery using the embedding γ. Moreover, the trace of this surgery is again T. The reader should look at a picture of T until he "sees this."

Thus the considerations above imply that for $j \leq n - k - 1$ the inclusion induces an isomorphisms

$$\pi_1(\Sigma) \xrightarrow{\cong} \pi_1(T) = 0,$$

and

$$H_j(\Sigma) \xrightarrow{\cong} H_j(T).$$

Combining this with the information above we conclude that for $r \leq n/2$ we have

$$\widetilde{H}_r(\Sigma) = 0$$

and

$$\pi_1(\Sigma) = 0.$$

By Poincaré duality this implies that $\widetilde{H}_r(\Sigma) = 0$ for $r \neq n$. Thus Σ is a homotopy sphere. We summarize:

Theorem 10.1. *Let M and the embedding $\beta \colon S^k D^{n-k} \hookrightarrow M$ be as above. Then surgery with β replaces M by a homotopy sphere Σ.*

10.3 Application to Knottings

The following section is not related to the Novikov Conjecture (except that we use Novikov's result about the homeomorphism invariance of the rational Pontrjagin classes). It gives a nice geometric application of the previous result to knottings.

A (generalized) *knot* is an r-dimensional submanifold $K \subseteq S^n$. Two knots (S^n, K) and (S^n, K') are called equivalent if there is a homeomorphism $f \colon S^n \to S^n$ mapping K to K', i.e., $f(K) = K'$. Of particular interest is the situation where K is itself homeomorphic to a sphere S^r. Such a knot is called a *spherical knot*. The standard embedding $S^r \subseteq \mathbb{R}^{r+1} \subseteq \mathbb{R}^{n+1}$ gives an embedding $S^r \subseteq S^n$ which is called the *unknot*. It is natural to ask for which r and n there are non-trivial knots. One should exclude examples which are non-trivially knotted for obvious reasons. Such a reason is, for example, if K has no trivial tubular neighborhood, i.e., an embedding $K \times D^{n-r} \hookrightarrow S^n$ extending the inclusion of K to S^n. If $r = n - 2$ one has many examples of knottings, in particular for $n = 3$, the classical knots. But the methods used there to distinguish from the unknot do not apply to $r > n - 2$.

After this introduction we return to the manifolds M as considered in the beginning of this chapter. Surgery on $S^k \times D^{n-k} \subseteq M$ leads to a homotopy sphere Σ. As a consequence of the h-cobordism theorem we have shown that Σ is homeomorphic to S^n if $n > 4$ (Theorem 7.4) (by a much a deeper result this is also true for $n = 4$ [94]). Since $\Sigma = (M - S^k \times D^{n-k}) \cup D^{k+1} \times S^{n-k-1}$, we have an embedding of $\{0\} \times S^{n-k-1}$ to Σ and if we choose a homeomorphism to S^n, we obtain a knotting of S^{n-k-1} in S^n with trivial tubular neighborhood (here $r = n - k$). It is natural to ask whether by this construction one can obtain knottings which are not equivalent to the unknot.

As an interesting example we consider the situation where $n = 6$ and $k = 2$. The situation is particularly simple, since for such a manifold M a 2-sphere representing a generator of $H_2(M)$ has trivial normal bundle if and only if the manifold is a spin manifold which is equivalent to $w_2(M) = 0$. Furthermore the normal bundle has then a canonical framing, since $\pi_2(SO(4)) = 0$. Thus there is a canonical knot associated to M, if M is spin, once we have distinguished a generator of $H_2(M)$.

One next proves that, if x is a generator of $H^2(M)$ (dual to a fixed generator of $H_2(M)$), then the corresponding knot is the unknot if and only if the expression $\langle x \cup p_1(M), [M] \rangle - 4 \cdot \langle x^3, [M] \rangle$ vanishes. This is shown by starting from the unknot and computing this expression for all manifolds obtained by surgery on the unknot (the different possibilities for this correspond to the different framings of the normal bundle of the standard embedding of S^3 to S^6). Since we work with topological manifolds we have to use Novikov's result about the homeomorphism invariance of the rational Pontrjagin classes.

Finally one needs to construct 6-manifolds M of the type under consideration, which are spin, such that $\langle x \cup p_1(M), [M] \rangle - 4 \cdot \langle x^3, [M] \rangle \neq 0$. Unfortunately we don't know a simple construction of such manifolds. The closest one can come with explicit examples is hyper-planes $V(d)$ of odd degree d in $\mathbb{CP}^4$, which have all the desired properties except that $H_3(V(d))$ is in general non-trivial. But by surgery on 3-spheres one can pass from $V(d)$ to $V'(d)$ with $H_3(V'(d)) = 0$ (the basic idea behind this will be explained in Chapter 14). These manifolds are closely related: $V(d) = V'(d)\natural_l(S^3 \times S^3)$. Thus

$$\langle x \cup p_1(V'(d)), [V'(d)]\rangle - 4 \cdot \langle x^3, [V'(d)]\rangle \ = \ \langle x \cup p_1(V(d)), [V(d)] > -4 < x^3, [V(d)]\rangle.$$

It is a good exercise (see Exercise 10.2) to compute this expression. The answer is

$$\langle x \cup p_1(V'(d)), [V'(d)]\rangle - 4 \cdot \langle x^3, [V'(d)]\rangle \ = \ d(5 - d^2) - 4d,$$

which is non-zero if $d > 1$. Moreover the value is different for different d's leading to the result that the corresponding knottings are pairwise non-equivalent. The existence of infinitely many knottings was originally proven by [106],

Chapter 11

Surgery Below the Middle Dimension II: Systematically

In the following chapters we give a brief introduction to surgery. This is a complicated theory and we cannot treat all aspects in these lecture notes. We recommend consulting other articles and books. For example [33], [129], [153], [203] and [248].

11.1 The Effect of Surgery in Homology and Homotopy

We recall that for simplicity's sake we assume that all manifolds are oriented.

Let W be a compact connected manifold of dimension n and

$$\varphi \colon S^k \times D^{n-k} \hookrightarrow \overset{\circ}{W}$$

an embedding. We say that

$$W' := W - \varphi(S^k \times \overset{\circ}{D}{}^{n-k}) \cup_{\varphi^{-1}|S^k \times S^{n-k-1}} D^{k+1} \times S^{n-k-1}$$

is obtained from W by *surgery via* φ.

As in the special case studied in the last chapter, we want to see how the homology of W' relates to the homology of W. But since W is in general not 1-connected, we are more interested in the homology groups of the universal coverings of W and W'. The reason is that if $\partial W = M_0 + M_1$ with M_i and W connected, then W is an h-cobordism if and only if for $i = 0, 1$:

$$i_* \colon \pi_1(M_i) \overset{\cong}{\to} \pi_1(W)$$

and if for all j:

$$i_* \colon H_j(\widetilde{M_i}) \overset{\cong}{\to} H_j(\widetilde{W}).$$

This follows again from the Whitehead and Hurewicz theorems (see Exercise 11.1).

In fact, using a similar argument as in our special case in the last chapter, one sees that it is enough to require instead of this condition that $H_j \widetilde{M_i} \to H_j \widetilde{W}$ is bijective for $j < \left[\frac{n}{2}\right]$ and that $H_{\left[\frac{n}{2}\right]} \widetilde{M_i} \to H_{\left[\frac{n}{2}\right]} \widetilde{W}$ is surjective.

Now we return to the situation above. Let W be a compact manifold of dimension n and $\varphi \colon S^k \times D^{n-k} \hookrightarrow \overset{\circ}{W}$ an embedding. Besides W', the result of surgery via φ, we look at the result of attaching a handle to $W \times I$ denoted by T.

We recall that $H_j(\widetilde{W})$ is a module over $\mathbb{Z}[\pi_1(W)]$, where elements of $\pi_1(W)$ act on $H_j(\widetilde{W})$ via deck transformation and similarly the fundamental groups act on the homology groups of the other manifolds $\widetilde{W'}$ and $\widetilde{T}$. It is a standard notation to use for $H_j(\widetilde{W})$ considered as module over $\mathbb{Z}[\pi_1(W)]$ the notation: $H_j(W; \mathbb{Z}[\pi_1(W)])$, *the homology of W twisted coefficients.*

The homotopy groups depend on the choice of a base point. We always choose the base point in W disjoint from the image of φ and use the "same" base point for T. Of course, if we compare W' with T we have to choose a base point in W'. We take the same base point as for W. To compare the homotopy groups of T with respect to the two base points coming from W and W' we have to give a path between them. For this we take of course the path given by $x_0 \times I$, if x_0 denotes the base point in W.

Proposition 11.1 (effect of surgery). *Let W be connected, $\varphi \colon S^k \times D^{n-k} \hookrightarrow \overset{\circ}{W}$ as above and $1 < k < n-1$. Then*

i) $T := W \times I \cup_\varphi D^{k+1} \times D^{n-k}$ *is a bordism relative boundary between W and W'.*

ii) $\pi_1(W) \to \pi_1(T)$ *and* $\pi_1(W') \to \pi_1(T)$ *are isomorphisms.*

iii) *If* $k < \frac{n-1}{2}$, *then for* $i < k$

$$H_i(W; \mathbb{Z}[\pi_1]) \to H_i(T; \mathbb{Z}[\pi_1])$$

$$H_i(W'; \mathbb{Z}[\pi_1]) \to H_i(T; \mathbb{Z}[\pi_1])$$

are isomorphisms and

$$H_k(W'; \mathbb{Z}[\pi_1]) \cong H_k(W; \mathbb{Z}[\pi_1]) / \mathbb{Z}[\pi_1]\, x,$$

where $\pi_1 = \pi_1(W) \cong \pi_1(T)$ *and* $x \in H_k(\widetilde{W})$ *is the image of* $\left[\varphi|_{S^k \times \{0\}}\right] \in \pi_k(W)$ $\cong \pi_k(\widetilde{W})$ *under the Hurewicz map* $\pi_k(W; \mathbb{Z}[\pi_1]) \to H_k(\widetilde{W}) = H_k(W; \mathbb{Z}[\pi_1])$.

iv) *If* $k = 1$ *then* $\pi_1(W') \cong \pi_1(T)$ *but* $\pi_1(T) \cong \pi_1(W)/N(x)$, *where $N(x)$ is the normal subgroup generated by x.*

Proof. The proof is similar to the proof of the special case in the last chapter. The statements i) and ii) are obvious and for the third statement we note that $\widetilde{T}$ is a bordism between $\widetilde{W}$ and $\widetilde{T'}$ and that

$$\widetilde{T} = \widetilde{W} \times I \cup_{\widetilde{\varphi}} (+(D^{k+1} \times D^{n-k})),$$

where $+D^{k+1} \times D^k$ is the sum of as many copies of $D^{k+1} \times D^{n-k}$ as the order of $\pi_1(W)$ is. In other words, it is the total space of the trivial $\pi_1(W)$-covering over $D^{k+1} \times D^{n-k}$. Since $k > 1$, the restriction of the universal covering of W to $\varphi(S^k \times D^{n-k})$ is trivial and $\widetilde{\varphi}$ is the lift of φ to the universal covering.

Similarly, $\widetilde{T} = \widetilde{W'} \times I \cup_{\widetilde{\varphi}} (+(D^{k+1} \times D^{n-k}))$ and by a Mayer–Vietoris argument as in the last chapter of the proof of iii) is achieved.

The last point iv) follows from the Seifert–van Kampen Theorem. $\qquad\square$

11.2 Surgery below the Middle Dimension

If $\partial W = M_0 + M_1$ and $\pi_1(M_i) \overset{\cong}{\to} \pi_1(W)$, it would be nice if we could use this proposition to change the homology of W inductively until the maps

$$H_j(M_j; \mathbb{Z}[\pi_1]) \to H_j(W; \mathbb{Z}[\pi_1])$$

are isomorphisms. But this is impossible for various reasons. We have to add more information to have a chance. The additional data are given by normal maps as defined in Chapter 2. We recall that for a connected CW-complex X, together with a stable vector bundle E over X, a normal map in (X, E) is a triple (M, f, α), where M is an n-dimensional smooth manifold, $f \colon M \to X$ a continuous map and α an isomorphism between f^*E and the normal bundle $\nu(M)$.

A particularly interesting example is where W is a bordism between two manifolds M_0 and M_1, and X be equal to M_0, and E the normal bundle of M_0, and (f, α) a normal map from W to (M_0, E), such that $f|_{M_0}$ and $f|_{M_1}$ are homotopy equivalences. If in addition $f \colon W \to M_0$ is a homotopy equivalence, then both inclusions $M_0 \to W$ and $M_1 \to W$ are homotopy equivalences and so W is an h-cobordism. If f is not a homotopy equivalence, we will try to replace it by a homotopy equivalence via surgeries.

This is "almost" possible. By this we mean that after surgeries we may assume that $\pi_1(M_i) \to \pi_1(W)$ is an isomorphism and $H_j(W, M_i; \mathbb{Z}[\pi_1]) = 0$ for $j < [\frac{m}{2}]$, where m is the dimension of W. (Note that if also $H_{[\frac{m}{2}]}(W, M_1; \mathbb{Z}[\pi_1]) = 0$, then W is an h-cobordism.) If we use the map to M_0 we can formulate a slightly stronger condition. Namely the two conditions follow from the condition that

$$f_* \colon \pi_1(W) \to \pi_1(M_0)$$

is an isomorphism, and

$$f_* \colon H_j(W; \mathbb{Z}[\pi_1]) \to H_j(M_0; \mathbb{Z}[\pi_1])$$

is an isomorphism for $j < r = [\frac{m}{2}]$ and surjective for $j = r = [\frac{m}{2}]$. A map f with this property is called a r-*equivalence*.

To formulate the corresponding result we have to define bordism rel. boundary between two normal maps. If (W, f, α) is a normal map as above we say that (W, f, α) is *normally bordant rel. boundary* to (W', f', α') if

- $\partial W = \partial W'$

- $f|_{\partial W} = f'|_{\partial W'}$

- $\alpha|_{\partial W} = \alpha'|_{\partial W}$

- there is a normal map $(g\colon T \to X, \beta)$ to (X, E) with $\partial T = W \cup_{\partial w = \partial W'} W'$, $g|_{\partial T} = f \cup f'$ and $\alpha|_{\partial T} = \alpha \cup \alpha'$.

Note that if (W, f, α) is normally bordant rel. boundary to (W', f', α'), then these two triples are normally bordant, but the converse is not true, since the boundary of W and W' are not necessarily kept unchanged by a normal bordism.

Proposition 11.2 (Surgery below the middle dimension). *Let (W, f, α) be a normal map from an m-dimensional compact manifold W to (X, E) where X is a CW-complex with finite skeleta, then (W, f, α) is normally bordant rel. boundary to a normal map (W', f', α') such that*

$$f'_* \colon \pi_1(W') \overset{\cong}{\to} \pi_1(X)$$

and

$$f'_* \colon H_j(W'; \mathbb{Z}[\pi_1]) \to H_j(X; \mathbb{Z}[\pi_1])$$

is an isomorphism for $j < [\frac{m}{2}]$ and surjective for $j = [\frac{m}{2}]$, in other words: $f'\colon W' \to X$ is a $[\frac{m}{2}]$-equivalence.

Proof. We proceed inductively and assume at the beginning that $f_* \colon \pi_1(W) \overset{\cong}{\to} \pi_1(X)$ and $f_* \colon H_j(W; \mathbb{Z}[\pi_1]) \to H_j(\widetilde{X}; \mathbb{Z}[\pi_1])$ is an isomorphism for $j < k$ and surjective for $j = k$, where $1 < k < [\frac{m}{2}]$. We replace W by W' which is normally bordant rel. boundary to W such that W' has the same property for $k + 1$.

Replacing $f\colon W \to X$ by the mapping cylinder $W \times I \cup_f X$, we can assume that f is an inclusion and consider the relative homology groups $H_j(X, W; \mathbb{Z}[\pi_1])$ and $\pi_j(\widetilde{X}, \widetilde{W}) \cong \pi_j(X, W)$ for $j > 1$. By assumption $H_j(X, W; \mathbb{Z}[\pi_1]) = \{0\}$ for $j \leq k$ and the Hurewicz theorem implies that $\pi_j(X, W) = \{0\}$ for $j \leq k$ and

$$\pi_{k+1}(X, W) \cong H_{k+1}(X, W; \mathbb{Z}[\pi_1]).$$

Combining this with the exact homology sequence of the pair

$$H_{k+1}(X, W; \mathbb{Z}[\pi_1]) \to H_k(W; \mathbb{Z}[\pi_1]) \overset{f_*}{\to} H_k(X; \mathbb{Z}[\pi_1]) \to 0$$

we conclude that the kernel of

$$f_* \colon H_k(W; \mathbb{Z}[\pi_1]) \to H_k(X; \mathbb{Z}[\pi_1])$$

is equal to the image of

$$\pi_{k+1}(X, W) \cong H_{k+1}(X, W; \mathbb{Z}[\pi_1]) \overset{d}{\to} H_k(W; \mathbb{Z}[\pi_1]).$$

Thus, all elements in the image of d are representable by map $g\colon S^k \to W$. Since all skeleta of X are finite and also W is homotopy equivalent to a finite CW-complex (W is compact), the $\mathbb{Z}[\pi_1(X)]$-module $\mathrm{im}(d) = \ker(H_k(W;\mathbb{Z}[\pi_1]) \to H_k(X;\mathbb{Z}[\pi_1]))$ is finitely generated. We choose a map $g\colon S^k \to W$ representing a generator and apply Whitney's embedding theorem to approximate it by an embedding ($k < \dim(W)/2$)

$$g\colon S^k \hookrightarrow \overset{\circ}{W}.$$

Since fg is homotopic to zero and $\nu(W)$ is isomorphic to f^*E the stable normal bundle of $g(S^k)$ in $\overset{\circ}{W}$ is trivial. Since the dimension of the normal bundle is $> k$, this implies that the normal bundle is trivial (see Exercise 11.2). Thus we can extend g to an embedding

$$\varphi\colon S^k \times D^{m-k} \hookrightarrow \overset{\circ}{W}.$$

This extension depends on the choice of a trivialization of the normal bundle.

We attach a handle to $\overset{\circ}{W}$ via g to obtain a bordism rel. boundary

$$T := W \times I \cup_\varphi D^{k+1} \times D^{m-k}$$

between W and

$$W' = (W - \varphi(S^k \times \overset{\circ}{D}{}^{m-k})) \cup D^{k+1} \times S^{m-k-1}.$$

By Proposition 11.1 the fundamental group of W and the homology groups of $\widetilde{W}$ agree with that of $\widetilde{W}'$ for degree $< k$ and

$$H_k(W';\mathbb{Z}[\pi_1]) \cong H_k(W;\mathbb{Z}[\pi_1])/\mathbb{Z}[\pi_1]\, g_*([S^k \times \{0\}]).$$

We further note that, since $f_* g_*([S^k])$ is zero in $\pi_k(X)$, the map $f\colon W \to X$ can be extended to a map $F\colon T \to X$ whose restriction to W' we denote by f'. Again using $f_* g_*([S^k]) = 0$, we conclude that the pullback of E to S^k is trivial. We would like to extend this trivialization to T. Since the normal bundle of $D^{k+1} \times D^{n-k}$ is trivial, we have a canonical (up to isotopy) trivialization on the handle. The two trivializations on $S^k \times D^{n-k}$ must not agree. But — as mentioned above — we have freedom in choosing the trivialization of the normal bundle of S^k and using this freedom we can achieve that we can extend the isomorphism α between $f^*(E)$ and $\nu(T)$ to an isomorphism between $F^*(E)$ and $\nu(T)$ whose restriction to W' we denote by α'.

We summarize these considerations:

For each element $z \in \ker(H_k(W;\mathbb{Z}[\pi_1]) \to H_k(X;\mathbb{Z}[\pi_1]))$ we can construct a normal cobordism T from (W, f, α) to (W', f', α') such that

i) $\pi_1(W') \overset{\cong}{\to} \pi_1(X)$

ii) $H_j(W'; \mathbb{Z}[\pi_1]) \overset{\cong}{\to} H_j(W; \mathbb{Z}[\pi_1])$ *for* $j < k$

iii) $H_k(W'; \mathbb{Z}[\pi_1]) \cong H_k((W; \mathbb{Z}[\pi_1])/\mathbb{Z}[\pi_1(W)]z$.

Since the kernel of $H_k(\widetilde{W}) \to H_k(\widetilde{X})$ is finitely generated, we can — after finitely many such constructions — replace the third condition by

iii) $H_k(\widetilde{W'}) \overset{\cong}{\to} H_k(\widetilde{X})$

To finish the inductive step, we must achieve that:

iv) $H_{k+1}(W'; \mathbb{Z}[\pi_1]) \to H_{k+1}(X; \mathbb{Z}[\pi_1])$ is surjective.

Using the assumptions i)–iii), the exact homology and homotopy sequence of the pair $(\widetilde{X}, \widetilde{W'})$ and the Hurewicz isomorphism, we see that the condition iv) is equivalent to:

$$\pi_{k+1}(W) \to \pi_{k+1}(X)$$

is surjective. Since X is a finite CW-complex, $\pi_{k+1}(X)$ is a finitely generated $\mathbb{Z}[\pi_1(X)]$-module. Let $y_1, \ldots, y_r$ be generators. We replace W' by

$$W' \natural_r (S^{k+1} \times S^{m-k-1})$$

and "extend" f' by a map which maps the p'th summand $S^{k+1} \times S^{m-k-1}$ to S^{k+1} and then to the p'th generator y_p, we can achieve condition iv).

It is easy to construct a normal map on $W'_{\natural_r}(S^{k+1} \times S^{m-k-1})$ to X which is normally bordant rel. boundary to W'. This finishes the inductive step.

Finally we have to get the induction started by achieving that W is connected and the map to X is an isomorphism on π_1. We make W connected by forming the connected sum of the components. There is an obvious normal bordism between these manifolds. Since $\pi_1(X)$ is finitely presented we can make the map a surjection on π_1 by forming the connected sum with $S^1 \times S^{m-1}$, and then we can kill the kernel by surgeries using iv) of Proposition 11.1. $\square$

11.3   Construction of Certain 6-Manifolds

We now return to the situation in chapter 0, where we indicated the construction of certain 6-manifolds. The starting point was a closed 4-dimensional spin-manifold M. Then we considered $T^2 \times M$ and wanted to replace this by a bordant 6-manifold $N(M)$ with $\pi_1(N(M)) \cong \pi_1(T^2 \times M) \cong \mathbb{Z} \oplus \mathbb{Z}$, and $\pi_2(N(M)) = 0$. Now we can add details to this construction. For this we consider $X := T^2 \times BSpin$, where $BSpin$ is the classifying space of the stable Spin group, the universal covering of SO. We equip X with the stable vector bundle obtained from pulling the stable vector bundle over BSO back under the map $T^2 \times BSpin \to BSpin \to BSO$. Now we note that, since M is a spin manifold, the normal Gauss map factors through $BSpin$, and the identity on T^2 together with this map gives a map $f : T^2 \times M \to X$, under which the stable vector bundle over X pulls back to the normal bundle of $T^2 \times M$.

Now we apply Proposition 11.2 and conclude that $(T^2 \times M, f)$ is normally bordant to $(N(M), f')$, where f' is a 3-equivalence. Since $\pi_2(X) = 0$ this implies that $\pi_1(N(M)) \cong \pi_1(T^2 \times BSpin) \cong \mathbb{Z} \oplus \mathbb{Z}$ and $\pi_2(N(M)) = 0$, as was desired in Chapter 0. It is not clear from the construction that $N(M)$ is well defined and we don't claim this.

Chapter 12

Surgery in the Middle Dimension I

12.1   Motivation for the Surgery Obstruction Groups

In this chapter we define (following Wall [248]) the L-groups $L_m^h(\pi_1(X))$. In the next chapter we associate under certain conditions to each normal bordism class (W, f, α) an element

$$\Theta(W, f, \alpha) \subset L_m^h(\pi_1(X)).$$

We are mainly interested in the case where X is a closed manifold and W is a bordism between manifolds M_0 and M_1 and $f|_{M_i}$ is a homotopy equivalence. Then the geometric meaning of Θ is demonstrated by the result that $\Theta(W, f, \alpha) = 0$ if and only if (W, f, α) is normally bordant rel. boundary to an h-cobordism (assuming that the dimension of W is larger than 4). If in addition we can control the Whitehead torsion in such a way that it vanishes, then we obtain an s-cobordism and, if $m > 5$, the s-cobordism theorem implies that M_0 is diffeomorphic to M_1. This gives the fundamental tool for classifying manifolds in dimension > 5.

To give a first feeling for the groups $L_m^h(\pi_1(X))$ and the invariant $\Theta(W, f, \alpha)$, we note that if $\pi_1(X) = 0$ and $m = 0 \mod 4$, then $L_m^h(0) \cong \mathbb{Z}$ and $\Theta(W, f, \alpha)$ corresponds to the signature of W under this isomorphism. This looks like a good starting point for the Novikov Conjecture in the sense that the higher signatures are a generalization of the signature and sit in $h_m(B\pi_1) = \bigoplus_k H_{m-4k}(B\pi_1) \otimes \mathbb{Q}$, an abelian group attached to a group π_1. Although — as we will see — the definition of the L-groups attached to a group π_1, $L_m^h(\pi_1)$, is very different from the definition of the homology groups of $B\pi_1$, the fact that these groups agree (after taking the tensor product with $\mathbb{Q}$) for π_1 trivial and that then $\Theta(W, f, \alpha)$ is given by the signature, leaves room for speculations concerning the relation of these groups for non-trivial π_1. These speculations will be supported even more when we see that for $m = 0 \mod 4$ the invariant Θ is a natural generalization of the signature.

12.2   Unimodular Hermitian Forms

We will begin with the groups $L^h_{4m}(\pi_1)$. The groups $L^h_{4m+2}(\pi_1)$ are defined in a similar way, as we will indicate. The odd-dimensional L-groups are defined in a different way, which we will explain. We abbreviate the group ring $\mathbb{Z}[\pi_1]$ by R. As we have seen earlier, R is equipped with an anti-involution — mapping $g \in \pi_1$ to g^{-1}. For a ring R with anti-involution we define an *unimodular hermitian form*. It is a finitely generated free R-module A (i.e., A is isomorphic to R^k for some k) together with a map

$$\lambda : A \times A \to R$$

such that

i) for each fixed $\lambda \in A$ the map

$$A \to R, \quad x \mapsto \lambda(x,y)$$

is linear,

ii) $\lambda(x,y) = \overline{\lambda(y,x)}$,

iii) the associated map

$$A \to A^*, \quad y \mapsto (x \mapsto \lambda(x,y))$$

is an isomorphism, where $A^* = Hom_R(A, R)$.

We also need the concept of a *quadratic refinement* of a hermitian unimodular form λ. This is a map

$$q \colon A \to R/\{a - \overline{a}| a \in R\}$$

such that

iv) $\lambda(x,x) = q(x) + \overline{q(x)} \in R$,

v) $q(x + y) = q(x) + q(y) + [\lambda(x,y)] \in R/\{a - \overline{a}\}$,

vi) $q(ax) = aq(x)\overline{a} \in R/\{a - \overline{a}\}$.

Here we note that iv) has to be interpreted as follows. Choose a representative $b \in R$ for $q(x)$ and consider $b + \overline{b}$. If we change b by adding some element $a - \overline{a}$, then $b + \overline{b}$ is replaced by $b + a - \overline{a} + \overline{b} + \overline{a} - a = b + \overline{b}$. Thus, although b is not well defined if only $[b] \in R_{\{a - \overline{a}\}}$ is given, the sum $b + \overline{b}$ is well defined in R, so that the equation iv) makes sense in R. For equation vi) we have to convince ourselves that, if $b \in R$ represents $q(x)$, then $[ab\overline{a}] \in R/\{a - \overline{a}\}$ is independent of the choice of the representative b, which the reader can easily check.

Here is a simple and important example, the *hyperbolic form* H, in which $A = R \oplus R$ with basis e and f and λ is given by $\lambda(e,e) = \lambda(f,f) = 0$ and $\lambda(e,f) = 1$. The quadratic refinement is given by $q(e) = q(f) = 0$.

We abbreviate "unimodular hermitian bilinear forms together with a quadratic refinement" to *"unimodular hermitian quadratic form"*.

12.3   The L-Groups in Dimensions $4m$

We say that two unimodular hermitian quadratic forms (A, λ, q) and (A', λ', q') are *isomorphic*, if there is an isomorphism $\varphi : A \to A'$ such that $\lambda'(\varphi(x), \varphi(y)) = \lambda(x, y)$ for all $x, y \in A$ and $q'(\varphi(x)) = q(x)$. We say that (A, λ) and (A', λ', q') are *stably isomorphic* if there are k and l such that

$$(A, \lambda) \perp H^k \cong (A', \lambda') \perp H^l.$$

Here, $\perp$ is the orthogonal sum and H^k stands for the orthogonal sum of k copies of H.

Definition 12.1. *The abelian group of stable isomorphic classes of unimodular hermitian quadratic forms over the group ring $R = \mathbb{Z}[G]$, where G is a group, is called*

$$L_{4m}^h(G).$$

We have to explain how we define the structure of an abelian group. The sum is given by the orthogonal sum

$$[(A, \lambda, q)] + [A', \lambda', q'] \; := \; [A \oplus A', \lambda \perp \lambda', q \perp q'].$$

The neutral element is given by the hyperbolic form H. One has to prove that one has an inverse for $[(A, \lambda, q)]$.

Lemma 12.2. *If (A, λ, q) is an unimodular hermitian quadratic form, then*

$$(A, \lambda, q) \perp (A, -\lambda, -q)$$

is isomorphic to H^k, where k is the rank of the free module A.

The proof is a consequence of the following useful algebraic result, which is a generalization of the argument for $R = \mathbb{Z}$ (G trivial):

Lemma 12.3. *Let (A, λ, q) be an unimodular hermitian quadratic form of rank $2k$. If there is a direct free summand $B \subseteq A$ of rank k such that*

$$\lambda|_{B \times B} = 0$$

$$q|_B = 0,$$

then (A, λ, q) is isomorphic to H^k.

Proof. We choose a basis $x_1, \ldots, x_k$ of B. Since $\lambda|_{B \times B = 0}$ and λ is unimodular, there exist $y_1, \ldots, y_k \in A$ such that

$$\lambda(x_i, y_j) = \delta_{ij}.$$

If $\lambda(y_i, y_j) = 0$ and $q(y_i) = 0$ for all i and j, the lemma follows since then we find an isomorphism mapping e_i, f_i in the i-th summand of H^k to x_i, y_i.

If $\lambda(y_i, y_j) = 0$ and $q(y_i) = 0$ is not fulfilled, we proceed inductively. We first consider the case $k = 1$, so that we have a basis x, y of A with $\lambda(x, x) = 0$ and $q(x) = 0$ and $\lambda(x, y) = 1$. Now, we replace y by $y' := y - ax$ where a is a representative of $q(y)$. Then again $\lambda(x, y') = 1$ and $\lambda(y', y') = \lambda(y - ax, y - ax) = \lambda(y, y) - (a + \bar{a}) = 0$, since $a + \bar{a} = \lambda(y, y)$. And $q(y') = q(y - ax) = q(y) - q(ax) - \lambda(y, ax) = q(y) - \bar{a} = [a - \bar{a}] = 0$.

Now the proof follows inductively by considering $A' := \langle x, y' \rangle^{\perp}$ and noting that $x_2, \ldots, x_k$ is contained in A'. Thus, by induction hypothesis, we can find in A', which has rank $2k - 2$, elements $y_2, \ldots, y_k$ such that our conditions are fulfilled. $\qquad\square$

Now we are ready to give the proof of Lemma 12.2.

Proof. We apply the last lemma by choosing an arbitrary basis $a_1, \ldots, a_k$ of A and considering in $A \oplus A$ the diagonal elements

$$x_i := (a_i, a_i).$$

Since $\lambda(x_i, x_j) = 0$ and $q(x_i) = 0$, the proof follows. $\qquad\square$

Thus $L_{4m}^h(G)$ is an abelian group.

The computation of these groups is very hard and unknown in general. Already the case where G is trivial is very interesting and has many geometric applications. Since the signature of a hyperbolic form is zero, the signature gives a homomorphism

$$\mathrm{sign}\colon L_{4m}^h(1) \to \mathbb{Z}.$$

This is an injective homomorphism and non-trivial (see Exercise 12.1). It is not surjective since one can rather easily show that the elements in the image are divisible by 2 (see Exercise 12.2). With more effort one can determine the image and show that it consists of all numbers divisible by 8 [170]. Thus we obtain the computation of $L_{4m}^h(1)$:

Theorem 12.4. *The signature gives an isomorphism*

$$\mathrm{sign}\colon L_{4m}^h(1) \to 8\mathbb{Z}.$$

12.4   The L-Groups in Other Dimensions

The groups $L_{4m+2}^h(G)$ are very similar to $L_{4m}^h(G)$. The only difference is that instead of hermitian forms, we use *skew-hermitian* forms. This means that in the definition of unimodular hermitian bilinear forms we replace condition ii) by

$$\lambda(y, x) = -\overline{\lambda, (x, y)}$$

and for the quadratic refinement we consider

$$q\colon A \to R_{\{a + \bar{a}\}}$$

and replace condition iv) by

$$\lambda(x, x) = q(x) = q(x) - \overline{q(x)}.$$

The hyperbolic form is now given by $R \oplus R$, $\lambda(e, f) = 1$, $\lambda(f, e) = -1$, $\lambda(e, e) = \lambda(f, f) = 0$, and $q(e) = q(f) = 0$. With this we define $L^h_{4m+2}(G)$ as the set of stable isomorphism classes of skew-hermitian unimodular quadratic forms over $\mathbb{Z}[G]$. Again this becomes a group via orthogonal sum.

Also here the computation is hard. For the trivial group one can rather easily show that the L-group $L^h_{4m+2}(1)$ s non-trivial (see Exercise 12.3). More generally one has an invariant which is closely related to the signature, the Arf invariant [170]. This is a homomorphism

$$\mathrm{Arf} \colon L^h_{4m+2}(1) \to \mathbb{Z}/2.$$

One can show

Theorem 12.5. *The Arf invariant gives an isomorphism*

$$\mathrm{Arf} \colon L^h_{4m+2}(1) \xrightarrow{\cong} \mathbb{Z}/2.$$

The definition of the odd L-groups $L^h_{2m+1}(G)$ is more complicated but we have already seen similar groups. Namely we have previously defined the K-groups $K_0(R)$ and $K_1(R)$. One can consider L^h_{4m} and L^h_{4m+2} as analogous to K_0 in the sense that instead of equivalence classes of modules, we consider equivalence classes of modules with quadratic forms.

In analogy to $K_1(R)$ which by definition is the abelianization of $GL(R)$, i.e., $GL(R)/[GL(R), GL(R)]$ we consider the isometries of the hyperbolic form H^k (with the hermitian form for defining L^h_{4m+1}, and the skew hermitian forms for defining L^h_{4m+3}). We denote the group of isometries by $U(H^k)$, where $R = \mathbb{Z}[G]$. As for $GL(R)$ we pass to the limit under the inclusions $U(H^k) \subseteq U(H^{k+1}) \subseteq \ldots$ and denote this limit by $U(R)$. The abelianization of $U(R)$ is almost $L^h_{4m+1}(G)$, we have to divide by an element of order 2, namely the isometry σ on H which interchanges e and f. Thus we define

$$L^h_{4m+1}(G) := U(\mathbb{Z}[G])/\langle [U(\mathbb{Z}[G]), U(\mathbb{Z}[G])], \sigma \rangle.$$

Similarly, if we start with the skew-hermitian hyperbolic form H and replace σ by the map which maps e to $-f$ and f to e, we obtain the groups

$$L^h_{4m+3}(G).$$

For a trivial group the odd dimensional L-groups are zero:

Theorem 12.6. *The groups $L^h_{2k+1}(0)$ are trivial.*

Indirectly this follows from the algebraic input in the geometric arguments in [129], for a purely algebraic proof see [248].

Although the following remark is completely obvious from the definition of the L-groups, it is important in the geometric context: The L-groups are 4-periodic, i.e.,

$$L^h_{m+4}(G) = L^h_m(G).$$

Chapter 13

Surgery in the Middle Dimension II

13.1 Equivariant Intersection Numbers

We recall that if (W, f, α) is a normal map to (X, E), where X is a connected CW-complex with finite skeleta and E is a stable vector bundle, this normal map is normally bordant to (W', f', α') such that

i) $f'_* \colon \pi_1(W') \overset{\cong}{\to} \pi_1(x)$

ii) $f'_* \colon H_j(W'; \mathbb{Z}[\pi_1]) \to H_j(X; \mathbb{Z}[\pi_1])$

is an isomorphism for $j < \left[\frac{m}{2}\right]$ and surjective for $j = \left[\frac{m}{2}\right]$, in other words, f' is a $[m/2]$-equivalence. Here $m = \dim(W)$, and we assume that $m > 4$.

We are mainly interested in the special case where W is a bordism between two closed manifolds M_0 and M_1 and $f \colon M_i \to X$ are homotopy equivalences. Then one can define an invariant

$$\Theta(W, f, \alpha) \in L^h_m(\pi_1(W))$$

such that

$$\Theta(W, f, \alpha) = 0$$

if and only if (W, f, α) is bordant to (W', f', α') such that W' is an h-cobordism between M_0 and M_1. We explain the meaning of the letter h in the notation of $L^h_m(\pi_1(W))$, it stresses that we investigate homotopy equivalences and h-cobordisms. If we study simple homotopy equivalences and s-cobordisms there are corresponding groups $L^s_m(\pi_1(W))$.

We firstly define (following Wall [248]) the invariant if $m = 2k = 0 \mod 4$ (under some additional conditions) and indicate the definition in the other cases at the end of this chapter. We begin with a definition of intersection numbers between elements in $H_{2k}(W; \mathbb{Z}[\pi_1])$. We actually do this only for elements which are represented by a map

$$g \colon M \to \widetilde{W},$$

where M is a closed oriented smooth manifold and g is a smooth map and $\widetilde{W}$ is the universal covering of W. Then if $g' \colon M' \to \widetilde{W}$ is another such data, we can consider the integral intersection number $g_*([M]) \circ g'_*([M']) \in \mathbb{Z}$.

Now we enrich this information by using the operation of $\pi_1(W)$ on $\widetilde{W}$ by deck transformations. Passing to the induced operation of $\pi_1(W)$ on $H_k(\widetilde{W})$, we can consider for homology classes α and β the expression

$$\lambda(\alpha, \beta) := \Sigma_{g \in \pi_1(W)}(\alpha \circ g_*(\beta))\beta \in \mathbb{Z}[\pi_1(W)],$$

the *equivariant intersection number*.

More or less by construction of the intersection numbers it is clear that they give a hermitian, if k is even, (skew-hermitian, if k is odd) bilinear form on the $\mathbb{Z}[\pi_1]$-module $H_k(W; \mathbb{Z}[\pi_1])$.

In general, $H_k(W; \mathbb{Z}[\pi_1])$ is not a free $\mathbb{Z}[\pi_1]$-module, but the kernel

$$K_k(W, f)$$

of $H_k(W; \mathbb{Z}[\pi_1]) \to H_k(X; \mathbb{Z}[\pi_1])$ is stably free, if W fulfills the assumption that $\partial W = M_0 + M_1$ and $f|_{M_i}$ is a homotopy equivalence, and $f \colon W \to X$ is a $2k$-equivalence (see Exercise 13.2).

13.2   Stably Free Modules

Here, stably free means that $K_k(W, f) \oplus \mathbb{Z}[\pi_1(W)]^n$ is free for some n. The argument uses the Poincaré–Lefschetz duality for $\widetilde{W}$ and the universal coefficient theorem relating cohomology (with compact support) to homology. Namely, if $\partial W = M_0 + M_1$, then there is an isomorphism

$$H_c^i(\widetilde{W}, \widetilde{M_1}) \xrightarrow{\cong} H_{2k-i}(\widetilde{W}, \widetilde{M_0}),$$

where H_c^i is cohomology with compact support. The universal coefficient theorem relates $H_c^i(\widetilde{W}, \widetilde{M_1})$ to $H_i(\widetilde{W}, \widetilde{M_1})$ and $H_{i-1}(\widetilde{W}, \widetilde{M_1})$. We don't want to go into details. For us it is enough to note that if both homology groups are trivial, then $H_c^i(\widetilde{W}, \widetilde{M_1})$ is zero.

Next we note that in our situation

$$H_j(\widetilde{W}, \widetilde{M_i}) = 0$$

for $0 \le j < k$. The reason is that both $H_j(\widetilde{W})$ and $H_j(\widetilde{M_1})$ are mapped isomorphically to $H_j(\widetilde{X})$ by assumption and so

$$H_j(\widetilde{M_i}) \xrightarrow{\cong} H_j(\widetilde{W})$$

implying the statement by the long exact sequence of a pair.

Combining this with Lefschetz duality and the universal coefficient theorem, we conclude that the only potentially non-trivial homology group of $(\widetilde{W}, \widetilde{M_1})$ is $H_k(\widetilde{W}, \widetilde{M_1})$. But if all homology groups, except perhaps one, are trivial, this homology group is a stably free $\mathbb{Z}[\pi_1(W)]$-module. This is an exercise, where one plays with exact sequences and the definitions, and which we recommended to the reader (see Exercise 13.1).

Finally, we note that the inclusion $H_k(\widetilde{W}) \to H_k(\widetilde{W}, \widetilde{M_1})$ induces an isomorphism

$$K_k(W, f) \xrightarrow{\cong} H_k(\widetilde{W}, \widetilde{M_1}).$$

The argument is given by diagram chasing in the diagram given by the exact sequences of $(\widetilde{X}, \widetilde{W})$ and $\widetilde{W}, \widetilde{M_1})$, and, since this diagram has the shape of a butterfly, it is called the Butterfly Lemma.

To have $K_k(W, f)$ stably free is not enough, we want it to be free. This can easily be achieved by taking the connected sum of W with enough copies of $S^k \times S^k$ and extending f to this connected sum by a map which is constant on $(S^k \times S^k - \overset{\circ}{D}{}^{2k})$. There is an obvious bordism between W and $W \sharp S^k \times S^k$ and so we can assume that $K_k(W, f)$ is a free $\mathbb{Z}[\pi_1(W)]$-module.

13.3 The Quadratic Refinement

To finish the construction of the even-dimensional surgery obstruction, we have to define a quadratic refinement of

$$\lambda \colon K_k(W, f) \times K_k(W, f) \to \mathbb{Z}[\pi_1(W)],$$

a map

$$q \colon K_k(W, f) \to \mathbb{Z}[\pi_1(W)]/_\sim,$$

a certain quotient of $\mathbb{Z}[\pi_1(W)]$.

To simplify the presentation, we make further restrictions and assume that $\dim(W) = 0 \mod 4$ and $\pi_1(W)$ has no element of order 2. These assumptions simplify the definition of the quadratic refinement. The first assumption implies that λ is a hermitian form. Using the second we will see that the quadratic refinement is completely determined by λ. The reason is that, if $R = \mathbb{Z}[\pi_1(W)]$, then $R/\{a - \overline{a}\} = \Sigma_{g^2=1}\mathbb{Z}g \oplus \Sigma_{\{g,g^{-1}\}_{g^2 \neq 1}}\mathbb{Z}$, a torsion free abelian group (see Exercise 13.2). But since $q(2x) = 2q(x)\overline{2} = 4q(x)$ and $q(x + x) = q(x) + q(x) + \lambda(x, x)$, we conclude that

$$2q(x) = \lambda(x, x) \in R/\{a - \overline{a}\}$$

and so q is determined by λ. Moreover, we will show that the elements represented by $\lambda(x, x)$ in $R/\{a - \overline{a}\}$ for x in $K_k(W, f)$ are all divisible by 2, and so we can define

$$q(x) := \frac{1}{2}\lambda(x, x).$$

To see that the elements represented by $\lambda(x, x)$ in $R/\{a-\overline{a}\}$ for x in $K_k(W, f)$ are all divisible by 2 we first consider those components of $\lambda(x, x)$ sitting in one of the summands $\mathbb{Z}g \oplus \mathbb{Z}^{-1}/g - g^{-1}$. They are represented by $(x \circ gx)g + (x \circ g^{-1}(x))g^{-1}$. Since g is orientation preserving on $\widetilde{W}$, we have $x \circ g^{-1}x = gx \circ g(g^{-1}x) = gx \circ x = x \circ gx$, since the form is hermitian, and so: $(x \circ gx)g + (x \circ g^{-1}(x))g^{-1} = (x \circ (x))(g + g^{-1})$. But in $\mathbb{Z}g + \mathbb{Z}g^{-1}/g - g^{-1}$ we have $[g - g^{-1}] = 0$ and so $[g + g^{-1}] = [g + g^{-1}] + [g - g^{-1}] = 2[g]$.

Finally we have to show that the component of the summand in $R/\{a - \overline{a}\}$ corresponding to 1, the neutral element in π_1, is even. But this summand is just given by the integral self intersection number

$$x \circ x = 0 \mod 2.$$

for $x \in K_k(W, f)$ To see this we note that $H_{k+1}(\widetilde{X}, \widetilde{W}) \to K_k(W, f)$ is surjective and, since f is a k-equivalence, the Hurewicz map $\pi_{k+1}(\widetilde{X}, \widetilde{W}) \to H_{k+1}(\widetilde{X}, \widetilde{W})$ is an isomorphism. Thus, all elements in $K_k(W, f)$ are representable by maps $g : S^k \to \widetilde{W}$, which go to zero in $\pi_k(X)$.

We approximate g by an embedding and so we assume that g is an embedding to $\widetilde{W}$. The self intersection number is equal to the Euler class of the normal bundle of this embedding and so mod 2 it is given by the k-th Stiefel–Whitney class w_k. But since $g_*([S^k])$ is zero in $\pi_k(X)$ and the normal bundle of W is the pullback of a bundle on X, the restriction of the normal bundle of $\widetilde{W}$ to the image of the embedding is a trivial bundle. Since the tangent bundle of S^k is stably trivial, we conclude that the stable normal bundle of the embedding in $\widetilde{W}$ is trivial and so $w_k = 0$ implying that

$$x \circ x = 0 \mod 2.$$

Thus we define our quadratic refinement as

$$q(X) := \left[\frac{\lambda(x, x)}{2}\right] \in R/\{a - \overline{a}\}.$$

To obtain an element in $L_{2k}^h(\pi_1(W))$, we have to show that the form λ on $K_k(W)$ is unimodular. This follows from Lefschetz duality. Namely, one has a pairing given by intersection numbers of relative homology classes

$$\lambda : H_k(W, M_0; \mathbb{Z}[\pi_1]) \times H_k(W, M_1; \mathbb{Z}[\pi_1]) \to \mathbb{Z}[\pi_1]$$

defined completely analogously as the pairing on $K_k(\widetilde{W})$. The Lefschetz duality says that this pairing is unimodular. But we have shown already that the inclusions $K_k(W) \to H_k(W; \mathbb{Z}[\pi_1])$ and $H_k(W; \mathbb{Z}[\pi_1]) \to H_k(W, M_i; \mathbb{Z}[\pi_1])$ give an isomorphism

$$j_i : K_k(W) \xrightarrow{\cong} H_k(W, M_i; \mathbb{Z}[\pi_1])$$

and by definition of the intersection numbers we have

$$\lambda(x, y) = \lambda(j_0(x), j_1(y)).$$

Thus
$$\lambda \colon K_k(W) \times K_k(W) \to \mathbb{Z}\left[\pi_1(W)\right]$$
is unimodular.

13.4 The Surgery Obstruction

Now we have all the information needed to define the surgery obstructions if $\dim(W) = 0 \mod 4$ and if $\pi_1(W)$ has no element of order 2. We summarize:

Suppose that $f \colon W \to X$ is as in the beginning of this chapter and $\partial W = M_0 + M_1$ and $f|_{M_i} \colon M_i \to X$ is a homotopy equivalence. Then the surgery obstruction is represented by
$$(K_k(W), \lambda, q),$$
where $K_k(W) = \ker\left(H_k(W; \mathbb{Z}[\pi_1]) \to H_k(X; \mathbb{Z}[\pi_1])\right)$ and λ is the restriction of the equivariant intersection form to $K_k(W)$. Here we assume that $K_k(W)$ is free. The quadratic refinement is

$$q(x) = \frac{\lambda(x,x)}{2} \in \mathbb{Z}\left[\pi_1(W)\right]/\{a - \bar{a}\}.$$

In the following chapter we will show that if we change (W, f, α) in its bordism class, the element represented by $(K_k(W), \lambda, q)$ in $L_m^h(\pi_1(W))$ will not be changed. We call it
$$\Theta(W, f, \alpha) \in L_m^h(\pi_1(W)),$$
the *surgery obstruction*.

A few words about the definition of the surgery obstruction in the other cases. If the dimension of W is $4k + 2$, we proceed as above, but even under the assumptions above the quadratic refinement is not determined by λ. This is also the case for dimension $W = 4k$, if the assumptions above are not fulfilled. In both cases one defines the quadratic refinement geometrically in terms of the self intersection points of an immersed sphere with trivial normal bundle. This will be indicated later (see Chapter 14), for details we refer to [248].

In the remaining cases, where the dimension of W is $2k + 1$ one chooses a set of generators of $K_k(W, f)$ and represents them by disjoint embeddings $\varphi_i \colon S^k \times D^{k+1} \to W$. We remove the interiors of these embeddings to obtain T. The boundary of T consists of $M_0 + M_1$ and the disjoint union S of $\varphi_i(S^k \times S^k)$. Now we pass to the universal covering of T and note that the homology of $\widetilde{S}$, the restriction of the universal covering to S, with its intersection form is a hyperbolic form. Now we consider the kernel of the map

$$H_k(\widetilde{S}) \to H_k(\widetilde{T}, \widetilde{M_0} + \widetilde{M_1}).$$

Poincaré–Lefschetz duality implies that this is a free direct summand of half rank, on which the intersection form and quadratic refinement of the hyperbolic form

vanish (see Exercise 13.3). Choose a basis $x_1, \ldots, x_r$ of this kernel. The considerations above in the proof of Lemma 12.3 imply that we can extend this by
elements $y_1, \ldots, y_k$ to a basis of the hyperbolic forms, such that again the intersection form and the quadratic refinement vanish for all elements y_i. Furthermore
$\lambda(x_i, y_j) = \delta_{i,j}$. Let A be the isometry of the hyperbolic form mapping the standard base elements e_i to x_i and f_j to y_j. This gives an element of $L^h_{2k+1}(\pi_1)$. One
shows that this is well defined and that this is the surgery obstruction.

Chapter 14

Surgery in the Middle Dimension III

14.1 Stable Diffeomorphism Classification

As announced at the end of the last section, we have to show that the surgery obstruction is a bordism invariant. Our proof of this result is different from the standard approach. It is a consequence of an improvement of the following theorem which is very useful in itself.

Theorem 14.1. (Stable classification of manifolds). *Suppose that* (W, f, α) *and* (W', f', α') *are 2k-dimensional normal maps with equal boundaries such that* f *and* f' *are k-equivalences (the maps are then called* normal $(k-1)$-smoothings *[133]). If* $k \geq 2$ *and* (W, f, α) *and* (W', f', α') *are bordant, then they are stably diffeomorphic.*

Here we call two $2k$-dimensional manifolds W and W' with same boundary $\partial W = \partial W'$ stably diffeomorphic if there are integers n and m such that

$$W \natural n(S^k \times S^k) \cong N \cong W' \natural m(S^k \times S^k),$$

where the restriction of the diffeomorphism to the boundary is the identity.

Proof. Applying surgery below the middle dimension to a bordism (T, g, β) between W and W' (rel. boundary), we can assume that $g : T \to X$ is a k-equivalence. This implies that there is a bordism T between W and W' (rel. boundary) such that the inclusions $W \to T$ and $W' \to T$ are isomorphisms on π_1 and the relative homotopy group $\pi_i(T, W)$ and $\pi_i(T, W')$ are trivial for $i < k$. Then the proof of the s-cobordism theorem implies that T has a handle decomposition starting from $W \times [0, \frac{1}{2}]$ by attaching to $W \times \{\frac{1}{2}\}$ only handles of index k firstly and then of index $k + 1$. If we stop after adding the handles of index k to $W \times \{\frac{1}{2}\}$ we call the resulting new boundary component N. The attaching maps of the k-handles define homotopy classes in $W \times \{\frac{1}{2}\}$ and this implies that the handles are trivially attached and so

$$N \cong W \natural n(S^k \times S^k),$$

where n is the number of k-handles.

Now we look at the other half of the bordism T obtained from $N \times \left[\frac{1}{2}\right]$ by attaching handles of index $k + 1$ only. Turning this bordism from the top to the bottom, we notice that it is obtained from $W' \times \left[0, \frac{1}{2}\right]$ by attaching handles of index k to $W' \times \{\frac{1}{2}\}$. The new boundary component is N again, and so — by the same considerations as before — diffeomorphic to $W' \natural m(S^k \times S^k)$, where m is the number of $(k + 1)$- handles (or k-handles seen from W'). Thus

$$W \natural n(S^k \times S^k) \cong N \cong W' \natural m(S^k \times S^k). \qquad \square$$

To see how useful this result is, we apply it to closed smooth 4-manifolds M with $w_2(\widetilde{M}) \neq 0$. Let M and M' be such closed 4-manifolds with isomorphic fundamental group π_1. Let K be $B\pi_1$. An isomorphism between $\pi_1(M)$ and π_1 and $\pi_1(M')$ and π_1 gives maps

$$f \colon M \to K$$

and

$$f' \colon M' \to K.$$

inducing an isomorphism on π_1.

We also consider the normal Gauss map $\nu \colon M \to BSO$ and $\nu' \colon M' \to BSO$ and call $X := K \times BSO$. Over X we consider the bundle E given by the pullback of the universal bundle over BSO. Then $(f \times \nu)^* E = \nu(M)$ and $(f' \times \nu)^* E = \nu(M')$. Thus we can consider the normal smoothings (M, f, id) and (M', f', id). By construction these are normal 1-smoothings. Here we use that since $w_2(\widetilde{M}) \neq 0$ the map ν is surjective on π_2. Our results say that M and M' are stably diffeomorphic if these triples are normally bordant. By standard considerations one shows that this is the case if and only if the signatures and the images of the fundamental classes in $H_4(K)$ are equal:

$$f_*([M]) = f'_*([M']) \in H_4(K) = H_4(\pi).$$

Thus we have proved

Theorem 14.2 (Stable classification of 4-manifolds). *Let M and M' be closed oriented 4-manifolds with $w_2(\widetilde{M}) \neq 0$ and $w_2(\widetilde{M'}) \neq 0$. Then M and M' are stably diffeomorphic if and only if their signatures and fundamental classes in $H_4(\pi)$ agree:*

$$\mathrm{sign}(M) \; = \; \mathrm{sign}(M')$$

and

$$f_*([M]) = f_*([M']) \in H_4(B\pi_1).$$

We note that this is a simple version of the very hard Theorem 3.1 by Donaldson and Freedman mentioned in Chapter 3.

Returning to the stable classification, we note that there is an obvious bordism between W and $W \sharp n(S^k \times S^k)$, namely $W \times [0,1] \natural n(D^{k+1} \times S^k)$, where $\natural$ is the connected sum along the boundary, which in $W \times [0,1]$ we take with respect to the component $M \times \{1\}$. We can extend the normal structure on W to this bordism in an obvious way by firstly projecting $W \times [0,1]$ to $W \times \{0\}$ and composing with f and extending to $D^{k+1} \times S^k$ by the constant map. This gives the map $\hat{f}$ to X. Secondly, since the normal bundle of $D^{k+1} \times S^k$ is trivial, we can extend α to $\hat{\alpha}$ on this bordism.

14.2 The Surgery Obstruction is a Bordism Invariant

From the proof of Theorem 14.1, which gives the existence of a diffeomorphism between $W \sharp n(S^k \times S^k)$ and $W' \sharp m(S^k \times S^k)$, it is not clear that this preserves the normal structures. But one can improve Theorem 14.1 so that the diffeomorphism preserves the normal structures [133] and using this information, it is obvious that the surgery obstruction is a bordism invariant. For, by construction of the normal structure on $W \sharp_m(S^k \times S^k)$ it is clear that the surgery construction on $W \sharp_m(S^k \times S^k)$ is

$$\Theta((W \sharp_m(S^k \times S^k), \hat{f}, \hat{\alpha}) = \Theta(W, f, \alpha) + H^m,$$

the orthogonal sum of $\Theta(W, f, \alpha)$ and the hyperbolic form.

Thus we obtain:

Theorem 14.3 (Bordism invariance of the surgery obstruction). *The surgery obstruction in $L_{2k}^h(\pi_1)$ is a bordism invariant.*

The proof of the corresponding result for the obstructions in L_{2k+1}^h is completely different and we refer to [248].

14.3 The Main Result

Now we come to the main result of this chapter: if the surgery obstruction vanishes, then W is bordant rel. boundary to an h-cobordism (if $\dim(W) > 5$). As before, we will show this only if $\dim(W)$ is $2k = 0 \mod 4$ (the proof in the case $W = 2 \mod 4$ is almost the same but for odd dimensions the proof is different).

We begin with the proof and suppose that (W, f, α) is a normal map to X equipped with a stable vector bundle E, where $\partial W = M_0 + M_1$ and $f|_{M_i}$ is a homotopy equivalence. We further assume that f is a k-equivalence, where $\dim(W) = 2k$ and k is even. Then $H_j(W, M_i; \mathbb{Z}[\pi_1]) = 0$ for $j \neq k$ and we want to replace W by W' such that these groups vanish, too, if $\Theta(W, f, \alpha) = 0$.

If $\Theta(W, f, \alpha) = 0$ then there exists an n such that $\Theta(W, f, \alpha) + H^n$ is isomorphic to H^m. As discussed before, we replace (W, f, α) by $(W \natural_n(S^k \times S^k), \hat{f}, \hat{\alpha})$ in such a way that $\Theta(W \natural_n(S^k \times S^k)) = \Theta(W) + H^n$. Using this we can assume that

$$\Theta(W, f, \alpha) \cong H^m.$$

Now we consider the canonical basis $e_1, f_1, e_2, f_2, \ldots, e_m, f_m$ of H^m with $\lambda(e_i, e_j) = \lambda(f_i, f_j) = 0$, $\lambda(e_j, f_j) = \delta_{i,j}$ and $q(e_i) = q(f_i) = 0$.

We give the same name to the images under the map from $K_k(W)$ to $H_k(W)$. The following lemma is a fundamental result in topology.

Lemma 14.4. *Let W be a $2k$-dimensional manifold and $k > 2$ and $\alpha \in \pi_k(W)$ an element such that $\alpha^* \nu(W)$ is trivial. Then α can be represented by an embedding $S^k \hookrightarrow W$ with trivial normal bundle if and only if $\lambda(\alpha, \alpha) \in \mathbb{Z}[\pi_1(W)]$ vanishes. Here, we assume that k is even and $\pi_1(W)$ has no elements of order 2.*

Remark 14.5. If $\pi_1(W)$ contains 2-torsion or if $\dim(W) = 2 \mod 4$ the vanishing of the quadratic refinement $q(x)$ in the definition of the surgery obstruction is an additional necessary and sufficient condition. The following indication of the proof of Lemma 14.4 will indirectly give a definition of the quadratic refinement in these cases.

Proof. Since even the 1-connected case is not completely obvious and the key idea is visible in this case, we firstly assume $\pi_1(W) = \{1\}$. If $\dim(W) > 4$, the Whitney embedding theorem implies that we can represent α by an embedding

$$f : S^k \to W.$$

If $\dim(W) = 4$ this is not true, since the Whitney trick will not work. The reader should be warned that the failure of the Whitney trick in dimension 4 is a highly non-trivial result. For certain fundamental groups one can repair the Whitney trick for topological embeddings by work of Freedman [94]. Approximately at the same time Donaldson [69] proved that even for simply connected 4-manifolds the Whitney trick does not work in the smooth category. Both results are very deep.

Returning to the situation above, in general the normal bundle of the embedding f will be non-trivial even if the stable normal bundle is trivial (consider for example the diagonal $\Delta S^k \subseteq S^k \times S^k$, whose normal bundle is isomorphic to the tangent bundle of S^k and this is only trivial if $k = 0, 1, 3, 7$ [8]).

If k is even, a stably trivial bundle over S^k is trivial if and only if its Euler class vanishes (see Exercise 14.1). But the Euler class is equal to the self intersection number of the 0-section [171], which by definition is $\lambda(\alpha, \alpha)$, where α is the homology class represented by S^k. This finishes the argument in the 1-connected case.

If W is not simply connected, the considerations above in the simply connected case imply, that we can represent α by an embedding $f : S^k \hookrightarrow \widetilde{W}$ into the universal covering $\widetilde{W}$ with trivial normal bundle if $\lambda(\alpha, \alpha) = 0$ (actually, we only need to know that the coefficient of 1 in $\lambda(\alpha, \alpha) = \sum_{g \in \pi_1} \alpha \circ g_*(\alpha)$ is zero). If $p: \widetilde{W} \to W$ is the projection, then pf is an immersion. By general position (Sard's theorem), we can assume that pf has only double points in which two branches intersect transversally. These double points are in some sense equivalent to the other coefficients of $\lambda(\alpha, \alpha)$. Namely, we fix a base point x_0 on $pf(S^k)$. Then we choose two paths which join x_0 with a double point y which reach the double point on the two different branches of the immersed sphere and avoid all other double points. If we compose these two paths starting with the first from x_0 to p using the second to return from p to x_0, we obtain a closed path. We denote the corresponding element in $\pi_1(W)$ by g_p. It is not difficult to show that

$$\lambda(\alpha, \alpha) = \sum_p g_p,$$

where the sum is taken over all double points p.

Thus, we see at least that if α can be represented by an embedded sphere with trivial normal bundle in W, the self intersection number $\lambda(\alpha, \alpha)$ vanishes.

We indicate the argument why the converse is true by considering the case of a single double point p. Then, if $\lambda(\alpha, \alpha) = 0$, we have $g_p = 0$ in $\pi_1(W)$. Thus, the curve g_p is the boundary of a map $D^2 \to W$. Since $\dim(W) > 4$ (this is a harmless looking condition but the only reason why surgery does not work in dimension 4, as we know from the results by Donaldson [69]), we can choose an embedding $D^2 \hookrightarrow W$ whose boundary is homotopic to g_p. Now, the Whitney trick comes into play and gives an isotopy between the immersion and an embedding. $\qquad\square$

14.4 Proof of the Main Theorem

Returning to the situation where we have a normal map (W, f, α) with $\Theta(W, f, \alpha) = 0$, we have found a basis $e_1, f_1, e_2, f_2, \ldots, e_m, f_m$ of $K_k(W)$ with $\lambda(e_i, e_j) = \lambda(f_i, f_j) = 0$, $\lambda(e_i, f_j) = \delta_{i,j}$ and $q(e_i) = q(f_i) = 0$. According to Lemma 14.4 there is an embedding

$$g: S^k \to W$$

representing e_1 with trivial normal bundle. Thus we can extend g to an embedding

$$g: S^k \times D^k \to W,$$

and we use this to do surgery on W:

$$W' := (W - g(S^k \times \overset{\circ}{D}{}^k)) \cup D^{k+1} \times S^{k-1}.$$

What is the effect on the homology groups of $\widetilde{W}$? For this we consider the following diagram of exact sequences of the pairs $(W, W - g(S^k \times \overset{\circ}{D}{}^k))$ and

$(W', W' - g(S^k \times \overset{0}{D}{}^k))$. We abbreviate $W - g(S^k \times \overset{0}{D}{}^k))$ by T:

$$H_{j+1}(W', T; \mathbb{Z}[\pi_1])$$
$$\downarrow$$
$$\to H_{j+1}(W, T; \mathbb{Z}[\pi_1]) \to \quad H_j(T; \mathbb{Z}[\pi_1]) \to \quad H_j(W; \mathbb{Z}[\pi_1]) \to$$
$$\downarrow$$
$$H_j(W'; \mathbb{Z}[\pi_1])$$
$$\downarrow$$

By excision

$$H_j(WT; \mathbb{Z}[\pi_1]) \cong H_j(S^k \times D^k, \partial; \mathbb{Z}[\pi_1]) = \left\{ \begin{array}{ll} 0, & \text{if} \quad j \neq k \\ \mathbb{Z}[\pi_1(W)], & \text{if} \quad j = k \end{array} \right.$$

and similarly

$$H_j(W, T; \mathbb{Z}[\pi_1]) \cong \left\{ \begin{array}{ll} 0, & \text{for} \quad j \neq k+1 \\ \mathbb{Z}[\pi_1(W)], & \text{for} \quad j = k+1 \end{array} \right. .$$

Thus for $j < k$ the homology groups of $\widetilde{W}$ and $\widetilde{W'}$ are isomorphic and for $j = k$ the diagram looks as follows:

$$H_{k+1}(W', T; \mathbb{Z}[\pi_1])$$
$$\downarrow$$
$$0 \to \quad H_k(T; \mathbb{Z}[\pi_1]) \to \quad H_k(W; \mathbb{Z}[\pi_1]) \to \quad H_k(W, T; \mathbb{Z}[\pi_1]) \to$$
$$\downarrow$$
$$H_k(W'; \mathbb{Z}[\pi_1])$$
$$\downarrow$$
$$0$$

The map

$$H_k(W; \mathbb{Z}[\pi_1]) \to H_k(W, T; \mathbb{Z}[\pi_1]) \cong \mathbb{Z}[\pi_1(x)]$$

has a geometric interpretation, it maps $\alpha \in H_k(W; \mathbb{Z}[\pi_1])$ to $\lambda(\alpha, e_1)$ (see Exercise 14.2, this is a nice exercise which helps one to understand the geometric meaning of intersection numbers). Since $\lambda(f_1, e_1) = 1$ and $\lambda(e_i, e_1) = \lambda(f_i, e_1) = 0$ for $i > 1$, the consequence is that the kernel of

$$H_k(W; \mathbb{Z}[\pi_1]) \to H_k(W, T; \mathbb{Z}[\pi_1])$$

contains the submodule generated by e_1 and e_i and f_i for $i > 1$. Moreover — more of less by definition of the maps in the diagram — the generator of $H_k(W, T; \mathbb{Z}[\pi_1])$ is given by a fibre $[\{x_0\} \times D^{k+1}, \partial]$ maps to e_1 in $H_k(W; \mathbb{Z}[\pi_1])$.

If one compares the diagram with the homology groups of $\widetilde{X}$, one concludes that

$$K_k(W') \cong H^{m-1},$$

where H^{m-1} geometrically corresponds to the elements $e_2, f_2, e_3, f_3, \ldots, e_m, f_m$ in $K_k(W)$.

Proceeding inductively, after a sequence of m surgeries we can replace W by W' by where

$$K_k(W') = \{0\}.$$

From this it is easy to show that $H_k(W', M_0; \mathbb{Z}[\pi_1])$ and $H_k(W', M_1; \mathbb{Z}[\pi_1])$ are zero and so W' is an h-cobordism.

Conversely, if W is an h-cobordism, then the group $K_K(W)$ is trivial and hence $\Theta(W, f, \alpha) = 0$. Thus, we have shown the main theorem of surgery in the case $\dim(W) = 0 \mod 4$ and $\pi_1(W)$ without 2-torsion.

A similar argument works once we have defined $\Theta(W, f, \alpha)$ in the general case and this leads to an important theorem of surgery due to Wall [248]

Theorem 14.6 (The surgery obstruction). *Let (W, f, α) be a normal map to a CW-complex X with finite skeleta equipped with a stable vector bundle E. We furthermore suppose that $\partial W = M_0 + M_1$ and $f|_{M_i}$ is a homotopy equivalence. Then W is normally bordant rel. boundary to an h-cobordism W' if and only if the surgery obstruction*

$$\Theta(W, f, \alpha) \in L_m^h(\pi_1(W))$$

vanishes. Here $m = \dim(W)$ is assumed to be larger than 4.

If f is a simple homotopy equivalence, then we obtain an obstruction

$$\Theta(W, f, \alpha) \in L_m^s(\pi_1(W))$$

which vanishes if and only if W is normally bordant to an s-cobordism.

This is a central result in Wall's surgery theory. It raises the obvious question, which elements in the L-groups can be realized as surgery obstructions. The answer is: all of them, and there is a comparatively simple construction giving a proof. The proof actually leads to a stronger result:

Theorem 14.7 (The surgery obstruction). *Let M_0 be a closed m-dimensional manifold with fundamental group π_1 and $m \geq 4$. Then there is a map r, the realization map from $L_{m+1}^h(\pi_1)$ to the set of normal bordisms (W, f, α), which has the following properties. If $x \in L_{m+1}^h(\pi_1)$ and $r(x) = (W, f, \alpha)$, then:*

  i) *The boundary of W is $M_0 + M_1$ for some closed manifold M_1.*

  ii) *The normal map goes to $(M_0, \nu(M_0))$.*

  iii) *$f|_{M_0} = \mathrm{id}$, and $f|_{M_1}$ is a homotopy equivalence.*

  iv) *$\Theta r(x) = x$, i.e., the surgery obstruction of the normal bordism associated to the image under r is the element itself.*

There is a similar result for the groups L^s, where we replace in iii) a homotopy equivalence by a simple homotopy equivalence.

For the — as mentioned above — not so hard proof of this result we refer to [248].

14.5   The Exact Surgery Sequence

We want to discuss some consequences of this theorem. We recall that in Chapter 9 we defined the set of homotopy smoothings $\mathcal{S}^h(M)$ of a closed manifold M. The realization map r gives a map

$$L_{m+1}^h(\pi_1) \to \mathcal{S}^h(M)$$

given by restricting $f\colon W \to M$ to M_1. On the other hand we have a map

$$\mathcal{S}^h(M) \to \bigoplus_E \Omega_m(M, E)/\operatorname{Aut}(E)$$

where E is the isomorphism class of a stable vector bundle over M and $\operatorname{Aut}(E)$ is the group of self isomorphism on E, which act on $\Omega_m(M, E)$ in the obvious way. This map is given by firstly associating to a homotopy equivalence $f\colon N \to M$ the bundle $E := g^*(\nu(N))$, where g is a homotopy inverse of f and the element $(N, f, \alpha) \in \Omega_m(M, E)$, where $\alpha\colon \nu(N) \to f^*(E)$ is an isomorphism given by identifying the two bundles via a homotopy between gf and id. Since this isomorphism is not unique we have to divide by $\operatorname{Aut}(E)$. Then the map $\mathcal{S}^h(M) \to \bigoplus_E \Omega_m(M, E)/\operatorname{Aut}(E)$ is given by the difference of this element and (M, id, id).

The composition of the two maps

$$L_{m+1}^h(\pi_1) \to \mathcal{S}^h(M) \to \bigoplus_E \Omega_m(M, E)/\operatorname{Aut}(E)$$

is zero, since if $x \in L_{m+1}^h(\pi_1)$ maps to $r(x) = (W, f, \alpha)$, then (W, f, α) is a zero bordism of the element associated to x by the composition map.

In turn, if $(N, f) \in \mathcal{S}^h(M)$ maps to zero in $\bigoplus_E \Omega_m(M, E)/\operatorname{Aut}(E)$ one can consider the surgery obstruction of a zero bordism to obtain an element $x \in L_{m+1}^h(\pi_1)$ and it turns out that $r(x) = (N, f)$. Thus the fibre over 0 of the map $\mathcal{S}^h(M) \to \bigoplus_E \Omega_m(M, E)/\operatorname{Aut}(E)$ is the image of r. If this is the case one says that the sequence

$$L_{m+1}^h(\pi_1) \to \mathcal{S}^h(M) \to \bigoplus_E \Omega_m(M, E)/\operatorname{Aut}(E)$$

is an exact sequence of sets. One can actually say more: The group $L_{m+1}^h(\pi_1)$ acts on $\mathcal{S}^h(M)$ and the orbit space injects into $\Omega_m(M, E)/\operatorname{Aut}(E)$.

The next obvious question is, what the image of the map

$$\mathcal{S}^h(M) \to \bigoplus_E \Omega_m(M, E)/\operatorname{Aut}(E)$$

is. There is a simple restriction for elements in this image: if (N, f, α) is in the image, the degree of f is 1. Thus we consider the subset

$$\bigoplus_E \Omega_m^1(M, E) := \{(N, f, \alpha) \in \bigoplus_E \Omega_m(M, E) \mid \operatorname{degree}(f) = 1\}.$$

The group $\mathrm{Aut}(E)$ acts on $\Omega^1_m(M, E)$ and we denote the orbit set by

$$\mathcal{N}(M) := \bigoplus_E (\Omega^1_m(M, E) / \mathrm{Aut}(E).$$

If (N, f, α) is in $\Omega^1_m(M, E) / \mathrm{Aut}(E)$, we can ask whether it is bordant to a homotopy equivalence. By surgery below the middle dimension we can almost achieve this and by a similar argument as in the situation where $N = W$ has two boundary components and the restriction of f to these boundary components are homotopy equivalences, one associates an element $\Theta(N, f, \alpha)$ to $(N, f, \alpha) \in L^h_m(\pi_1)$ with the property that this element is zero if and only if (N, f) is bordant to a homotopy equivalence. This leads to the *exact surgery sequence* [248]:

Theorem 14.8 (The Sullivan–Wall surgery exact sequence). *Let M be a closed connected manifold of dimension $m > 4$ with fundamental group π_1. Then there is an exact sequence of sets:*

$$L^h_{m+1}(\pi_1) \to \mathcal{S}^h(M) \to \mathcal{N}(M) \to L^h_m(\pi_1).$$

Similarly, if we replace homotopy equivalences by simple homotopy equivalences and $\mathcal{S}^h$ by $\mathcal{S}^s$ and L^h by L^s we obtain a corresponding exact sequence of sets:

$$L^s_{m+1}(\pi_1) \to \mathcal{S}^s(M) \to \mathcal{N}(M) \to L^s_m(\pi_1).$$

As noted before the left part of the sequence can be replaced by a stronger the statement, the group acts on the structure set and the orbit space injects into the normal bordism classes $\mathcal{N}(M)$.

We note that one has the same result for the topological structure sets $\mathcal{S}^h_{\mathrm{TOP}}(M)$ and $\mathcal{S}^s_{\mathrm{TOP}}(M)$. The only difference is that one has to replace the bordism set $\mathcal{N}(M)$ by the corresponding set $\mathcal{N}_{\mathrm{TOP}}(M)$ of topological manifolds and the stable vector bundle E by a stable topological vector bundle. We note that one can equip the topological structure set with a group structure that the whole sequence becomes an exact sequence of abelian groups. This is not known in the smooth case. Moreover one can identify the topological structure set of a manifold with the homotopy fiber of the surgery assembly map interpreted as maps of spectra (see [200, Chapter 18]).

14.6 Stable Classification of Certain 6-Manifolds

In this section we give the proof of Theorem 0.6 from Chapter 0. The proof is not relevant for the Novikov Conjecture. But on the one hand it is a nice application of surgery, a central tool in the study of the Novikov Conjecture, and on the other hand Theorem 0.6 leads to an interesting application of the Novikov Conjecture (Corollary 0.7).

Proof. We firstly determine the normal 2-type of N. The normal 2-type of a smooth oriented manifold is a fibration $\pi\colon B \to BSO$ with the property that the homotopy groups of the fibre vanish in dimension > 2 and that the normal Gauss map $\nu\colon N \to BSO$ admits a lift $\bar{\nu}\colon N \to B$, which is a 3-equivalence.

The normal 2-type depends on the behavior of the second Stiefel–Whitney class. We begin with the case $w_2(N) = 0$. Then we define $B := T \times BSpin$, where T is the 2-torus. and $BSpin$ is the classifying space of the spino group $Spin$, the universal cover of SO. The fibration over BSO is simply the composition $\pi_0 := pp_2$, where $p\colon BSpin \to BSO$ is the fibration induced from the homomorphism $Spin \to SO$. If $w_2(N) = 0$, then the normal Gauss map admits a lift $\hat{\nu}\colon N \to BSpin$ and we obtain our 3-equivalence $\bar{\nu} := f \times \hat{\nu}\colon N \to B = T \times BSpin$. Here $f\colon N \to T$ is the classifying map of the universal covering, a map inducing an isomorphism on the fundamental group.

If $w_2(N) \neq 0$ we make a small modification. We choose an oriented vector bundle E over T such that $f^* w_2(E) = w_2(N)$. Then we define $B = T \times BSpin$ as before but we obtain the fibration in a different way. Namely we consider the classifying map $g\colon T \to BSO$ of E. Then we take the composition $\oplus (g \times p_s)\colon B \to BSO \times BSO \to BSO$, where $\oplus\colon BSO \times BSO \to BSO$ is the map given by the Whitney sum. Finally we replace this map by a fibration [255] to obtain our normal 2-type $\pi\colon B \to BSO$. Once we have defined the fibration this way it is not difficult to show that we can find a 3-equivalence $\bar{\nu}\colon N \to B$ such that $\nu = \pi\bar{\nu}$.

Now we apply Theorem 14.1 and conclude that the stable diffeomorphism type is determined by the bordism class of $(N, \bar{\nu})$. If $w_2(N) = 0$ the corresponding bordism group is $\Omega_6^{Spin}(T)$. Since the spin bordism groups are $\mathbb{Z}$ in dimension 4 detected by p_1 and 0 in dimension 5 and 6, a simple computation with the Mayer–Vietoris sequence implies that our bordism group is isomorphic to $\mathbb{Z}$ and is detected by our invariant $\langle x \cup p_1(N), [N] \rangle$ (see Exercise 14.3). With a little bit more input one shows that the same result holds for the other bordism group, where $w_2(N) \neq 0$. $\square$

Chapter 15

An Assembly Map

15.1 More on the Definition of the Assembly Map

In Chapter 9 we explained the relation between the Novikov Conjecture and surgery and indicated (following [248], 17H) assembly maps (see (9.4))

$$A_m^G \colon h_m(BG) \ \to \ L_m^h(G) \otimes_{\mathbb{Z}} \mathbb{Q},$$

and (see (9.10))

$$\widehat{A}_m^G \colon \widehat{h}_m(BG) \ \to \ L_m(G) \otimes \mathbb{Q}$$

where $h_m(X) = \bigoplus_{i \in \mathbb{Z}} H_{m+4i}(X_i; \mathbb{Q})$ and $\widehat{h}_m(X) = \bigoplus_{i \in \mathbb{Z}, i \geq 0} H_{m-4i}(X_i; \mathbb{Q})$. We give now more details of the construction of the map

$$\overline{\overline{A}}_m^G \colon \Omega_m(BG) \ \to \ L_m(G)$$

introduced in (9.7) which was one of the key ingredients in the construction of A_m^G and $\widehat{A}_m^G$.

Let M be a closed oriented manifold with fundamental group H. We further want to suppose that M is connected. Later we want to generalize to the situation where M is not connected, which is relevant if we want to compare the assembly map of a disjoint union with the assembly maps of the individual manifolds. The solution is very simple, if M is not connected we take the connected sum of the components. We note that the disjoint union is bordant to the connected sum, and that the assembly map will be a bordism invariant. This justifies considering only connected manifolds.

Let $K = \{[x_0, x_1, x_2, x_3] \in \mathbb{CP}^3 \mid x_0^4 + x_1^4 + x_2^4 + x_3^4 = 0\}$ be the Kummer surface. Let Q be K with two disjoint open disks $\overset{\circ}{D}{}^4$ deleted, so that $\partial Q = S^3 + S^3 = N_0 + N_1$. The manifold Q stably parallelizable. We choose a trivialization of the stable normal bundle of Q. The identity maps $N_i \to S^3$ extend to a map $g\colon Q \to S^3$ (see Exercise 15.1). Now we consider

$$W := M \times Q$$

and

$$f\colon W \to M \times S^3$$

which is the product of $\mathrm{id}\colon M \to M$ with g. The trivialization of $\nu(Q)$ induces an isomorphism

$$\alpha\colon f^*\nu(M \times S^3) \overset{\cong}{\to} \nu(W).$$

Thus

$$(W, f, \alpha)$$

is a normal map to $X := M \times S^3$ equipped with the stable vector bundle $E := \nu(M \times S^3)$. Thus we can consider the surgery obstruction

$$\Theta(W, f, \alpha) \in L^h_{m+4}(\pi_1(M \times S^3)) = L^h_m(H)$$

where H is the fundamental group of M.

A map $h\colon M \to BG$ is up to homotopy the same as a map

$$h\colon H \to G,$$

and we define

$$\overline{\overline{A}}^G_m([M, h]) := h_*\Theta(W, f, \alpha).$$

Here h_* is the map from $L^h_m(H) \to L^h_m(G)$ obtained by taking the tensor product with $\mathbb{Z}[G]$ considered as $\mathbb{Z}[H]$-module via h.

The following result contains the basic properties of the invariant $\overline{\overline{A}}^G_m([M, h])$:

Theorem 15.1.

i) *The assembly map induces a homomorphism from $\Omega_m(K(G, 1))$ to $L^h_m(G)$.*

ii) *If $f : N \to M$ is a homotopy equivalence, then $\overline{\overline{A}}^G_m([N, hf]) = \overline{\overline{A}}^G_m([M, f])$.*

iii) *If we replace M by the product of M with a closed simply connected manifold V, the invariant is zero unless the dimension of V is $4k$, in which case the invariant changes by multiplying with the signature of V.*

Idea of proof: The detailed proof of these results is too long for the purpose of the seminar.

The proof of i) is more or less standard and in spirit similar to the proof of the bordism invariance of the ordinary signature.

Wall derives ii) from the following formula. Let $f : M \to N$, with normal structure α in a bundle E over N, be a normal map between closed oriented manifolds. Then ([248, page 263])

$$16 \cdot \theta(M, f, \alpha) \;=\; \theta(W_M, f_M, \alpha_M) - \theta(W_N, f_N, \alpha_N), \qquad (15.2)$$

where (W_M, f_M, α_M) and (W_N, f_N, α_N) are the normal maps for M and N constructed above. Now, if $f : M \to N$ is a homotopy equivalence we can find a bundle E over N pulling back to the normal bundle of M and so we can add to f normal data. Since the left side vanishes if f is a homotopy equivalence, we conclude the statement. Wall's proof of this formula (15.2) is based on two lemmas: a product formula for the surgery obstruction for a product with a closed manifold (Lemma 1) and a composition formula for the surgery obstruction (Lemma 2) [248, page 264]. This composition formula is not correct. A counterexample is given by a normal map $f \colon T^2 \to S^2$ realizing the non-trivial Arf invariant and a normal map $g \colon T^2 \to T^2$ whose underlying map is the identity and whose normal structure is chosen in such a way that the Arf invariant of the composition $f \circ g \colon T^2 \to S^2$ is zero. Wall's formula would imply that the surgery obstruction of $f \circ g$ is non-trivial since the surgery obstruction of g is zero because g is a homotopy equivalence. For a proof of (15.2) we refer to Ranicki [197, Proposition 6.6], where he defines algebraic bordism groups to give a different approach to the surgery obstructions.

The proof of iii) is again more or less standard and in spirit similar to the proof of the multiplicativity of the ordinary signature. Again the best approach is via Ranicki's theory. We note that for the proof of the Novikov conjecture for finitely generated free abelian groups in the next chapter we only use properties i) and ii). $\qquad\square$

The properties i) and iii) imply that $\overline{\overline{A}}{}^G_m$ induces homomorphisms

$$A^G_m \colon h_m(BG) \to L^h_m(G) \otimes_{\mathbb{Z}} \mathbb{Q},$$

and (see (9.10))

$$\widehat{A}^G_m \colon \widehat{h}_m(BG) \to L_m(G) \otimes \mathbb{Q}.$$

We summarize:

Theorem 15.3. *The invariant* $\overline{\overline{A}}{}^G_m$ *induces homomorphisms*

$$A^G_m \colon h_m(BG) \to L^h_m(G) \otimes_{\mathbb{Z}} \mathbb{Q},$$

and (see (9.10))

$$\widehat{A}^G_m \colon \widehat{h}_m(BG) \to L_m(G) \otimes \mathbb{Q},$$

and if $f : N \to M$ *is a homotopy equivalence, then*

$$A^G_m(N, hf) = A^G_m(M, h)$$

and

$$\widehat{A}^G_m(N, hf) = \widehat{A}^G_m(M, h).$$

15.2   The Surgery Version of the Novikov Conjecture

We have already mentioned in Section 9.2 the following result whose proof we
want to sketch (see also Lemma 23.2 and Remark 23.8

Proposition 15.4.   (1) *The Novikov Conjecture* 1.2 *for the group* G *is equivalent*
to the rational injectivity of the assembly map

$$\widehat{A}_m^G : \widehat{h}_m(BG) \;\to\; L_m(G) \otimes \mathbb{Q}$$

introduced in (9.10).

(2) *The Novikov Conjecture* 1.2 *for the group* G *follows from the rational injec-*
tivity of the assembly map

$$A_m^G : h_m(BG) \;\to\; L_m(G) \otimes \mathbb{Q}$$

introduced in (9.4).

Proof. (1) We have already explained that the Novikov Conjecture 1.2 for the
group G is equivalent to the triviality of the map

$$\mathrm{sign}^G : \mathcal{S}^h(M) \;\to\; \widehat{h}_m(BG; \mathbb{Q})$$

defined in (9.3) for all closed manifolds M with $G = \pi_1(M)$. Let m be the dimension of M. The composition of the map above with the map

$$\widehat{A}_m^G : \widehat{h}_m(BG) \;\to\; L_m(G) \otimes \mathbb{Q}.$$

can be identified with the composition

$$\mathcal{S}^h(M) \to \mathcal{N}(M) \to L_m^h(\pi_1(M)) \to L_m^h(\pi_1(M)) \otimes_{\mathbb{Z}} \mathbb{Q}$$

where the first two maps appear in the exact surgery sequence (see Theorem 14.8)
which implies that this composition is trivial. Hence $\widehat{A}_m^G \circ \mathrm{sign}^G = 0$. Thus the
injectivity of $\widehat{A}_m^G$ implies the triviality of sign^G.

Using the exact surgery sequence and a further analysis of $\mathcal{N}(M)$ one can
show that the kernel of $\widehat{A}_m^G$ is the image of sign^G. Hence the triviality of sign^G
implies the injectivity of $\widehat{A}_m^G$.

(2) follows from assertion (1) since A_m^G composed with the inclusion $\widehat{h}_m(BG) \to$
$h_m(BG)$ is $\widehat{A}_m^G$. $\square$

Chapter 16

The Novikov Conjecture for $\mathbb{Z}^n$

16.1 The Idea of the Proof

In this section we want to prove

Theorem 16.1. *The Novikov Conjecture holds for $\mathbb{Z}^n$.*

Because of Proposition 15.4 this result is equivalent to the next theorem

Theorem 16.2. *The assembly*

$$\widehat{A}_m^G \colon \widehat{h}_m(BG) \to L_m^h(G) \otimes_{\mathbb{Z}} \mathbb{Q}$$

defined in (9.10) is injective for $G = \mathbb{Z}^n$ and every $m \in \mathbb{Z}$.

The strategy of the proof is the following. We consider the composition of the homomorphism given by the higher signature $\text{sign}_G \colon \Omega_m(BG) \to \widehat{h}_m(BG)$ and the assembly map $\widehat{A}_m^G$. Since, after taking the tensor product with $\mathbb{Q}$, the map sign^G is surjective, we are finished if we can show: for all $[M, g] \in \Omega_m(BG)$ with $\widehat{A}_m^G \circ \text{sign}^G([M, f]) = 0$ in $L_m^h(\mathbb{Z}^n) \otimes \mathbb{Q}$ we have $\text{sign}^G(M, f) = 0$. This statement is open for general groups.

A different proof of Theorem 16.1 and Theorem 16.2 based on the Shaneson splitting is outlined in Remark 21.7

16.2 Reduction to Mapping Tori

To give a first feeling for the geometric input of the assembly map, we prove a reformulation of the Novikov Conjecture in terms of mapping tori. Let $\varphi \colon N \to N$ be an orientation preserving diffeomorphism. The *mapping torus* is the manifold $N_\varphi := N \times [0, 1]/\sim_\varphi$, where $\sim_\varphi$ is the relation obtained by identifying $(x, 1)$ with $(\varphi(x), 0)$. This is a smooth fibre bundle over S^1 with fibre N. A *fibre homotopy equivalence* to $N \times S^1$ is a homotopy equivalence $h\colon N_\varphi \to N \times S^1$ which commutes with the projection to S^1. In particular this is a homotopy equivalence and, if $f\colon N \to BG$ is a map, the Novikov conjecture would imply that the higher signatures of (N_φ, fp_1h) are those of $(N \times S^1, fp_1)$, which are 0 since $(N \times D^2, fp_1)$ is a zero bordism. Thus the Novikov Conjecture would imply

$$\text{sign}^G(N_\varphi, fp_1h) = 0.$$

The Novikov Conjecture follows from this special case:

Theorem 16.3 (Reduction to mapping tori). *If for all diffeomorphisms* $\varphi\colon N \to N$, *fibre homotopy equivalences* $h\colon N_\varphi \to N \times S^1$ *and maps* $f\colon N \to BG$ *the higher signatures* $\text{sign}_G(N_\varphi, fp_1h)$ *vanishes, then the Novikov Conjecture holds for* G.

Moreover it is enough to find for each generator of $\widehat{h}_m(BG)$ *an element* $[M, f] \in \Omega_m(BG)$ *in the preimage of this generator under the map* $\Omega_m(BG) \to \widehat{h}_m(BG)$ *induced by* sign^G *such that for any diffeomorphism* $\varphi\colon M \times S^3 \to M \times S^3$ *and fibre homotopy equivalence* $h\colon (M \times S^3)_\varphi \to M \times S^3 \times S^1$ *the higher signatures* $\text{sign}_G((M \times S^3)_\varphi, fp_1h)$ *vanishes.*

Proof. To prove the Novikov Conjecture if the assumptions hold, we consider (M, f) and suppose $\widehat{A}_m^G \circ \text{sign}^G([M, f]) = 0$. Since the Novikov Conjecture is trivial for manifolds of dimension < 4 we assume that $M := \dim(M) \geq 4$. Then by a sequence of surgeries we can assume that f is an isomorphism on the fundamental groups. For this we consider the normal bordism class in $BG \times BSO$ given by f and the normal Gauss map and apply Proposition 11.2 to make this normal map a 2-equivalence. Since $\pi_1(BSO)$ is trivial, the statement follows.

Next we note that our definition of the assembly map in $L_m^h(G)$ actually gives an element in the Wall group $L_m^s(G)$ (it is the surgery obstruction of a normal map with boundary whose restriction to the boundary components is a simple homotopy equivalence, even a homeomorphism). Since the map from $L_m^s(G)$ to $L_m^h(G)$ is an isomorphism after taking the tensor product with $\mathbb{Q}$ (this follows from the Rothenberg sequence of (21.3)), we can work with L^s instead of L^h.

By definition

$$\widehat{A}_m^G \circ \text{sign}^G([M, f]) \; = \; f_*(\Theta(M \times Q, \text{id} \times g, \text{id} \times \alpha)) \; \in \; L_{m+4}^s(G) \otimes \mathbb{Q}.$$

After passing from (M, f) to a multiple we can assume that

$$f_*(\Theta(M \times Q, \text{id} \times g, \text{id} \times \alpha)) \; = \; 0 \quad \text{in } L_m^s(G)$$

or, since f induces an isomorphism on π_1:

$$\Theta(M \times Q, \text{id} \times g, \text{id} \times \alpha) \; = \; 0 \quad \text{in } L_m^s(G).$$

We conclude from Theorem 14.6 that $(M \times Q, \text{id} \times g, \text{id} \times \alpha)$ is normally bordant rel. boundary to an s-cobordism W between the two boundary components which are both $M \times S^3$. The normal structure of this bordism rel. boundary between $M \times Q$ and W is irrelevant and we only keep the map to $M \times S^3$. By the s-cobordism theorem W is diffeomorphic to $M \times S^3 \times I$, where we can choose the diffeomorphism to be the identity on one end. The diffeomorphism on the other end is denoted by φ. The map on the bordism gives a map from W to $M \times S^3$, which is on both end the identity, or if we use the diffeomorphism to identify W with $M \times S^3 \times I$, a homotopy between id and φ.

Now we consider a useful bordism relation between cutting and pasting of manifolds and the mapping torus. Let V_1 and V_2 be smooth manifolds with same boundary P (with opposite orientations induced from V_1 and V_2 and φ and ψ be orientation preserving diffeomorphisms on the boundary P. Then we obtain a bordism between $(V_1 \cup_\varphi V_2) + (V_1 \cup_\psi V_2)$ and the mapping torus of the composition $P_{\varphi\psi}$ by considering $V_1 \times [0, 1] + V_2 \times [0, 1]$ and identifying for $x \in \partial V_1$ the points $(\varphi(x), t)$ for $0 \leq t \leq 1/3$ with (x, t) and for $1/3 \leq t \leq 1$ with $(\psi(x), t)$. Applying this to $M \times Q \cup W = M \times Q \cup_{\mathrm{id} + \varphi} M \times S^3 \times [0, 1]$ we obtain a bordism between $M \times Q \cup_{\mathrm{id} + \varphi} M \times S^3 \times [0, 1] + M \times Q \cup_{\mathrm{id} + \mathrm{id}} M \times S^3 \times [0, 1]$ and $(M \times S^3)_\varphi$. The homotopy between φ and id gives a fibre homotopy equivalence from $(M \times S^3)_\varphi$ to $M \times S^3 \times S^1$ and composing this map with the projection to M and further with the map from M to BG we obtain a map from the mapping torus $(M \times S^3)_\varphi$ to BG. By construction of the bordism between $M \times Q \cup_{\mathrm{id} + \varphi} M \times S^3 \times [0, 1]$ and $M \times Q \cup_{\mathrm{id} + \mathrm{id}} M \times S^3 \times [0, 1] + (M \times S^3)_\varphi$ the map extends to this bordism in such a way that the restriction to $M \times Q \cup_{\mathrm{id} + \mathrm{id}} M \times S^3 \times [0, 1]$ is the projection to M composed with the map to BG and similarly the map on the other boundary component is the composition of our normal map with the map to BG. It is easy to see (see Exercise 16.1) that the higher signatures of $M \times Q \cup_{\mathrm{id} + \mathrm{id}} M \times S^3 \times [0, 1]$ equipped with the map above vanish if and only if the higher signatures of (M, f) vanish (construct a bordism between this manifold and $M \times K_3$ over BG). Since $M \times Q \cup W = M \times Q \cup_{\mathrm{id} + \varphi} M \times S^3 \times [0, 1]$ is zero bordant (over BG) we conclude that if the higher signatures of our mapping torus vanish, then the higher signatures of $M \times Q \cup_{\mathrm{id} + \mathrm{id}} M \times S^3 \times [0, 1]$ vanish, finishing the argument. $\quad\square$

16.3 The Proof for Rank 1

Now, we apply this result to prove the Novikov Conjecture for $\mathbb{Z}$ and give the proof of Theorem 16.2 in the case $n = 1$.

Proof. The homology groups $\widehat{h}_m(S^1)$ are zero except for $m = 0, 1 \mod 4$ where they are $\mathbb{Z}$. The case $m = 0$ is trivial, since then the higher signature is equivalent to the ordinary signature. If $m = 1 \mod 4$ the higher signature is equivalent to the signature of a regular value of the map $M \to S^1$ representing the cohomology class for which we compute the higher signature.

Now we apply Theorem 16.3 and consider a diffeomorphism on φ on $S^1 \times P$ for some simply connected manifold P together with a homotopy between φ and id.

After perhaps taking the product with $\mathbb{CP}^4$ (or any other manifold with signature non-trivial) we can assume that $\dim(P) > 4$. The homotopy gives a fibre homotopy equivalence from the mapping torus to $S^1 \times S^1 \times P$. We compose this map with the projection to the second S^1 factor (the factor of the fibre $S^1 \times P$). The higher signature associated to a generator of $H^1(S^1)$ is the signature of the preimage of a regular value of this map. This map to S^1 corresponds to a cohomology class in the mapping torus, whose restriction to the fibre $S^1 \times P$ is a generator. If we consider another class in H^1 of the mapping torus which also restricts to a generator of the fibre then the signature of the preimage of a regular value of a smooth map representing this class is unchanged (see Exercise 16.2 and Exercise 16.3). Thus it is enough to show that the signature of the fibre of some map from the mapping torus to S^1, whose restriction to the fibre is a generator of H^1, is trivial.

After passing to a finite covering we will show that the diffeomorphism is isotopic to one which preserves $\{x\} \times P$, for some point x in S^1. Then the fibre of the map from the corresponding mapping torus to S^1 is the mapping torus of the restriction of this diffeomorphism to P. The signature of the total space of a bundle over S^1 is trivial, as one can see for example from the Novikov additivity of the signature [10]. Since the higher signatures are multiplicative under finite coverings we conclude that the higher signature of our original mapping torus vanishes. This finishes the argument.

To prove the existence of the desired isotopy we note that up to isotopy we can assume that φ fixes a base point x_0. Now we pass from $S^1 \times P$ to the universal covering $\mathbb{R} \times P$ and consider the lifted diffeomorphism $\hat\varphi$. Here we construct the universal covering as usual as the space of homotopy classes rel. boundary of paths α starting from the base point x_0. Then the lifted diffeomorphism maps $[\alpha]$ to $[\varphi\alpha]$.

We note that $\mathbb{R} \times P$ is the union of $(-\infty, 0] \times P$ and $[0, \infty) \times P$, two manifolds with boundary $0 \times P$ and each of them has a connected end, in the first case $-\infty \times P$ and in the second case $\infty \times P$. Here we say that a manifold W has end L, where L is a manifold, if there is an a *closed* subset E homeomorphic to $[0, 1) \times L$. Now we apply $\hat\varphi$ to this situation and obtain a decomposition of $\mathbb{R} \times P$ into $\hat\varphi((-\infty, 0] \times P)$ and $\hat\varphi([0, \infty) \times P)$. Since $\hat\varphi$ is orientation preserving both ends are preserved (we say that a homeomorphism $f \colon W \to W'$ preserves the ends L and L', if there are closed subsets E and E' as in the definition of an end, which are mapped to each other), implying that $\hat\varphi((-\infty, 0] \times P)$ has end $-\infty \times P$ and $\hat\varphi([0, \infty) \times P)$ has end $\infty \times P$.

Since $0 \times P$ is compact, there is an integer k such that $\hat\varphi(0 \times P)$ is contained in $[-k, k] \times P$ and $\hat\varphi(0 \times P)$ decomposes $[-k, k] \times P$ into two compact manifolds A_- and A_+ intersecting in $\hat\varphi(0 \times P)$. The boundary of A_- is $-k \times A + \hat\varphi(0 \times P)$ and the boundary of A_+ is $k \times P + \hat\varphi(0 \times P)$. The inclusion of A_+ to $\hat\varphi([0, \infty) \times P)$ is a homotopy equivalence and A_+ is an h-cobordism (see Exercise 16.4). (Similarly, A_- is also an h-cobordism.)

Using the h-cobordism theorem we choose a diffeomorphism ρ from $\hat\varphi(0 \times P) \times [0, 1]$ to A_1 whose restriction to $\hat\varphi(0 \times P)$ is the identity. Then $(x, t) \mapsto \rho(\hat\varphi(x), t)$ is an isotopy between $\hat\varphi|_{0 \times P}$ and a diffeomorphism to $k \times P$. Now we pass to $\mathbb{R} \times P$

divided by $2k\mathbb{Z}$, i.e., we identify (t,x) with $(t+2k,x)$. The resulting space is again $S^1 \times P$, the $2k$-fold covering over $S^1 \times P$. The isotopy above induces an isotopy between the restriction to a fibre $0 \times P$ of the diffeomorphism induced by $\hat{\varphi}$ on $\mathbb{R}/2k\mathbb{Z} \times P$ and an embedding preserving the fibre. We embed this isotopy into a diffeotopy and compose it with our diffeomorphism induced by $\hat{\varphi}$ on $\mathbb{R}/2k\mathbb{Z} \times P$ to obtain the desired isotopy between the diffeomorphism induced by $\hat{\varphi}$ on $\mathbb{R}/2k\mathbb{Z} \times P$ and a diffeomorphism preserving a fibre.

This finishes our proof of the surgery version of the Novikov Conjecture for $\mathbb{Z}$, i.e., the proof of Theorem 16.2 in the case $n = 1$. $\square$

This idea of reducing a problem about manifolds with fundamental group $\mathbb{Z}$ to the simply connected case is well known in the literature and was applied by Browder and Levine [35] to determine those manifolds with fundamental group $\mathbb{Z}$ which are total spaces of fibre bundles over S^1. This idea was generalized and exploited many times. In particular one can try to do inductive arguments for surgery problems, where the fundamental group G has a normal subgroup H with G/H isomorphic to $\mathbb{Z}$, assuming that one can solve problems over H. Shaneson exploited this idea in his analysis of certain L-groups [221]. For a systematic early treatment of these ideas see Wall's book [248, chapter 12B]. In the following we extend our proof for $\mathbb{Z}$ to $\mathbb{Z}^n$ using inductive an argument.

16.4 The Generalization to Higher Rank

Finally we indicate the proof of Theorem 16.2 for arbitrary n. The idea is to work inductively. If f is a diffeomorphism on $T^n \times P$ with P a 1-connected manifold, one can isotope it so that it preserves $T^{n-1} \times P$. The argument is more or less the same as for $n = 1$. One considers the induced diffeomorphism on $\mathbb{R} \times T^{n-1} \times P$. The image of $\{0\} \times T^{n-1} \times P$ is contained in $[-k,k] \times T^{n-1} \times P$ and decomposes it into two parts W_1 and W_2. Then both W_i are again h-cobordisms (see Exercise 16.4) and so, since the Whitehead group vanishes, they are products. As in the case $n = 1$ this leads to the desired isotopy. The rest of the argument is as in the case $n = 1$.

Acknowledgment

During the seminar I (K.) gave a proof of the Novikov Conjecture for finitely generated free abelian groups, which I myself found too simple. Some of the participants found a gap in my argument. What can be saved is explained in this chapter. I would like to thank the participants, in particular Diarmuid Crowley, for discussing my argument so carefully and explaining the gap.

Chapter 17

Poincaré Duality and Algebraic L-Groups

In this chapter we explain the notion of Poincaré duality geometrically and algebraically, and we discuss how a chain complex version of bordism yields a definition of symmetric and quadratic algebraic L-groups for rings with involutions.

17.1 Poincaré duality

In this section we introduce the fundamental definitions of Poincaré spaces and Poincaré duality.

First we need to recall some algebraic facts. A *ring with involution* is an associative ring with unit R together with an *involution* $^-\colon R \to R$, $r \mapsto \bar{r}$, i.e., a map satisfying $\bar{\bar{r}} = r$, $\overline{r+s} = \bar{r} + \bar{s}$, $\overline{rs} = \bar{s}\,\bar{r}$, and $\bar{1} = 1$ for all $r, s \in R$. For us the main example will be the group ring AG for a commutative associative ring with unit A and a group G equipped with a homomorphism $w\colon G \to \{\pm 1\}$. The so-called *w-twisted involution* takes $\sum_{g \in G} a_g \cdot g$ to $\sum_{g \in G} w(g) \cdot a_g \cdot g^{-1}$. Now let M be a left R-module. Then $M^* := \hom_R(M, R)$ carries a canonical right R-module structure given by $(fr)(m) = f(m) \cdot r$ for a homomorphism of left R-modules $f\colon M \to R$ and $m \in M$. The involution allows us to turn every right module into a left module and viceversa; in particular we can view $M^* = \hom_R(M, R)$ as a left R-module, namely define rf for $r \in R$ and $f \in M^*$ by $(rf)(m) := f(m) \cdot \bar{r}$ for all $m \in M$. There is a natural homomorphism of left R-modules $M \to M^{**}$, which sends $m \in M$ to the homomorphism $M^* \to R$, $f \mapsto \overline{f(m)}$. If M is finitely generated projective, then so is M^*, and $M \to M^{**}$ is an isomorphism. Given a chain complex of left R-modules C_* and an integer $n \in \mathbb{Z}$, we define its *n-th dual chain complex* C^{n-*} to be the chain complex of left R-modules whose p-th chain module is $\hom_R(C_{n-p}, R)$ and whose p-th differential is given by

$$(-1)^{n-p}(d_{n-p+1})^* \colon (C^{n-*})_p = (C_{n-p})^* \to (C^{n-*})_{p-1} = (C_{n-p+1})^*.$$

Then $C^{n-*} = \Sigma^n C^{-*}$, where C^{-*} stands for C^{0-*}, and the n-suspension Σ^n of a chain complex B_* is defined as the chain complex $(\Sigma^n B_*)_p = B_{p-n}$, $d_{\Sigma B} = (-1)^n d_B$. Recall also that for two R-chain complexes C_* and D_* one can define a chain complex of abelian groups $\hom_R(C_*, D_*)$ whose p-th chain group is

$$\hom_R(C_*, D_*)_p := \prod_{k \in \mathbb{Z}} \hom_R(C_{k-p}, D_k)$$

and whose p-th differential sends $f \in \hom_R(C_{k-p}, D_k)$ to $d_D f - (-1)^p f d_C \in \hom_R(C_*, D_*)_{p-1}$, where d_C and d_D denote the differentials of C and D respectively. Observe that the group of 0-cycles of $\hom_R(C_*, D_*)$ is precisely the group of R-chain maps from C to D and that the 0-boundaries correspond to chain homotopies, and in general $H_n(\hom_R(C_*, D_*))$ is the group $[\Sigma^n C_*, D_*]_R$ of R-chain homotopy classes of R-chain maps from $\Sigma^n C$ to D.

Now consider a connected finite CW-complex X with fundamental group π and a group homomorphism $w \colon \pi \to \{\pm 1\}$. Equip the integral group ring $\mathbb{Z}\pi$ with the w-twisted involution. Denote by $C_*(\widetilde{X})$ the cellular $\mathbb{Z}\pi$-chain complex of the universal covering of X. Recall that this is a free $\mathbb{Z}\pi$-chain complex, and that the cellular structure on X determines a cellular $\mathbb{Z}\pi$-basis on it such that each basis element corresponds to a cell in X. This basis is not unique but its equivalence class depends only on the CW-structure of X (see Section 8.2). The product $\widetilde{X} \times \widetilde{X}$ equipped with the diagonal π-action is again a π-CW-complex. The diagonal map $D \colon \widetilde{X} \to \widetilde{X} \times \widetilde{X}$ sending $\widetilde{x}$ to $(\widetilde{x}, \widetilde{x})$ is π-equivariant but not cellular. By the equivariant cellular approximation Theorem (see for instance [152, Theorem 2.1 on page 32]) there is up to cellular π-homotopy precisely one cellular π-map $\overline{D} \colon \widetilde{X} \to \widetilde{X} \times \widetilde{X}$ which is π-homotopic to D. It induces a $\mathbb{Z}\pi$-chain map unique up to $\mathbb{Z}\pi$-chain homotopy

$$C_*(\overline{D}) \colon C_*(\widetilde{X}) \to C_*(\widetilde{X} \times \widetilde{X}). \tag{17.1}$$

There is a natural isomorphism of based free $\mathbb{Z}\pi$-chain complexes

$$i_* \colon C_*(\widetilde{X}) \otimes_{\mathbb{Z}} C_*(\widetilde{X}) \xrightarrow{\cong} C_*(\widetilde{X} \times \widetilde{X}). \tag{17.2}$$

Denote by $\mathbb{Z}^w$ the $\mathbb{Z}\pi$-module whose underlying abelian group is $\mathbb{Z}$ and on which $g \in G$ acts by $w(g) \cdot \mathrm{id}$. Given two chain complexes C_* and D_* of projective $\mathbb{Z}\pi$-modules we obtain a natural $\mathbb{Z}$-chain map unique up to $\mathbb{Z}$-chain homotopy

$$s \colon \mathbb{Z}^w \otimes_{\mathbb{Z}\pi} (C_* \otimes_{\mathbb{Z}} D_*) \to \hom_{\mathbb{Z}\pi}(C^{-*}, D_*) \tag{17.3}$$

by sending $1 \otimes x \otimes y \in \mathbb{Z} \otimes C_p \otimes D_q$ to

$$s(1 \otimes x \otimes y) \colon \hom_{\mathbb{Z}\pi}(C_p, \mathbb{Z}\pi) \to D_q, \quad (\varphi \colon C_p \to \mathbb{Z}\pi) \mapsto \overline{\varphi(x)} \cdot y.$$

The composite of the chain map (17.3) for $C_* = D_* = C_*(\widetilde{X})$, the inverse of the chain map (17.2) and the chain map (17.1) yields a $\mathbb{Z}$-chain map

$$\mathbb{Z}^w \otimes_{\mathbb{Z}\pi} C_*(\widetilde{X}) \to \hom_{\mathbb{Z}\pi}(C^{-*}(\widetilde{X}), C_*(\widetilde{X})).$$

Recall that the n-th homology of $\hom_{\mathbb{Z}\pi}(C^{-*}(\widetilde{X}), C_*(\widetilde{X}))$ is the abelian group $[C^{n-*}(\widetilde{X}), C_*(\widetilde{X})]_{\mathbb{Z}\pi}$ of $\mathbb{Z}\pi$-chain homotopy classes of $\mathbb{Z}\pi$-chain maps from $C^{n-*}(\widetilde{X})$ to $C_*(\widetilde{X})$. Define $H_n(X; \mathbb{Z}^w) := H_n(\mathbb{Z}^w \otimes_{\mathbb{Z}\pi} C_*(\widetilde{X}))$. Taking the n-th homology group yields an homomorphism of abelian groups

$$\cap : H_n(X; \mathbb{Z}^w) \to [C^{n-*}(\widetilde{X}), C_*(\widetilde{X})]_{\mathbb{Z}\pi} \tag{17.4}$$

which sends a class $x \in H_n(X; \mathbb{Z}^w) = H_n(\mathbb{Z}^w \otimes_{\mathbb{Z}\pi} C_*(\widetilde{X}))$ to the $\mathbb{Z}\pi$-chain homotopy class of a $\mathbb{Z}\pi$-chain map denoted by $? \cap x : C^{n-*}(\widetilde{X}) \to C_*(\widetilde{X})$.

Definition 17.5 (Poincaré complex). *A connected finite n-dimensional Poincaré complex is a connected finite CW-complex X of dimension n together with a group homomorphism $w = w_1(X) : \pi_1(X) \to \{\pm 1\}$ called* orientation homomorphism *and an element $[X] \in H_n(X; \mathbb{Z}^w)$ called fundamental class such that the $\mathbb{Z}\pi$-chain map $? \cap [X] : C^{n-*}(\widetilde{X}) \to C_*(\widetilde{X})$ is a $\mathbb{Z}\pi$-chain homotopy equivalence. We will call it the* Poincaré $\mathbb{Z}\pi$-chain homotopy equivalence.

Obviously there are two possible choices for $[X]$, since it has to be a generator of the infinite cyclic group $H_n(X, \mathbb{Z}^w) \cong H^0(X; \mathbb{Z}) \cong \mathbb{Z}$. A choice of $[X]$ will be part of the structure of a Poincaré complex.

Remark 17.6 (Uniqueness of the orientation homomorphism). The orientation homomorphism $w : \pi_1(X) \to \{\pm 1\}$ is uniquely determined by the homotopy type of X, as can be seen using the following argument. Denote by $C^{n-*}(\widetilde{X})_{\text{untw}}$ the n-dual $\mathbb{Z}\pi$-chain complex of $C_*(\widetilde{X})$ defined using the untwisted involution on $\mathbb{Z}\pi$. The $\mathbb{Z}\pi$-module given by its n-th homology $H_n(C^{n-*}(\widetilde{X})_{\text{untw}})$ depends only on the homotopy type of X. If X carries the structure of a Poincaré complex with respect to $w : \pi_1(X) \to \{\pm 1\}$, then the Poincaré $\mathbb{Z}\pi$-chain homotopy equivalence induces a $\mathbb{Z}\pi$-isomorphism $H_n(C^{n-*}(\widetilde{X})_{\text{untw}}) \cong \mathbb{Z}^w$. Thus we rediscover w from $H_n(C^{n-*}(\widetilde{X})_{\text{untw}})$ since the $\mathbb{Z}\pi$-isomorphism type of $\mathbb{Z}^w$ determines w.

Remark 17.7 (Poincaré duality on (co)homology with coefficients). Suppose that X is a Poincaré complex with respect to the trivial orientation homomorphism. Definition 17.5 implies that Poincaré duality holds for any G-covering $\overline{X} \to X$ and for all possible coefficient systems. In particular we get a $\mathbb{Z}$-chain homotopy equivalence

$$\mathbb{Z} \otimes_{\mathbb{Z}\pi} (? \cap [X]) : \mathbb{Z} \otimes_{\mathbb{Z}\pi} C^{n-*}(\widetilde{X}) = C^{n-*}(X) \to \mathbb{Z} \otimes_{\mathbb{Z}\pi} C_*(\widetilde{X}) = C_*(X),$$

which induces for any commutative ring R an R-isomorphism on (co)homology with R-coefficients

$$? \cap [X] : H^{n-*}(X; R) \xrightarrow{\cong} H_*(X; R). \tag{17.8}$$

Definition 17.9 (Simple Poincaré complex). *A connected finite n-dimensional Poincaré complex X is called* simple *if the Whitehead torsion (see (6.6)) of the $\mathbb{Z}\pi$-chain homotopy equivalence of finite based free $\mathbb{Z}\pi$-chain complexes*

$$? \cap [X] \colon C^{n-*}(\widetilde{X}) \to C_*(\widetilde{X})$$

vanishes.

Theorem 17.10 (Simple Poincaré structures on closed manifolds). *Let M be a connected closed n-dimensional manifold. Then M carries the structure of a connected finite simple n-dimensional Poincaré complex.*

Proof. For a proof we refer for instance to [248, Theorem 2.1 on page 23]. □

Remark 17.11 (Poincaré duality as obstruction). Theorem 17.10 yields the first obstruction for a topological space X to be homotopy equivalent to a connected closed n-dimensional manifold. Namely, X must be homotopy equivalent to a connected finite simple n-dimensional Poincaré complex.

Remark 17.12 (Dual cell decomposition). Here is an explanation of the proof of Theorem 17.10 in terms of dual cell decompositions.

Any closed manifold admits a smooth triangulation $h\colon K \to M$, i.e., a finite simplicial complex K together with a homeomorphism $h\colon K \to M$ which restricted to any simplex is a smooth C^∞-embedding. In particular M is homeomorphic to a finite CW-complex. Any two such smooth triangulations admit a common subdivision. Fix such a triangulation K. Denote by K' its barycentric subdivision. The vertices of K' are the barycenters $\widehat{\sigma^r}$ of simplices σ^r in K. A p-simplex in K' is given by a sequence $\widehat{\sigma^{i_0}} \widehat{\sigma^{i_1}} \ldots \widehat{\sigma^{i_p}}$, where σ^{i_j} is a proper face of $\sigma^{i_{j+1}}$. Now we define the dual CW-complex K^* as follows. It is not a simplicial complex but shares the property of a simplicial complex that all attaching maps are embeddings. Each p-simplex σ in K determines an $(n-p)$-dimensional cell σ^* of K^*, which is the union of all simplices in K' which begin with $\widehat{\sigma^p}$. So K has as many p-simplices as K^* has $(n-p)$-cells. One calls K^* the *dual cell decomposition*.

The cap product with the fundamental cycle, which is given by the sum of the n-dimensional simplices, yields an isomorphism of $\mathbb{Z}\pi$-chain complexes $C^{n-*}(\widetilde{K^*}) \to C_*(\widetilde{K})$. It preserves the cellular $\mathbb{Z}\pi$-bases and so in particular its Whitehead torsion is trivial. Since K' is a common subdivision of K and K^*, there are canonical $\mathbb{Z}\pi$-chain homotopy equivalences $C_*(\widetilde{K'}) \to C_*(\widetilde{K})$ and $C_*(\widetilde{K'}) \to C_*(\widetilde{K^*})$ which have trivial Whitehead torsion. Thus we can write the $\mathbb{Z}\pi$-chain map $? \cap [M] \colon C^{n-*}(\widetilde{K'}) \to C_*(\widetilde{K'})$ as a composite of three simple $\mathbb{Z}\pi$-chain homotopy equivalences. Hence it is a simple $\mathbb{Z}\pi$-chain homotopy equivalence.

Remark 17.13 (Dual handlebody decomposition). From a Morse theoretic point of view Poincaré duality corresponds to the *dual handlebody decomposition*, which we explain next.

Suppose that the smooth compact manifold W is obtained from $\partial_0 W \times [0,1]$ by attaching one q-handle (φ^q), i.e., $W = \partial_0 W \times [0,1] + (\varphi^q)$. Then we can interchange the role of $\partial_0 W$ and $\partial_1 W$ and try to built W from $\partial_1 W$ by handles. It turns out that W can be written as

$$W = \partial_1 W \times [0,1] + (\psi^{n-q}); \tag{17.14}$$

this can be seen as follows.

Let M be the manifold with boundary $S^{q-1} \times S^{n-1-q}$ obtained from $\partial_0 W$ by removing the interior of $\varphi^q(S^{q-1} \times D^{n-q})$. We get

$$\begin{aligned}
W &\cong \partial_0 W \times [0,1] \cup_{S^{q-1} \times D^{n-q}} D^q \times D^{n-q} \\
&= M \times [0,1] \cup_{S^{q-1} \times S^{n-2-q} \times [0,1]} \\
&\qquad \left(S^{q-1} \times D^{n-1-q} \times [0,1] \cup_{S^{q-1} \times D^{n-q} \times \{1\}} D^q \times D^{n-q} \right).
\end{aligned}$$

Inside $S^{q-1} \times D^{n-1-q} \times [0,1] \cup_{S^{q-1} \times D^{n-q} \times \{1\}} D^q \times D^{n-q}$ we have the submanifolds

$$X := S^{q-1} \times 1/2 \cdot D^{n-1-q} \times [0,1] \cup_{S^{q-1} \times 1/2 \cdot D^{n-q} \times \{1\}} D^q \times 1/2 \cdot D^{n-q},$$

$$Y := S^{q-1} \times 1/2 \cdot S^{n-1-q} \times [0,1] \cup_{S^{q-1} \times 1/2 \cdot S^{n-q} \times \{1\}} D^q \times 1/2 \cdot S^{n-q}.$$

The pair (X, Y) is diffeomorphic to $(D^q \times D^{n-q}, D^q \times S^{n-1-q})$, i.e., it is a handle of index $(n-q)$. Let N be obtained from W by removing the interior of X. Then W is obtained from N by adding an $(n-q)$-handle, the so-called *dual handle*. One easily checks that N is diffeomorphic to $\partial_1 W \times [0,1]$ relative $\partial_1 W \times \{1\}$. Thus (17.14) follows.

Remark 17.15 (The Hodge–de Rham Theorem). From an analytic point of view Poincaré duality can be explained as follows. Let M be a connected closed oriented Riemannian manifold. Let $(\Omega^*(M), d^*)$ be the de Rham complex of smooth p-forms on M. The p-th Laplacian is defined by $\Delta_p = (d^p)^* d^p + d^{p-1} (d^{p-1})^* \colon \Omega^p(M) \to \Omega^p(M)$, where $(d^p)^*$ is the adjoint of the p-th differential d^p. The kernel of the p-th Laplacian is the space $\mathcal{H}^p(M)$ of harmonic p-forms o M. The Hodge–de Rham Theorem yields an isomorphism

$$A^p \colon \mathcal{H}^p(M) \xrightarrow{\cong} H^p(M; \mathbb{R}) \tag{17.16}$$

from the space of harmonic p-forms to the singular cohomology of M with coefficients in $\mathbb{R}$. Let $[M]_{\mathbb{R}} \in H^n(M; \mathbb{R})$ be the fundamental cohomology class with $\mathbb{R}$-coefficients, which is characterized by the property $\langle [M]_{\mathbb{R}}, i_*([M]) \rangle = 1$, where $\langle \, , \, \rangle$ is the Kronecker product and $i_* \colon H_n(M; \mathbb{Z}) \to H_n(M; \mathbb{R})$ the change of rings homomorphism. Then A^n sends the volume form $d\text{vol}$ to the class $\frac{1}{\text{vol}(M)} \cdot [M]_{\mathbb{R}}$. The Hodge-star operator $* \colon \Omega^{n-p}(M) \to \Omega^p(M)$ induces an isomorphism

$$* \colon \mathcal{H}^{n-p}(M) \xrightarrow{\cong} \mathcal{H}^p(M). \tag{17.17}$$

We obtain from (17.16) and (17.17) an isomorphism

$$H^{n-p}(M; \mathbb{R}) \xrightarrow{\cong} H^p(M; \mathbb{R}).$$

This is the analytic version of the Poincaré duality. It is equivalent to the claim that the bilinear pairing

$$P^p \colon \mathcal{H}^p(M) \otimes_{\mathbb{R}} \mathcal{H}^{n-p}(M) \to \mathbb{R}, \quad (\omega, \eta) \mapsto \int_M \omega \wedge \eta \qquad (17.18)$$

is non-degenerate. Recall that for any commutative ring R with unit we have the intersection pairing

$$I^p \colon H^p(M; R) \otimes_R H^{n-p}(M; R) \to R, \quad (x, y) \mapsto \langle x \cup y, i_*[M] \rangle, \qquad (17.19)$$

where i_* is the change of coefficients map associated to $\mathbb{Z} \to R$. The fact that the intersection pairing is non-degenerate is for a field R equivalent to the bijectivity of the homomorphism $? \cap [X] \colon H^{n-*}(X; R) \to H_*(X; R)$ of (17.8). If we take $R = \mathbb{R}$, then the pairings (17.18) and (17.19) agree under the Hodge–de Rham isomorphism (17.16).

We close this section by discussing without proofs the extension of the material above to pairs. Consider a pair of finite CW-complexes (Y, X) with a group homomorphism $w(Y) \colon \pi_1(Y) \to \{\pm 1\}$. Let $\widetilde{Y}$ be the universal covering of Y and $\widetilde{X}$ the induced $\pi_1(Y)$-covering of X. The construction of (17.4) can be generalized in order to get a group homomorphism

$$\cap \colon H_{n+1}\big(Y, X; \mathbb{Z}^{w(Y)}\big) \to [C^{n+1-*}(\widetilde{Y}, \widetilde{X}), C_*(\widetilde{Y})]_{\mathbb{Z}[\pi_1(Y)]}.$$

Definition 17.20 (Poincaré pair). *A finite $(n + 1)$-dimensional Poincaré pair is a pair (Y, X) of finite CW-complexes, such that X is a finite n-dimensional Poincaré complex, together with a group homomorphism $w(Y) \colon \pi_1(Y) \to \{\pm 1\}$ such that $w(X)$ factors as $\pi_1(X) \to \pi_1(Y) \xrightarrow{w(Y)} \{\pm 1\}$, and with an element $[Y, X] \in H_{n+1}\big(Y, X; \mathbb{Z}^{w(Y)}\big)$ such that*

$$? \cap [Y, X] \colon C^{n+1-*}(\widetilde{Y}, \widetilde{X}) \to C_*(\widetilde{Y})$$

is a $\mathbb{Z}[\pi_1(Y)]$-chain homotopy equivalence and such that

$$\partial[Y, X] = [X] \in H_n\big(X, \mathbb{Z}^{w(X)}\big).$$

Theorem 17.21 (Poincaré pair structures on manifolds with boundary). *Let W be an $(n+1)$-dimensional manifold with boundary M. Then the pair (W, M) carries the structure of a finite $(n + 1)$-dimensional Poincaré pair.*

17.2   Algebraic L-groups

Now we want to illustrate how to define an algebraic version of Poincaré complexes in the world of chain complexes of modules over a ring with involution. The idea, due to Mishchenko [173] and especially to Ranicki [196], [197], is to mimic the geometric topological phenomena discussed in the preceding Section 17.1. We have seen that a connected closed n-dimensional manifold with fundamental group π yields a $\mathbb{Z}\pi$-chain homotopy equivalence

$$\varphi_0 =? \cap [M] \colon C^{n-*}(\widetilde{M}) \to C_*(\widetilde{M}).$$

This should be the prototype of a non-degenerate symmetric bilinear form on the chain complex $C_*(\widetilde{M})$. So far we have not discussed the symmetry properties of this map, but for example we already know that if $n = 4k$ the induced map $H^{2k}(M;\mathbb{R}) \otimes_{\mathbb{R}} H^{2k}(M;\mathbb{R}) \to \mathbb{R}$ of (17.19) is a non-degenerate symmetric bilinear form. In general the $\mathbb{Z}\pi$-chain map $\varphi_0 =? \cap [M]$ itself is not symmetric, i.e., it is not equal to φ_0^*, but it turns out that φ is coherently homotopic to φ^*, see (17.25) below. Morally, φ_0 is not a fixed point for the involution on $\mathrm{hom}_{\mathbb{Z}\pi}(C^{n-*}(\widetilde{M}), C_*(\widetilde{M}))$, but only a homotopy fixed point. In order to make these ideas precise we need to develop some algebraic language.

Let us first revisit the notions of symmetric and quadratic forms over modules. Assume for the rest of this chapter that R is a ring with involution. Let M be a finitely generated free R-module. We can use the identification $M \cong M^{**}$ to regard the abelian group $\mathrm{hom}_R(M^*, M)$ as a $\mathbb{Z}[\mathbb{Z}/2]$-module. Here $\mathbb{Z}/2 = \langle T \mid T^2 = 1 \rangle$ denotes the cyclic group of order 2, and for an R-homomorphism $f \colon M \to M^*$ we set $Tf := f^* \colon M^{**} \cong M \to M^*$. Then a symmetric bilinear form on M is nothing but an element of

$$Q^0(M) := \mathrm{hom}_R(M^*, M)^{\mathbb{Z}/2} = \mathrm{hom}_{\mathbb{Z}[\mathbb{Z}/2]}(\mathbb{Z}, \mathrm{hom}_R(M^*, M)),$$

the fixed points (algebraically known as invariants) of $\mathrm{hom}_R(M^*, M)$. A symmetric bilinear form is non-degenerate if it is an isomorphism. Similarly, a quadratic form on M is an element of

$$Q_0(M) := \mathrm{hom}_R(M^*, M)_{\mathbb{Z}/2} = \mathbb{Z} \otimes_{\mathbb{Z}[\mathbb{Z}/2]} \mathrm{hom}_R(M^*, M),$$

the orbits (algebraically known as coinvariants) of $\mathrm{hom}_R(M^*, M)$. There is a symmetrization (or norm) homomorphism $1 + T \colon Q_0(M) \to Q^0(M)$, and a quadratic form is non-degenerate if its symmetrization is an isomorphism. Notice that the groups Q^0 and Q_0 are covariantly functorial and satisfy the sum formulas

$$Q^0(M \oplus N) \cong Q^0(M) \oplus Q^0(N) \oplus \mathrm{hom}_R(M^*, N),$$
$$Q_0(M \oplus N) \cong Q_0(M) \oplus Q_0(N) \oplus \mathrm{hom}_R(M^*, N).$$

Now we want to generalize these definitions to chain complexes of finitely generated free R-modules, in such a way that the yet-to-be-defined "Q-groups" of symmetric and quadratic forms on a chain complex satisfy homotopy invariance and sum formulas similar to the ones above. In order to achieve homotopy invariance when we pass from modules to chain complexes, a natural idea is to consider homotopy fixed points (algebraically known as hypercohomology) and homotopy orbits (hyperhomology) instead of ordinary fixed points (invariants) and orbits (coinvariants). Recall that if X is a $\mathbb{Z}/2$-space, then its homotopy fixed points are defined as $X^{h\mathbb{Z}/2} := \mathrm{map}_{\mathbb{Z}/2}(E\mathbb{Z}/2, X)$, where $E\mathbb{Z}/2 = \widetilde{B\mathbb{Z}/2}$ is the free contractible $\mathbb{Z}/2$-space — whereas the fixed points $X^{\mathbb{Z}/2}$ can be identified with $\mathrm{map}_{\mathbb{Z}/2}(\{\bullet\}, X)$, and the $\mathbb{Z}/2$-equivariant maps $\mathbb{Z}/2 \to E\mathbb{Z}/2 \to \{\bullet\}$ yield maps $X^{\mathbb{Z}/2} \to X^{h\mathbb{Z}/2} \to X$. Similarly the homotopy orbits (also known as the Borel construction) of X are defined as $X_{h\mathbb{Z}/2} := E\mathbb{Z}/2 \times_{\mathbb{Z}/2} X$, and there are maps $X \to X_{h\mathbb{Z}/2} \to \mathbb{Z}/2 \backslash X$. This discussion should motivate the following definitions.

Let W_* be the standard free $\mathbb{Z}[\mathbb{Z}/2]$-resolution of the trivial $\mathbb{Z}[\mathbb{Z}/2]$-module $\mathbb{Z}$.

$$W_* : \qquad \cdots \longrightarrow \mathbb{Z}[\mathbb{Z}/2] \xrightarrow{1-T} \mathbb{Z}[\mathbb{Z}/2] \xrightarrow{1+T} \mathbb{Z}[\mathbb{Z}/2] \xrightarrow{1-T} \mathbb{Z}[\mathbb{Z}/2]$$
$$\downarrow$$
$$\mathbb{Z}$$

Now let C_* be a chain complex of finitely generated free R-modules. There is a natural homomorphism of a R-chain complexes $C_* \to (C^{-*})^{-*}$, which sends $m \in C_p$ to the homomorphism $(C_p)^* \to R$, $f \mapsto \overline{(-1)^p f(m)}$. Using this identification $(C^{-*})^{-*} \cong C_*$ we can regard $\mathrm{hom}_R(C^{-*}, C_*)$ as a chain complex of $\mathbb{Z}[\mathbb{Z}/2]$-modules. For any integer $n \in \mathbb{Z}$ define the so-called Q-*groups* as

$$Q^n(C_*) \; := \; H_n(\mathrm{hom}_{\mathbb{Z}[\mathbb{Z}/2]}(W_*, \mathrm{hom}_R(C^{-*}, C_*))),$$
$$Q_n(C_*) \; := \; H_n(W_* \otimes_{\mathbb{Z}[\mathbb{Z}/2]} \mathrm{hom}_R(C^{-*}, C_*)).$$

These groups should be seen as the groups of symmetric respectively quadratic forms on the chain complex C_*. They are covariantly functorial in C_*.

Remark 17.22 (Homotopy invariance of the Q-groups). The fundamental property of the Q-groups defined above is their homotopy invariance: If $f : C_* \to C_*'$ is an R-chain homotopy equivalence, then $Q^n(f)$ and $Q_n(f)$ are isomorphisms. For a proof we refer to [196, Proposition 1.1.(i)].

Note that there are natural group homomorphisms

$$Q_n(C_*) \xrightarrow{1+T} Q^n(C_*) \xrightarrow{\mathrm{ev}} [C^{n-*}, C_*]_R,$$

where ev is induced by the inclusion into W_* of the $\mathbb{Z}[\mathbb{Z}/2]$-chain complex concentrated in degree zero defined by the module $\mathbb{Z}[\mathbb{Z}/2]$ itself. Also note that if 2 is invertible in R, i.e., if R is a $\mathbb{Z}[\frac{1}{2}]$-algebra, then the so-called symmetrization homomorphism $1 + T$ induces an isomorphism $Q_n(C_*) \cong Q^n(C_*)$, see Exercises 17.2 and 17.3.

More explicitly, any element $\varphi \in Q^n(C_*)$ is represented by an n-chain in $\mathrm{hom}_{\mathbb{Z}[\mathbb{Z}/2]}(W_*, \mathrm{hom}_R(C^{n-*}, C_*))$ (unique modulo boundaries), and such an n-chain is given by a sequence of chains

$$\left(\varphi_s \in \hom_R(C^{n-*}, C_*)_s = \prod_{r \in \mathbb{Z}} \hom_R((C_{n+s-r})^*, C_r) \right)_{s \geq 0}$$

such that for all $s \geq 0$ and all $r \in \mathbb{Z}$

$$d_C \varphi_s - (-1)^{n+s} \varphi_s d_{C^{-*}} = (-1)^n (\varphi_{s-1} + (-1)^s T \varphi_{s-1}) : C^{n+s-r-1} \to C_r$$

(where $\varphi_{-1} = 0$). In particular this means that φ_0 is a chain map (whose chain homotopy class is nothing but $\mathrm{ev}(\varphi)$), φ_1 is a chain homotopy between φ_0 and φ_0^*, and so on.... An analogous formula for $Q_n(C_*)$ is available, too.

Remark 17.23 (Sum formula for the Q-groups). There are direct sum decompositions of abelian groups

$$Q^n(C \oplus D) \cong Q^n(C) \oplus Q^n(D) \oplus [C^{n-*}, D_*]_R,$$
$$Q_n(C \oplus D) \cong Q_n(C) \oplus Q_n(D) \oplus [C^{n-*}, D_*]_R,$$

which are compatible with the natural symmetrization homomorphism $1 + T$. The homomorphism $1 + T : [C^{n-*}, D_*]_R \to [C^{n-*}, D_*]_R$ is an isomorphism. Compare [196, Proposition 1.4.(i)].

Definition 17.24 (Algebraic Poincaré complexes). *Let R be a ring with involution and $n \in \mathbb{Z}$ be any integer.*

An n-dimensional symmetric algebraic Poincaré complex over R is a bounded chain complex C_* of finitely generated free R-modules together with an element $\varphi \in Q^n(C_*)$ such that $\mathrm{ev}(\varphi) \in [C^{n-*}, C_*]_R$ is (the R-chain homotopy class of) an R-chain homotopy equivalence.

An n-dimensional quadratic algebraic Poincaré complex over R is a bounded chain complex C_* of finitely generated free R-modules together with an element $\varphi \in Q_n(C_*)$ such that $\mathrm{ev}((1+T)\varphi) \in [C^{n-*}, C_*]_R$ is (the R-chain homotopy class of) an R-chain homotopy equivalence.

Notice that in this definition we do not require that the chain complex C_* is n-dimensional, but only that it is bounded. This agrees with [200] but not with [196], [197].

One could develop an analogous theory in the category of finitely generated projective R-modules, as well as a theory of simple algebraic Poincaré complexes in the category of finitely generated free and based R-modules — see also Remark 17.32 below.

Using the sum formula of Remark 17.23 above one can define the sum of two n-dimensional symmetric algebraic Poincaré complexes $(C, \varphi \in Q^n(C))$, $(C', \varphi' \in Q^n(C'))$ as the n-dimensional symmetric algebraic Poincaré complex

$$\left(C \oplus C', \varphi \oplus \varphi' \in Q^n(C) \oplus Q^n(C') \subset Q^n(C \oplus C') \right).$$

The inverse of (C, φ) is defined as $(C, -\varphi)$. Analogously for quadratic algebraic Poincaré complexes.

This definition is motivated by the following facts, that we state without proofs. Using the theory of acyclic models it is possible to factorize the map (17.4) as

$$H_n(X; \mathbb{Z}^w) \to Q^n(C_*(\widetilde{X})) \xrightarrow{\text{ev}} [C^{n-*}(\widetilde{X}), C_*(\widetilde{X})]_{\mathbb{Z}\pi}, \qquad (17.25)$$

see [197, Section 1]. Then one has the following result:

Proposition 17.26. *Let X be a connected finite n-dimensional Poincaré complex. Then $C_*(\widetilde{X})$ is an n-dimensional symmetric algebraic Poincaré complex over the ring $\mathbb{Z}[\pi_1(Y)]$, with the involution twisted by the orientation homomorphism of X.*

Proof. For a proof we refer to [197, Proposition 2.1]. $\square$

Now we want to sketch how to define pairs and cobordisms of algebraic Poincaré complexes. Let $f \colon C_* \to D_*$ be an R-chain map of chain complexes of finitely generated free R-modules. Consider the induced $\mathbb{Z}[\mathbb{Z}/2]$-chain map

$$\hom_R(f^{-*}, f) \colon \hom_R(C^{-*}, C_*) \to \hom_R(D^{-*}, D_*)$$

and take its mapping cone $C(\hom_R(f^{-*}, f))$. Define the *relative Q-groups*

$$Q^n(f) := H_n(\hom_{\mathbb{Z}[\mathbb{Z}/2]}(W, C(\hom_R(f^{-*}, f_*)))),$$
$$Q_n(f) := H_n(W \otimes_{\mathbb{Z}[\mathbb{Z}/2]} C(\hom_R(f^{-*}, f_*))).$$

One then gets a ladder of long exact sequences

$$
\begin{array}{ccccccccc}
\cdots & \longrightarrow & Q^{n+1}(f) & \xrightarrow{\partial} & Q^n(C_*) & \longrightarrow & Q^n(D_*) & \longrightarrow & Q^n(f) & \longrightarrow \cdots \\
 & & \big\uparrow{\scriptstyle 1+T} & & \big\uparrow{\scriptstyle 1+T} & & \big\uparrow{\scriptstyle 1+T} & & \big\uparrow{\scriptstyle 1+T} & \\
\cdots & \longrightarrow & Q_{n+1}(f) & \xrightarrow{\partial} & Q_n(C_*) & \longrightarrow & Q_n(D_*) & \longrightarrow & Q_n(f) & \longrightarrow \cdots
\end{array}
$$

see [196, Proposition 3.1]. Note that there is a natural group homomorphism

$$Q^{n+1}(f) \xrightarrow{\text{ev}} [C(f)^{n+1-*}, D_*]_R,$$

see Exercise 17.3.

Definition 17.27 (Algebraic Poincaré pair). *An $(n + 1)$-dimensional symmetric (quadratic) algebraic Poincaré pair over a ring with involution R is an R-chain map $f \colon C_* \to D_*$ of bounded chain complex C_*, D_* of finitely generated free R-modules together with an element $\psi \in Q^{n+1}(f)$ (respectively, $\psi \in Q_{n+1}(f)$) such that $\text{ev}(\psi) \in [C(f)^{n+1-*}, D_*]_R$ (respectively, $\text{ev}((1 + T)\psi))$ is (the R-chain homotopy class of) an R-chain homotopy equivalence.*

Note that in this case C_* together with $\partial(\psi)$ is an n-dimensional algebraic Poincaré complex, which is called the *boundary* of the pair.

Proposition 17.28. *Let (Y, X) be a finite $(n+1)$-dimensional Poincaré pair. Then $C_*(\widetilde{X}) \to C_*(\widetilde{Y})$ is an $(n+1)$-dimensional symmetric algebraic Poincaré pair over the ring $\mathbb{Z}[\pi_1(Y)]$, with the involution twisted by the orientation homomorphism of Y.*

Proof. For a proof we refer to [197, Proposition 6.2]. $\qquad\square$

Definition 17.29 (Cobordism of algebraic Poincaré complexes). *Two n-dimensional symmetric algebraic Poincaré complexes $(C, \varphi \in Q^n(C))$, $(C', \varphi' \in Q^n(C'))$ over a ring with involution R are* cobordant *if there exists an $(n+1)$-dimensional symmetric algebraic Poincaré pair $((f, f'): C \oplus C' \to D, \psi \in Q^{n+1}(f, f'))$ called* cobordism *such that $\partial(\psi) = \varphi \oplus -\varphi' \oplus 0 \in Q^n(C \oplus C') \cong Q^n(C) \oplus Q^n(C') \oplus [C^{n-*}, C'_*]_R$.*

Cobordism of quadratic algebraic Poincaré complexes is defined completely analogously.

Theorem 17.30 (Algebraic cobordism). *Let R be a ring with involution and $n \in \mathbb{Z}$ be any integer.*

(1) *Let $(C, \varphi \in Q^n(C))$ and $(C', \varphi' \in Q^n(C'))$ be n-dimensional symmetric algebraic Poincaré complexes over a ring with involution R, and let $f: C \to C'$ be an R-chain homotopy equivalence such that $Q^n(f)(\varphi) = \varphi'$. Then (C, φ) and (C', φ') are cobordant.*

(2) *Cobordism is an equivalence relation on the class of n-dimensional symmetric algebraic Poincaré complexes.*

Analogous statements hold for quadratic algebraic Poincaré complexes.

Proof. See [196, Proposition 3.2]. $\qquad\square$

Definition 17.31 (Algebraic L-groups). *Let R be a ring with involution and $n \in \mathbb{Z}$ be any integer.*

The n-th symmetric algebraic L-group $L^n_h(R)$ of R is defined as the set of cobordism classes of n-dimensional symmetric algebraic Poincaré complexes, with addition and inverses given by the corresponding operations for symmetric algebraic Poincaré complexes.

The n-th quadratic algebraic L-group $L^h_n(R)$ of R is defined analogously using n-dimensional quadratic algebraic Poincaré complexes.

Notice that the natural transformation $1 + T: Q_n \to Q^n$ defines a group homomorphism $s: L^h_n(R) \to L^n_h(R)$, called symmetrization.

Remark 17.32 (Decorations in L-theory — Rothenberg sequences). The decoration h in the notation $L^n_h(R)$, $L^h_n(R)$ stands for free, and refers to the fact that we are working in the category of finitely generated free R-modules. Completely analogously one can define groups $L^n_p(R)$, $L^p_n(R)$ by working in the category of finitely generated projective R-modules. Working with finitely generated free and based R-modules and simple algebraic Poincaré complexes one can define groups

$L_s^n(R)$, $L_n^s(R)$. All these groups are related by the long exact *Rothenberg sequences* (see [196, Proposition 9.1])

$$\ldots \to L_h^n(R) \to L_p^n(R) \to \widehat{H}^n(\mathbb{Z}/2; \widehat{K}_0(R)) \to L_h^{n-1}(R) \to \ldots,$$

$$\ldots \to L_s^n(R) \to L_h^n(R) \to \widehat{H}^n(\mathbb{Z}/2; \widehat{K}_1(R)) \to L_s^{n-1}(R) \to \ldots,$$

and analogously for quadratic algebraic *L*-theory — just replace the symmetric groups with the quadratic ones in the sequences above. Here $\widehat{H}^n(\mathbb{Z}/2; \widehat{K}_i(R))$ denotes the Tate cohomology of $\mathbb{Z}/2$ with coefficients in the $\mathbb{Z}[\mathbb{Z}/2]$-module $\widehat{K}_i(R)$. Notice that the Tate cohomology groups are 2-periodic and that they are 2-torsion. In particular one sees that after tensoring with $\mathbb{Z}\left[\frac{1}{2}\right]$ the algebraic *L*-groups corresponding to different decorations are all isomorphic.

Theorem 17.33 (Properties of the algebraic *L*-groups). *Let R be a ring with involution.*

(1) *The algebraic L-groups are 4-periodic:*

$$L_h^{n+4}(R) \cong L_h^n(R) \quad and \quad L_{n+4}^h(R) \cong L_n^h(R).$$

(2) *If 2 is invertible in R, i.e., if R is a $\mathbb{Z}\left[\frac{1}{2}\right]$-algebra, then the symmetrization homomorphism is an isomorphism $L_n^h(R) \cong L_h^n(R)$.*

(3) *For any R the symmetrization homomorphism induces an isomorphism*

$$L_n^h(R) \otimes_{\mathbb{Z}} \mathbb{Z}\left[\tfrac{1}{2}\right] \cong L_h^n(R) \otimes_{\mathbb{Z}} \mathbb{Z}\left[\tfrac{1}{2}\right].$$

(4) *For any $n \geq 0$ the groups $L_n^h(\mathbb{Z}G)$ of Definition 17.31 above are isomorphic to the L-groups $L_m(G)$ introduced in Sections 12.3 and 12.4.*

Proof. The claims (4) and (3) are proved in [196, Propositions 5.1 and 5.2, and Proposition 8.2 respectively], but using the "old" definition of *L*-theory, the one that requires the underlying chain complex of an n-dimensional algebraic Poincaré complex to be itself n-dimensional. See also [200, Examples 1.11 and 3.18], where the definitions are the ones adopted here. With the latter definitions the 4-periodicity of *L*-theory (1) is a very easy exercise, compare [200, Proposition 1.10] and Exercise 17.4. Finally, (2) follows easily from Exercise 17.2. $\square$

We conclude this chapter by putting together most of the things we have seen in order to describe the very important construction known as symmetric signature. Let X be a connected topological space with fundamental group π, and let M be an oriented closed n-dimensional manifold equipped with a so-called reference map $u\colon M \to X$. Then from Theorems 17.10 and 17.26 we get that $\mathbb{Z}\pi \otimes_{\mathbb{Z}\pi_1(M)} C_*(\widetilde{M})$ is an n-dimensional symmetric algebraic Poincaré complex over $\mathbb{Z}\pi$. We call the class it represents in symmetric *L*-theory the *symmetric signature* and denote it by

$$\sigma(M, u) \quad \in \quad L^n(\mathbb{Z}G). \tag{17.34}$$

Combining Theorems 17.21, 17.28 and 17.30(1) we obtain the following result.

Theorem 17.35 (Symmetric signature). *Let X be a connected topological space with fundamental group π. Then the symmetric signature defines for any $n \geq 0$ a group homomorphism*

$$\sigma(X)\colon \Omega_n(X) \to L_h^n(\mathbb{Z}\pi), \quad [M, u] \mapsto \sigma(M, u)$$

which is homotopy invariant in the following sense: Let $[M, u], [M', u'] \in \Omega_n(X)$ and suppose that there exists an oriented homotopy equivalence $f\colon M \to M'$ such that $u'f$ is homotopic to u; then $\sigma(M, u) = \sigma(M', u') \in L_h^n(\mathbb{Z}\pi)$.

Remark 17.36 (The surgery obstruction and symmetric signatures). Consider a normal map of closed oriented m-dimensional manifolds given by a map $f\colon M \to N$ of degree one and a stable bundle isomorphism $\alpha\colon f^*E \to \nu_M$. Let

$$\Theta(f, \alpha) \in L_m^h(\pi_1(N))$$

be its surgery obstruction. There is a canonical map

$$L_m^h(\pi_1(N)) \to L_h^m(\pi_1(N))$$

which is an isomorphism after inverting 2. It sends the surgery obstruction to the difference

$$\sigma(N, \mathrm{id}_N) - \sigma(M, f)$$

as proven in [197]. Notice that this difference is independent of the bundle map α, whereas $\Theta(f, \alpha) \in L_m^h(\pi_1(N))$ does depend on α in general.

Chapter 18

Spectra

In this section we give a brief and elementary introduction to spectra mentioning examples such as K- and L-theory spectra. We will also introduce as illustration for spectra the Thom spectrum of a stable vector bundle. The best motivation for them is to consider in detail the bordism groups associated to stable vector bundles. Therefore we give some extended version of the material which has already appeared in Chapter 2.

To understand the remaining chapters is suffices to go through the basics of Sections 18.1, 18.4, 18.5 and 18.6.

18.1   Basic Notions about Spectra

We will work in the category of compactly generated spaces (see [229], [255, I.4]). So space means compactly generated space and all constructions like mapping spaces and products are to be understood in this category. We will always assume that the inclusion of the base point into a pointed space is a cofibration and that maps between pointed spaces preserve the base point.

Working in the category of compactly generated spaces has several technical advantages, for instance the exponential map

$$\exp\colon \operatorname{map}(X \times Y, Z) \ \stackrel{\cong}{\longrightarrow} \ \operatorname{map}(X, \operatorname{map}(Y, Z)) \tag{18.1}$$

is a natural homeomorphism for all spaces and the product of two CW-complexes is always a CW-complex again, no additional assumptions like locally compact are needed. A reader who is not familiar with this category may simply ignore this technicality.

We define the category of spectra SPECTRA as follows. A *spectrum* $\mathbf{E} = \{(E(n), \sigma(n)) \mid n \in \mathbb{Z}\}$ is a sequence of pointed spaces $\{E(n) \mid n \in \mathbb{Z}\}$ together with pointed maps (called *structure maps*) $\sigma(n)\colon E(n) \wedge S^1 \to E(n+1)$. A *map of spectra* (sometimes also called function in the literature) $\mathbf{f}\colon \mathbf{E} \to \mathbf{E}'$ is a sequence of maps of pointed spaces $f(n)\colon E(n) \to E'(n)$ that are compatible with the structure maps $\sigma(n)$, i.e., we have $f(n+1) \circ \sigma(n) = \sigma'(n) \circ (f(n) \wedge \mathrm{id}_{S^1})$ for all n. This should not be confused with the notion of map of spectra in the stable homotopy category (see [3, III.2]). The *homotopy groups of a spectrum* are defined as

$$\pi_n(\mathbf{E}) := \mathrm{colim}_{k \to \infty} \pi_{n+k}(E(k)),$$

where the maps in this system are given by the composition

$$\pi_{p+k}(E(k)) \to \pi_{p+k+1}(E(k) \wedge S^1) \xrightarrow{\sigma(k)_*} \pi_{p+k+1}(E(k+1))$$

of the suspension homomorphism and the homomorphism induced by the structure map. Notice that the homotopy groups $\pi_n(\mathbf{E})$ are abelian for $n \in \mathbb{Z}$ and can be non-trivial also for negative $n \in \mathbb{Z}$. If $\pi_n(\mathbf{E}) = 0$ for $n \leq -1$, we call $\mathbf{E}$ *connective*, otherwise it is called *non-connective*.

A *weak equivalence of spectra* is a map $\mathbf{f}\colon \mathbf{E} \to \mathbf{F}$ inducing an isomorphism on all homotopy groups.

If $X = (X, x)$ and $Y = (Y, y)$ are pointed spaces, define their *wedge,* or sometimes also called one-point-union, to be the pointed space

$$X \vee Y := X \times \{y\} \cup \{x\} \times Y \subseteq X \times Y$$

with the subspace topology and their *smash product* to be the pointed space

$$X \wedge Y := X \times Y / X \vee Y,$$

where in both cases we choose the obvious base point. The exponential map yields in the category of pointed spaces a natural homeomorphism

$$\mathrm{map}(X \wedge Y, Z) \xrightarrow{\cong} \mathrm{map}(X, \mathrm{map}(Y, Z)). \tag{18.2}$$

For a pointed space X let $\Omega X := \mathrm{map}(S^1, X)$ be its associated *loop space*. We get from the exponential map above in the special case $X = S^1$ a natural homeomorphism

$$\mathrm{map}(S^1 \wedge X, Y) \xrightarrow{\cong} \mathrm{map}(X; \Omega Y).$$

We call the image of a map $S^1 \wedge X \to Y$ under this bijection its adjoint $X \to \Omega Y$. A spectrum $\mathbf{E}$ is called Ω-*spectrum* if the adjoint of each structure map $\bar{\sigma}(n)\colon E(n) \to \Omega E(n+1)$ is a weak homotopy equivalence of spaces. We denote by Ω-SPECTRA the corresponding full subcategory of SPECTRA.

Given a family of spectra $\{\mathbf{E}_i \mid i \in I\}$, we define its *product* to be the spectrum $\prod_{i \in I} \mathbf{E}$ whose n-th underlying space is $\prod_{i \in I} E(n)_i$. The structure maps are given by the product of the adjoints of individual structure maps taking into account that Ω commutes with products.

Lemma 18.3. *Let $\{\mathbf{E}_i \mid i \in I\}$ be a family of Ω-spectra, where I is an arbitrary index set. Then the canonical map induced by the various projections* $\mathrm{pr}_i \colon \prod_{i \in I} \mathbf{E}_i \to \mathbf{E}_i$

$$\pi_n \left(\prod_{i \in I} \mathbf{E}_i \right) \xrightarrow{\cong} \prod_{i \in I} \pi_n(\mathbf{E}_i)$$

is bijective for all $n \in \mathbb{Z}$.

Lemma 18.3 does not hold without the assumption that each $\mathbf{E}_i$ is an Ω-spectrum.

Example 18.4 (Suspension spectrum). Let X be a pointed space. Define its *suspension spectrum* $\Sigma^\infty X$ to be the spectrum whose n-th space is the one-point-space for $n \leq -1$ and is the n-fold suspension $S^n \wedge X = S^1 \wedge S^1 \wedge \ldots \wedge S^1 \wedge X$ for $n \geq 0$. The structure maps are given by the identity. The suspension spectrum is not an Ω-spectrum.

The n-th homotopy group $\pi_n(\Sigma^\infty X)$ of the suspension spectrum is also called the *n-th stable homotopy group* $\pi_n^s(X)$ of the pointed space X.

As a special case we get the sphere spectrum $\mathbf{S} = \Sigma^\infty S^0$ whose n-th space is S^n for $n \geq 0$ and the one-point space for $n \leq 1$.

Example 18.5 (Eilenberg–MacLane spectrum). Let G be an abelian group. There is an in G natural construction of the so-called associated *Eilenberg–MacLane spectrum* $\mathbf{H}_G$. Its n-th space is a point for $n \leq 0$ and an *Eilenberg–MacLane space of type* (G, n) for $n \geq 1$, i.e., a pointed space $K(G, n)$ whose homotopy groups are all trivial except in dimension n where it is isomorphic to G.

Example 18.6 (Topological K-theory spectrum). Let BU be the classifying space of the unitary group U. Fix a base point in BU. Bott periodicity yields a pointed homotopy equivalence

$$\beta \colon \mathbb{Z} \times BU \to \Omega^2(\mathbb{Z} \times BU).$$

Define an Ω-spectrum $\mathbf{K}^{\mathrm{top}}$ by the sequence of spaces which is $\mathbb{Z} \times BU$ in even dimensions and $\Omega(\mathbb{Z} \times BU)$ in odd dimensions. The adjoints of the structure maps are given by the identity on $\Omega(\mathbb{Z} \times BU)$ and by the homotopy equivalence β above. It is called the *topological K-theory spectrum*. It satisfies $\pi_{2n+1}(\mathbf{K}^{\mathrm{top}}) = 0$ and $\pi_{2n}(\mathbf{K}^{\mathrm{top}}) = \mathbb{Z}$ for $n \in \mathbb{Z}$.

18.2 Homotopy Pushouts and Homotopy Pullbacks for Spaces

In this section we give some basic properties of homotopy pushouts and homotopy pullbacks of spaces. Given maps $i_1 \colon X_0 \to X_1$ and $i_2 \colon X_0 \to X_2$, the *homotopy pushout* is the quotient space of $X_1 \amalg X_0 \times [0, 1] \amalg X_2$ obtained by identifying $(x_0, 0) = i_1(x_0)$ and $(x_0, 1) = i_2(x_0)$ for all $x_0 \in X_0$. There are natural inclusions

$j_k \colon X_k \to$ hopushout for $k = 1, 2$ which are induced from the obvious inclusions $X_k \to X_1 \amalg X_0 \times [0, 1] \amalg X_2$. The diagram

$$
\begin{array}{ccc}
X_0 & \xrightarrow{\ i_1\ } & X_1 \\
{\scriptstyle i_2}\big\downarrow & & \big\downarrow{\scriptstyle j_1} \\
X_2 & \xrightarrow[\ j_2\ ]{} & \text{hopushout}
\end{array}
$$

is not commutative, but there is an explicit homotopy $h \colon X_0 \times [0, 1] \to$ hopushout from $j_1 \circ i_1$ to $j_2 \circ i_2$ which is the restriction of the canonical projection map $X_1 \amalg X_0 \times [0, 1] \amalg X_2 \to$ hopushout to $X_0 \times [0, 1]$. The maps j_1 and j_2 are cofibrations. The diagram above together with the homotopy h has the following universal property. For every space Z, every pair of maps $f_1 \colon X_1 \to Z$, $f_2 \colon X_2 \to Z$ and every homotopy $g \colon X_0 \times [0, 1] \to Z$ from $f_1 \circ i_1$ to $f_2 \circ j_2$ there is precisely one map $F \colon$ hopushout $\to Z$ such that $g = F \circ h$ and $F \circ j_k = f_k$ for $k = 1, 2$. There is an obvious map from the homotopy pushout to the pushout of $X_1 \xleftarrow{\ i_1\ } X_0 \xrightarrow{\ i_2\ } X_2$

$$
\varphi \colon \text{hopushout} \to \text{pushout} .
$$

It is a homotopy equivalence if one of the maps i_1 and i_2 is a cofibration.

Remark 18.7 (Homotopy invariance of the homotopy pushout). The main advantage of the homotopy pushout in comparison with the pushout is the following property called *homotopy invariance*. Consider the commutative diagram

$$
\begin{array}{ccc}
X_1 & \xleftarrow{\ i_1\ } X_0 \xrightarrow{\ i_2\ } & X_2 \\
{\scriptstyle f_2}\big\downarrow \quad\quad {\scriptstyle f_0}\big\downarrow \quad\quad {\scriptstyle f_2}\big\downarrow & & \\
Y_1 & \xleftarrow{\ k_1\ } Y_0 \xrightarrow{\ k_2\ } & Y_2
\end{array} .
$$

Let hopushout$(f_i) \colon$ hopushout$_X \to$ hopushout$_Y$ be the map induced between the homotopy pushouts of the two rows. Let pushout$(f_i) \colon$ pushout$_X \to$ pushout$_Y$ be the map induced between the pushouts of the two rows. Suppose that f_1, f_0 and f_2 are homotopy equivalences or weak homotopy equivalences respectively. Then hopushout(f_i) is a homotopy equivalence or weak homotopy equivalence respectively. The corresponding statement is false for pushout(f_i).

Dually one can define homotopy pullbacks, namely, invert all arrows, replace coproducts by products, replace subspaces by quotient spaces and replace $- \times [0, 1]$ by map$([0, 1], -)$. More precisely, given maps $i_1 \colon X_1 \to X_0$ and $i_2 \colon X_2 \to X_0$, the *homotopy pullback* is the subspace of $X_1 \times \text{map}([0, 1], X_0) \times X_2$ consisting of triples (x_1, w, x_2) satisfying $i_1(x_1) = w(0)$ and $i_2(x_2) = w(1)$. There are natural maps

j_k: hopullback $\to X_k$ for $k = 1, 2$. The diagram

$$
\begin{array}{ccc}
\text{hopullback} & \xrightarrow{\ j_1\ } & X_1 \\
{\scriptstyle j_2}\downarrow & & \downarrow{\scriptstyle i_1} \\
X_2 & \xrightarrow[\ i_2\ ]{} & X_0
\end{array}
$$

is not commutative, but there is an explicit homotopy h: hopullback $\times [0,1] \to X_0$ from $i_1 \circ j_1$ to $i_2 \circ j_2$. The maps j_1 and j_2 are fibrations. The diagram above together with the homotopy h has the following universal property. For each space Z, maps $f_1 \colon Z \to X_1$, $f_2 \colon Z \to X_2$ and homotopy $g \colon Z \times [0,1] \to X_0$ from $i_1 \circ f_1$ to $i_2 \circ f_2$ there is precisely one map $F \colon Z \to$ hopullback such that $g = h \circ F \times \mathrm{id}_{[0,1]}$ and $j_k \circ F = f_k$ for $k = 1, 2$. There is an obvious map from the pullback to the homotopy pullback of $X_1 \xleftarrow{i_1} X_0 \xrightarrow{i_2} X_2$,

$$
\psi \colon \text{pullback} \to \text{hopullback}.
$$

It is a homotopy equivalence if one of the maps i_1 and i_2 is a fibration.

Remark 18.8 (Homotopy invariance of the homotopy pullbacks). Consider the commutative diagram

$$
\begin{array}{ccccc}
X_1 & \xrightarrow{\ i_1\ } & X_0 & \xleftarrow{\ i_2\ } & X_2 \\
{\scriptstyle f_1}\downarrow & & {\scriptstyle f_0}\downarrow & & \downarrow{\scriptstyle f_2} \\
Y_1 & \xrightarrow[\ k_1\ ]{} & Y_0 & \xleftarrow[\ k_2\ ]{} & Y_2
\end{array}
\quad .
$$

Let hopullback(f_i): hopullback$_X \to$ hopullback$_Y$ be the map induced between the homotopy pullbacks of the two rows. Let pullback(f_i): pullback$_X \to$ pullback$_Y$ be the map induced between the pullbacks of the two rows. Suppose that f_1, f_0 and f_2 are homotopy equivalences or weak homotopy equivalences respectively. Then hopullback(f_i) is a homotopy equivalence or weak homotopy equivalence respectively. The corresponding statement is false for pullback(f_i).

Consider a commutative diagram of spaces

$$
\begin{array}{ccc}
X_0 & \xrightarrow{\ i_1\ } & X_1 \\
{\scriptstyle i_2}\downarrow & & \downarrow{\scriptstyle j_1} \\
X_2 & \xrightarrow[\ j_2\ ]{} & X
\end{array}
\quad .
$$

There are canonical maps

$$
a \colon X_0 \ \to \ \text{hopullback}(X_1 \xrightarrow{j_1} X \xleftarrow{j_2} X_2);
$$
$$
b \colon \text{hopushout}(X_1 \xleftarrow{i_1} X_0 \xrightarrow{i_2} X_2) \ \to \ X.
$$

We call the square *homotopy cartesian* if a is a weak equivalence and *homotopy cocartesian* if b is a weak equivalence.

Theorem 18.9 (Excision Theorem of Blakers–Massey). *Assume that the square*

$$
\begin{array}{ccc}
X_0 & \xrightarrow{\ i_1\ } & X_1 \\
{\scriptstyle i_2}\big\downarrow & & \big\downarrow{\scriptstyle j_1} \\
X_2 & \xrightarrow[\ 2\ ]{} & X
\end{array}
$$

is homotopy cocartesian and that i_1 is k_1-connected and i_2 is k_2-connected. Then the map $a\colon X_0 \to \mathrm{hopullback}(X_1 \xrightarrow{j_1} X \xleftarrow{j_2} X_2)$ is at least $(k_1 + k_2 - 1)$-connected.

Proof. See for example [99, Section 2, in particular Theorem 2.3], [238, Satz 14.1 on page 178]. □

18.3   Homotopy Pushouts and Homotopy Pullbacks for Spectra

Next we deal with spectra. Consider a commutative square $D_{\mathbf E}$ of spectra

$$
\begin{array}{ccc}
\mathbf{E}_0 & \xrightarrow{\ \mathbf{i}_1\ } & \mathbf{E}_1 \\
{\scriptstyle \mathbf{i}_2}\big\downarrow & & \big\downarrow{\scriptstyle \mathbf{j}_1} \\
\mathbf{E}_2 & \xrightarrow[\ \mathbf{j}_2\ ]{} & \mathbf{E}
\end{array}
$$

We denote by $\mathrm{hopullback}(\mathbf{E}_1 \xrightarrow{\mathbf{j}_1} \mathbf{E} \xleftarrow{\mathbf{j}_2} \mathbf{E}_2)$ the levelwise homotopy pullback spectrum and by $\mathrm{hopushout}(\mathbf{E}_1 \xleftarrow{\mathbf{i}_1} \mathbf{E}_0 \xrightarrow{\mathbf{i}_2} \mathbf{E}_2)$ the levelwise homotopy pushout spectrum, i.e., the k-th spaces are given by the homotopy pullback respectively the homotopy pushout of the corresponding diagrams of pointed spaces. For the structure maps use the fact that homotopy pullbacks commute with Ω and homotopy pushouts commute with $S^1 \wedge -$ up to natural homeomorphisms. There are canonical maps of spectra

$$
\begin{aligned}
\mathbf{a}\colon \mathbf{E}_0 \ &\to\ \mathrm{hopullback}(\mathbf{E}_1 \xrightarrow{\mathbf{j}_1} \mathbf{E} \xleftarrow{\mathbf{j}_2} \mathbf{E}_2), \\
\mathbf{b}\colon \mathrm{hopushout}(\mathbf{E}_1 \xleftarrow{\mathbf{i}_1} \mathbf{E}_0 \xrightarrow{\mathbf{i}_1} \mathbf{E}_2) \ &\to\ \mathbf{E}.
\end{aligned}
$$

We call the square $D_{\mathbf E}$ *homotopy cartesian* if $\mathbf{a}$ is a weak equivalence of spectra and *homotopy cocartesian* if $\mathbf{b}$ is a weak equivalence.

Theorem 18.10 (Homotopy cartesian and homotopy cocartesian is the same for spectra). *A commutative square $D_{\mathbf E}$ of spectra is homotopy cocartesian if and only* *if it is homotopy cartesian. In this case there is a long exact natural Mayer–Vietoris sequence*

$$\ldots \xrightarrow{\partial_{n+1}} \pi_n(\mathbf{E}_0) \xrightarrow{\pi_n(i_1) \oplus \pi_n(i_2)} \pi_n(\mathbf{E}_1) \oplus \pi_n(\mathbf{E}_2)$$

$$\xrightarrow{\pi_n(j_1) - \pi_n(j_2)} \pi_n(\mathbf{E}) \xrightarrow{\partial_n} \pi_{n-1}(\mathbf{E}_0) \xrightarrow{\pi_{n-1}(i_1) \oplus \pi_{n-1}(i_2)} \ldots$$

Proof. This is a consequence of the Excision Theorem of Blakers–Massey 18.9. The point is that taking the k-fold suspension raises the connectivity of i_1 and i_2 appearing in the Excision Theorem of Blakers–Massey 18.9 by k, whence the connectivity of the map a in the Excision Theorem of Blakers–Massey 18.9 is raised by $2k$.

We give the definition of the boundary map. There is a canonical map $c_m \colon \Omega E(m) \to \text{hopullback}(\mathbf{E}_1 \xrightarrow{j_1} \mathbf{E} \xleftarrow{j_2} \mathbf{E}_2)(m)$ for each $m \in \mathbb{Z}$. They induce maps

$$\pi_{n+1}(c_m) \colon \pi_{n+1}(\mathbf{E}(m)) \xrightarrow{\cong} \pi_n(\Omega E(m)) \to \pi_n\left(\text{hopullback}(\mathbf{E}_1 \xrightarrow{j_1} \mathbf{E} \xleftarrow{j_2} \mathbf{E}_2)(m)\right)$$

and thus homomorphisms

$$d_n \colon \pi_{n+1}(\mathbf{E}) \to \pi_n\left(\text{hopullback}(\mathbf{E}_1 \xrightarrow{j_1} \mathbf{E} \xleftarrow{j_2} \mathbf{E}_2)\right).$$

The map ∂_{n+1} is its composition with the inverse of the isomorphism

$$\pi_n(\mathbf{a}) \colon \pi_n(\mathbf{E}_0) \xrightarrow{\cong} \pi_n\left(\text{hopullback}(\mathbf{E}_1 \xrightarrow{j_1} \mathbf{E} \xleftarrow{j_2} \mathbf{E}_2)\right).$$

More details can be found for instance in [161, Lemma 2.5]. $\qquad\square$

18.4 (Co-)Homology Theories Associated to Spectra

Let $\mathbf{E}$ be a spectrum with structure maps $\sigma(n) \colon S^1 \wedge E(n) \to E(n+1)$. It defines a (generalized) homology theory $H_*(-; \mathbf{E})$ for the category of CW-pairs as follows.

Given a pointed space X, define the *smash product* of X and $\mathbf{E}$ to be the spectrum $X \wedge \mathbf{E}$ whose n-th space is $X \wedge E(n)$. The structure maps are given by

$$S^1 \wedge X \wedge E(n) \xrightarrow{\cong} X \wedge S^1 \wedge E(n) \xrightarrow{\text{id}_X \wedge \sigma(n)} X \wedge E(n+1).$$

Define the *reduced mapping cone* of a pointed space $X = (X, x)$ by $X \times [0,1]/X \times \{0\} \cup \{x\} \times [0,1]$. Given a space X, let X_+ be the pointed space obtained from X by adding a disjoint base point. Define for a pair (X, A) of CW-complexes

$$H_n(X, A; \mathbf{E}) \;=\; \pi_n\left((X_+ \cup_{A_+} \text{cone}(A_+)) \wedge \mathbf{E}\right). \tag{18.11}$$

Lemma 18.12. *We obtain a generalized homology theory $H_*(-;\mathbf{E})$ indexed by $\mathbb{Z}$ on the category of pairs of CW-complexes which satisfies the disjoint union axiom, i.e., for each family of pairs of CW-complexes (X_i, A_i) for $i \in I$ the map induced by the various canonical inclusions*

$$\bigoplus_{i \in I} H_n(X_i, A_i; \mathbf{E}) \xrightarrow{\cong} H_n\left(\coprod_{i \in I}(X_i, A_i); \mathbf{E}\right)$$

is bijective for all $n \in \mathbb{Z}$.

Proof. We explain why we can associate a natural long exact Mayer–Vietoris sequence to a pushout

$$
\begin{array}{ccc}
X_0 & \xrightarrow{\ i_1\ } & X_1 \\
{\scriptstyle i_2}\downarrow & & \downarrow{\scriptstyle j_1} \\
X_2 & \xrightarrow[\ j_2\ ]{} & X
\end{array}
$$

such that i_0 is an inclusion of CW-complexes, i_1 is cellular and X carries the induced CW-structure. If we take the smash product with $\mathbf{E}$, we obtain a commutative square of spectra

$$
\begin{array}{ccc}
X_0 \wedge \mathbf{E} & \xrightarrow{\ i_1 \wedge \mathrm{id}_{\mathbf{E}}\ } & X_1 \wedge \mathbf{E} \\
{\scriptstyle i_2 \wedge \mathrm{id}_{\mathbf{E}}}\downarrow & & \downarrow{\scriptstyle j_1 \wedge \mathrm{id}_{\mathbf{E}}} \\
X_2 \wedge \mathbf{E} & \xrightarrow[\ j_2 \wedge \mathrm{id}_{\mathbf{E}}\ ]{} & X \wedge \mathbf{E}
\end{array}
$$

such that in each dimension n the underlying commutative of spaces is a pushout with the top arrow a cofibration and hence a homotopy pushout of spaces. Hence the commutative diagram of spectra above is a homotopy pushout and now we can apply Lemma 18.10. $\qquad\square$

Given a pointed space X, the *mapping space spectrum* $\mathrm{map}(X; \mathbf{E})$ has as n-th space $\mathrm{map}(X; E(n))$. The definition of the n-th structure map involves the canonical map of pointed spaces (which is not a homeomorphism in general)

$$\mathrm{map}(X, E(n)) \wedge S^1 \to \mathrm{map}(X, E(n) \wedge S^1),$$

which assigns to $\varphi \wedge z$ the map from X to $E(n) \wedge S^1$ sending $x \in X$ to $\varphi(x) \wedge z \in E(n) \wedge S^1$.

Define for a pair (X, A) of CW-complexes and an Ω-spectrum $\mathbf{E}$

$$H^n(X, A; \mathbf{E}) \;=\; \pi_n\left(\mathrm{map}\left(X_+ \cup_{A_+} \mathrm{cone}(A_+), \mathbf{E}\right)\right). \qquad (18.13)$$

Lemma 18.14. *Given an Ω-spectrum, we obtain a generalized cohomology theory $H^*(-;\mathbf{E})$ indexed by $\mathbb{Z}$ on the category of pairs of CW-complexes which satisfies the disjoint union axiom, i.e., for each family of pairs of CW-complexes (X_i, A_i) for $i \in I$ the map induced by the various canonical inclusions*

$$H^n\left(\coprod_{i \in I}(X_i, A_i); \mathbf{E}\right) \xrightarrow{\cong} \prod_{i \in I} H^n(X_i, A_i; \mathbf{E})$$

is bijective for all $n \in \mathbb{Z}$.

Proof. The disjoint union axiom is satisfied because of Lemma 18.3. Here we need the assumption that $\mathbf{E}$ is an Ω-spectrum what is not required in Lemma 18.12. $\square$

Example 18.15 (Singular (co-)homology). If we apply the construction above to the Eilenberg–MacLane spectrum $\mathbf{H}_G$ associated to an abelian group G of Example 18.5, then $H_*(X, A; \mathbf{H}_G)$ and $H^*(X, A; \mathbf{H}_G)$ agree with the singular homology and cohomology with coefficients in G.

Example 18.16 (Topological K-theory). If we apply the construction above to the topological K-theory spectrum $\mathbf{K}^{\mathrm{top}}$ of Example 18.6 and a finite pair of CW-complexes (X, A), then $H^*(X, A; \mathbf{K}^{\mathrm{top}})$ agrees with topological K-theory $K^*(X, A)$ defined in terms of vector bundles (see for instance [8]).

Example 18.17 (Stable homotopy theory). If we apply the construction above to the sphere spectrum $\Sigma^\infty S^0$, the associated (co-)homology theory is denoted by $\pi_n^s(X_+, A_+) := H_n(X, A; \Sigma^\infty S^0)$ and $\pi_s^n(X_+, A_+) := H^n(X, A; \Sigma^\infty S^0)$ and called the *stable (co-)homotopy of the pair* (X, A).

18.5 K-Theory and L-Theory Spectra

Let RINGS be the category of associative rings with unit. An *involution* on R is a map $R \to R$, $r \mapsto \overline{r}$ satisfying $\overline{1} = 1$, $\overline{x+y} = \overline{x}+\overline{y}$ and $\overline{x \cdot y} = \overline{y} \cdot \overline{x}$ for all $x, y \in R$. Let RINGS^{inv} be the category of rings with involution. Let C^*-ALGEBRAS be the category of C^*-algebras. There are classical functors for $j \in -\infty \amalg \{j \in \mathbb{Z} \mid j \leq 2\}$

$$\mathbf{K}: \text{RINGS} \quad \to \quad \text{SPECTRA}; \tag{18.18}$$

$$\mathbf{L}^{(j)}: \text{RINGS}^{\mathrm{inv}} \quad \to \quad \text{SPECTRA}; \tag{18.19}$$

$$\mathbf{K}^{\mathrm{top}}: C^*\text{-ALGEBRAS} \quad \to \quad \text{SPECTRA}. \tag{18.20}$$

The construction of such a non-connective algebraic K-theory functor (18.18) goes back to Gersten [97] and Wagoner [243]. The spectrum for quadratic algebraic L-theory (18.19) is constructed by Ranicki in [200]. In a more geometric formulation it goes back to Quinn [193]. In the topological K-theory case (18.20) a construction using Bott periodicity for C^*-algebras can easily be derived from the Kuiper–Mingo Theorem (see [216, Section 2.2]). The homotopy groups of these spectra give the algebraic K-groups of Quillen (in high dimensions) and of Bass (in negative dimensions), the decorated quadratic L-groups and the topological K-groups of C^*-algebras.

In all three cases we need the non-connective versions of the spectra, i.e., the homotopy groups in negative dimensions are non-trivial in general.

Now let us fix a coefficient ring R. Then sending a group G to the group ring RG yields a functor GROUPS $\to$ RINGS, where GROUPS denotes the category of groups. If R comes with an involution $R \to R, r \mapsto \bar{r}$, we get a functor GROUPS $\to$ RINGS^{inv}, if we equip RG with the involution sending $\sum_{g \in G} r_g \cdot g$ to $\sum_{g \in G} \bar{r}_g \cdot g^{-1}$. Let GROUPS^{inj} be the category of groups with injective group homomorphisms as morphisms. Taking the reduced group C^*-algebra defines a functor C_r^*: GROUPS$^{\text{inj}} \to C^*$-ALGEBRAS. The composition of these functors with the functors (18.18), (18.19) and (18.20) above yields functors

$$\mathbf{K}R(-)\colon \text{GROUPS} \quad \to \quad \text{SPECTRA;} \tag{18.21}$$

$$\mathbf{L}^{\langle j \rangle}R(-)\colon \text{GROUPS} \quad \to \quad \text{SPECTRA;} \tag{18.22}$$

$$\mathbf{K}^{\text{top}}C_r^*(-)\colon \text{GROUPS}^{\text{inj}} \quad \to \quad \text{SPECTRA.} \tag{18.23}$$

They satisfy

$$\begin{aligned}
\pi_n(\mathbf{K}R(G)) &= K_n(RG); \\
\pi_n(\mathbf{L}^{\langle j \rangle}R(G)) &= L_n^{\langle j \rangle}(RG); \\
\pi_n(\mathbf{K}^{\text{top}}C_r^*(G)) &= K_n(C_r^*(G)),
\end{aligned}$$

for all groups G and $n \in \mathbb{Z}$.

A category is called *small* if the morphisms form a set. A *groupoid* is a small category, all of whose morphisms are isomorphisms. Let GROUPOIDS be the category of groupoids with functors of groupoids as morphisms. Let GROUPOIDS^{inj} be the subcategory of GROUPOIDS which has the same objects and whose morphisms consist of those functors $F\colon \mathcal{G}_0 \to \mathcal{G}_1$ which are *faithful*, i.e., for any two objects x, y in $\mathcal{G}_0$ the induced map $\mathrm{mor}_{\mathcal{G}_0}(x, y) \to \mathrm{mor}_{\mathcal{G}_1}(F(x), F(y))$ is injective. The next result essentially says that these functors above can be extended from groups to groupoids.

Theorem 18.24 (K- and L-Theory Spectra over Groupoids). *Let R be a ring (with involution). There exist covariant functors*

$$\mathbf{K}_R\colon \text{GROUPOIDS} \quad \to \quad \text{SPECTRA;} \tag{18.25}$$

$$\mathbf{L}_R^{\langle j \rangle}\colon \text{GROUPOIDS} \quad \to \quad \text{SPECTRA;} \tag{18.26}$$

$$\mathbf{K}^{\text{top}}\colon \text{GROUPOIDS}^{\text{inj}} \quad \to \quad \text{SPECTRA,} \tag{18.27}$$

with the following properties:

(1) *If $F\colon \mathcal{G}_0 \to \mathcal{G}_1$ is an equivalence of groupoids, then the induced maps $\mathbf{K}_R(F)$, $\mathbf{L}_R^{\langle j \rangle}(F)$ and $\mathbf{K}^{\text{top}}(F)$ are weak equivalences of spectra.*

(2) *Let* I: GROUPS $\to$ GROUPOIDS *be the functor sending a group G to the groupoid which has precisely one object and G as set of morphisms. This functor induces a functor* GROUPS$^{\mathrm{inj}} \to$ GROUPOIDS^{inv}.

There are natural transformations from $\mathbf{K}R(-)$ *to* $\mathbf{K}_R \circ I$, *from* $\mathbf{L}^{\langle j \rangle} R(-)$ *to* $\mathbf{L}_R^{\langle j \rangle} \circ I^{\mathrm{inv}}$ *and from* $\mathbf{K}C_r^*(-)$ *to* $\mathbf{K}^{\mathrm{top}} \circ I$ *such that the evaluation of each of these natural transformations at a given group is an equivalence of spectra.*

(3) *For every group G and all $n \in \mathbb{Z}$ we have*

$$\pi_n(\mathbf{K}_R \circ I(G)) = K_n(RG);$$
$$\pi_n(\mathbf{L}_R^{\langle j \rangle} \circ I^{\mathrm{inv}}(G)) = L_n^{\langle j \rangle}(RG);$$
$$\pi_n(\mathbf{K}^{\mathrm{top}} \circ I(G)) = K_n(C_r^*(G)).$$

Proof. We only sketch the strategy of the proof. More details can be found in [64, Section 2].

Let $\mathcal{G}$ be a groupoid. Similar to the group ring RG one can define an R-linear category $R\mathcal{G}$ by taking the free R-modules over the morphism sets of $\mathcal{G}$. Composition of morphisms is extended R-linearly. By formally adding finite direct sums one obtains an additive category $R\mathcal{G}_\oplus$. Pedersen–Weibel [186] (compare also [46]) define a non-connective algebraic K-theory functor which digests additive categories and can hence be applied to $R\mathcal{G}_\oplus$. For the comparison result one uses that for every ring R (in particular for RG) the Pedersen–Weibel functor applied to $R_\oplus$ (a small model for the category of finitely generated free R-modules) yields the non-connective K-theory of the ring R and that it sends equivalences of additive categories to equivalences of spectra. In the L-theory case $R\mathcal{G}_\oplus$ inherits an involution and one applies the construction of [200, Example 13.6 on page 139] to obtain the $L^{\langle 1 \rangle} = L^h$-version. Analogously one can construct the $L^{\langle 2 \rangle} = L^s$-version. The versions for $j \leq 1$ can be obtained by a construction which is analogous to the Pedersen–Weibel construction for K-theory, compare [47, Section 4]. In the C^*-case one obtains from $\mathcal{G}$ a C^*-category $C_r^*(\mathcal{G})$ and assigns to a C^*-category its topological K-theory Ω-spectrum. There is a construction of the topological K-theory spectrum of a C^*-category in [64, Section 2]. However the construction given there depends on two statements, which appeared in [93, Proposition 1 and Proposition 3], and those statements are incorrect, as already pointed out by Thomason in [236]. The construction in [64, Section 2] can easily be fixed but instead we recommend the reader to look at the more recent construction of Joachim [118]. $\square$

18.6 The Chern Character for Homology Theories

The following construction is due to Dold [67].

Theorem 18.28 (Chern character for homology theory). *Consider a (non-equivariant) homology theory $\mathcal{H}_*$ with values in R-modules for $\mathbb{Q} \subset R$ which satisfies the disjoint union axiom. Then there is a natural (non-equivariant) equivalence of homology theories called* Chern character

$$\mathrm{ch}_n(X)\colon \bigoplus_{p+q=n} H_p(X; \mathcal{H}_q(*)) \xrightarrow{\cong} \mathcal{H}_n(X).$$

Proof. For a CW-complex X the Chern character $\mathrm{ch}_n(X)$ is given by the composite

$$\mathrm{ch}_n\colon \bigoplus_{p+q=n} H_p(X; \mathcal{H}_q(*)) \xrightarrow{\oplus_{p+q=n} \alpha_{p,q}^{-1}} \bigoplus_{p+q=n} H_p(X; R) \otimes_R \mathcal{H}_q(*)$$

$$\xrightarrow{\oplus_{p+q=n} \mathrm{hur}^{-1} \otimes \mathrm{id}} \bigoplus_{p+q=n} \pi_p^s(X_+, *) \otimes_{\mathbb{Z}} R \otimes_R \mathcal{H}_q(*) \xrightarrow{\oplus_{p+q=n} D_{p,q}} \mathcal{H}_n(X).$$

Here the canonical maps $\alpha_{p,q}$ are bijective, since any R-module is flat over $\mathbb{Z}$ because of the assumption $\mathbb{Q} \subseteq R$. The second bijective map comes from the various Hurewicz homomorphisms (see (18.43)). The map $D_{p,q}$ is defined as follows. For an element $a \otimes b \in \pi_p^s(X_+, *) \otimes_{\mathbb{Z}} \mathcal{H}_q(*)$ choose a representative $f\colon S^{p+k} \to S^k \wedge X_+$ of a. Define $D_{p,q}(a \otimes b)$ to be the image of b under the composite

$$\mathcal{H}_q(*) \xrightarrow{\sigma} \mathcal{H}_{p+q+k}(S^{p+k}, *) \xrightarrow{\mathcal{H}_{p+q+k}(f)} \mathcal{H}_{p+q+k}(S^k \wedge X_+, *) \xrightarrow{\sigma^{-1}} \mathcal{H}_{p+q}(X),$$

where σ denotes the suspension isomorphism. This map turns out to be a transformation of homology theories and induces an isomorphism for $X = \{\bullet\}$. Hence it is a natural equivalence of homology theories. $\square$

An equivariant Chern character for equivariant homology theories is constructed in [154].

18.7   The Bordism Group Associated to a Vector Bundle

In this section we define the bordism group associated to a stable vector bundle over a space.

Let (M, i) be an embedding $i\colon M^n \to \mathbb{R}^{n+k}$ of a closed n-dimensional manifold M into $\mathbb{R}^{n+k}$. Notice that $T\mathbb{R}^{n+k}$ comes with an explicit trivialization $\mathbb{R}^{n+k} \times \mathbb{R}^{n+k} \xrightarrow{\cong} T\mathbb{R}^{n+k}$ and the standard Euclidean inner product induces a Riemannian metric on $T\mathbb{R}^{n+k}$. Denote by $\nu(M) = \nu(i)$ the *normal bundle* of i which is the orthogonal complement of TM in $i^*T\mathbb{R}^{n+k}$.

Fix a space X together with an oriented k-dimensional vector bundle $\xi\colon E \to X$. Next we define the (pointed) *oriented bordism set* $\Omega_n(\xi)$. An element in it is represented by a quadruple $(M, i, f, \overline{f})$ which consists of a closed oriented n-dimensional manifold M, an embedding $i\colon M \to \mathbb{R}^{n+k}$, a map $f\colon M \to X$ and an orientation preserving bundle map $\overline{f}\colon \nu(M) \to \xi$ covering f. Notice that an orientation on M is the same as an orientation on $\nu(M)$. The base point is given by the class of $M = \emptyset$.

We briefly explain what a bordism $(W, I, F, \overline{F})$ from one such quadruple $(M_0, i_0, f_0, \overline{f_0})$ to another quadruple $(M_1, i_1, f_1, \overline{f_1})$ is ignoring the orientations. We need a compact $(n+1)$-dimensional manifold W together with a map $F\colon W \to X \times [0, 1]$. Its boundary ∂W is written as a disjoint sum $\partial_0 W \coprod \partial_1 W$ such that F maps $\partial_0 W$ to $X \times \{0\}$ and $\partial_1 W$ to $X \times \{1\}$. There is an embedding $I\colon W \to \mathbb{R}^{n+k} \times [0, 1]$ such that $I^{-1}(\mathbb{R}^{n+k} \times \{j\}) = \partial_j W$ holds for $j = 0, 1$ and W meets $\mathbb{R}^{n+k} \times \{j\}$ for $j = 0, 1$ transversally. We require a bundle map $(\overline{F}, F)\colon \nu(W) \to \xi \times [0, 1]$. Moreover for $j = 0, 1$ there is a diffeomorphism $U_j\colon \mathbb{R}^{n+k} \to \mathbb{R}^{n+k} \times \{j\}$ which maps M_j to $\partial_j W$. It satisfies $F \circ U_j|_{M_j \times \{j\}} = f_j$. Notice that U_j induces a bundle map $\nu(U_j)\colon \nu(M_j) \to \nu(W)$ covering $U_j|_{M_j}$. The composition of $\overline{F}$ with $\nu(U_j)$ is required to be $\overline{f_j}$.

18.8 The Thom Space of a Vector Bundle

For a vector bundle $\xi\colon E \to X$ with Riemannian metric define its *disk bundle* $p_{DE}\colon DE \to X$ by $DE = \{v \in E \mid ||v|| \leq 1\}$ and its *sphere bundle* $p_{SE}\colon SE \to X$ by $SE = \{v \in E \mid ||v|| = 1\}$, where p_{DE} and p_{SE} are the restrictions of p. Its *Thom space* $\mathrm{Th}(\xi)$ is defined by DE/SE. It has a preferred base point $\infty = SE/SE$. The Thom space can be defined without choice of a Riemannian metric as follows. Put $\mathrm{Th}(\xi) = E \cup \{\infty\}$ for some extra point ∞. Equip $\mathrm{Th}(\xi)$ with the topology for which $E \subseteq \mathrm{Th}(E)$ is an open subset and a basis of open neighborhoods for ∞ is given by the complements of closed subsets $A \subseteq E$ for which $A \cap E_x$ is compact for each fiber E_x. If X is compact, E is locally compact and $\mathrm{Th}(\xi)$ is the one-point-compactification of E. The advantage of this definition is that any bundle map $(\overline{f}, f)\colon \xi_0 \to \xi_1$ of vector bundles $\xi_0\colon E_0 \to X_0$ and $\xi_1\colon E_1 \to X_1$ induces canonically a map $\mathrm{Th}(\overline{f})\colon \mathrm{Th}(\xi_0) \to \mathrm{Th}(\xi_1)$. Notice that we require that $\overline{f}$ induces a bijective map on each fiber. Denote by $\mathbb{R}^k$ the trivial vector bundle with fiber $\mathbb{R}^k$. We mention that there are homeomorphisms

$$\mathrm{Th}(\xi \times \eta) \;\cong\; \mathrm{Th}(\xi) \wedge \mathrm{Th}(\eta); \tag{18.29}$$

$$\mathrm{Th}(\xi \oplus \underline{\mathbb{R}^k}) \;\cong\; S^k \wedge \mathrm{Th}(\xi), \tag{18.30}$$

18.9 The Pontrjagin–Thom Construction

Let $(N(M), \partial N(M))$ be a tubular neighborhood of M. Recall that there is a diffeomorphism

$$u\colon (D\nu(M), S\nu(M)) \to (N(M), \partial N(M))$$

which is up to isotopy relative M uniquely determined by the property that its restriction to M is i and its differential at M is $\varepsilon \cdot \mathrm{id}$ for small $\varepsilon > 0$ under the canonical identification $T(D\nu(M))|_M = TM \oplus \nu(M) = i^*T\mathbb{R}^{n+k}$. The *collapse map*

$$c \colon S^{n+k} = \mathbb{R}^{n+k} \coprod \{\infty\} \quad \to \quad \mathrm{Th}(\nu(M)) \tag{18.31}$$

is the pointed map which is given by the diffeomorphism u^{-1} on the interior of $N(M)$ and sends the complement of the interior of $N(M)$ to the preferred base point ∞. The homology group $H_{n+k}(\mathrm{Th}(TM)) \cong H_{n+k}(N(M), \partial N(M))$ is infinite cyclic, since $N(M)$ is a compact orientable (n+k)-dimensional manifold with boundary $\partial N(M)$. The Hurewicz homomorphism $h \colon \pi_{n+k}(Th(TM)) \to H_{n+k}(\mathrm{Th}(TM))$ sends the class $[c]$ of c to a generator. This follows from the fact that any point in the interior of $N(M)$ is a regular value of c and has precisely one point in his preimage.

Theorem 18.32 (Pontrjagin–Thom Construction). *Let $\xi \colon E \to X$ be an oriented k-dimensional vector bundle over a CW-complex X. Then the map*

$$P_n(\xi) \colon \Omega_n(\xi) \xrightarrow{\cong} \pi_{n+k}(\mathrm{Th}(\xi)),$$

which sends the class of $(M, i, f, \overline{f})$ to the class of the composite

$$S^{n+k} \xrightarrow{c} \mathrm{Th}(\nu(M)) \xrightarrow{\mathrm{Th}(\overline{f})} \mathrm{Th}(\xi)$$

is a well-defined bijection and natural in ξ.

Proof. The details can be found in [32, Satz 3.1 on page 28, Satz 4.9 on page 35,]. The basic idea becomes clear after we have explained the construction of the inverse for a finite CW-complex X. Consider a pointed map $(S^{n+k}, \infty) \to (\mathrm{Th}(\xi), \infty)$. We can change f up to homotopy relative $\{\infty\}$ such that f becomes transverse to X. Notice that transversality makes sense although X is not a manifold, one needs only the fact that X is the zero-section in a vector bundle. Put $M = f^{-1}(X)$. The transversality construction yields a bundle map $\overline{f} \colon \nu(M) \to \xi$ covering $f|_M$. Let $i \colon M \to \mathbb{R}^{n+k} = S^{n+k} - \{\infty\}$ be the inclusion. Then the inverse of $P_n(\xi)$ sends the class of f to the class of $(M, i, f|_M, \overline{f})$. $\square$

18.10 The Stable Version of the Pontrjagin–Thom Construction

In the sequel we will denote for a finite-dimensional vector space V by $\underline{V}$ the trivial bundle with fiber V. Suppose we are given a sequence of inclusions of CW-complexes

$$X_0 \xrightarrow{j_0} X_1 \xrightarrow{j_1} X_2 \xrightarrow{j_2} \cdots$$

together with bundle maps

$$\overline{j_k} \colon \xi_k \oplus \mathbb{R} \to \xi_{k+1}$$

covering the map $j_k \colon X_k \to X_{k+1}$, where each vector bundle ξ_k has dimension k. Such a system ξ_* is called a *stable vector bundle*. If each bundle ξ_k is oriented and each bundle map $\overline{j_k}$ respects the orientations we call it an *oriented stable vector bundle*. We obtain a system of maps of pointed sets

$$\Omega_n(\xi_0) \xrightarrow{\Omega_n(\overline{j_0})} \Omega_n(\xi_1) \xrightarrow{\Omega_n(\overline{j_1})} \Omega_n(\xi_2) \xrightarrow{\Omega_n(\overline{j_2})} \dots$$

Let $\operatorname{colim}_{k\to\infty} \Omega_k(\xi_k)$ be the colimit of this system of pointed sets which is a priori a pointed set. But it inherits the structure of an abelian group by taking the disjoint union and possibly composing the given embedding $M \to \mathbb{R}^{n+k}$ with the canonical inclusion $\mathbb{R}^{n+k} \to \mathbb{R}^{n+k+l}$. The construction of an inverse is based on composing with the isomorphism

$$\operatorname{id}_{\gamma_k} \oplus - \operatorname{id}_{\mathbb{R}} \colon \gamma_k \oplus \mathbb{R} \to \gamma_k \oplus \mathbb{R}.$$

We call

$$\Omega_n(\xi_*) := \operatorname{colim}_{k\to\infty} \Omega_n(\xi_k) \tag{18.33}$$

the *(n-th oriented) bordism group of the oriented stable vector bundle* ξ_*. We also see a sequence of spaces $\mathrm{Th}(\gamma_k)$ together with maps

$$\mathrm{Th}(\overline{i_k}) \colon S^1 \wedge \mathrm{Th}(\gamma_k) = \mathrm{Th}(\gamma_k \oplus \mathbb{R}) \to \mathrm{Th}(\gamma_{k+1}).$$

They induce homomorphisms

$$s_k \colon \pi_{n+k}(\mathrm{Th}(\gamma_k)) \to \pi_{n+k+1}(S^1 \wedge \mathrm{Th}(\gamma_k)) \xrightarrow{\pi_{n+k}(\mathrm{Th}(\overline{i_k}))} \pi_{n+k+1}(\mathrm{Th}(\gamma_{k+1})),$$

where the first map is the suspension homomorphism. Let

$$\pi_n^s(\mathrm{Th}(\xi_*)) = \operatorname{colim}_{k\to\infty} \pi_{n+k}(\mathrm{Th}(\gamma_k)) \tag{18.34}$$

be the colimit of the directed system

$$\dots \xrightarrow{s_{k-1}} \pi_{n+k}(\mathrm{Th}(\gamma_k)) \xrightarrow{s_k} \pi_{n+k+1}(\mathrm{Th}(\gamma_{k+1})) \xrightarrow{s_{k+1}} \dots.$$

We get from Theorem 18.32 a bijection

$$P_k \colon \operatorname{colim}_{k\to\infty} \Omega_n(\gamma_k) \xrightarrow{\cong} \operatorname{colim}_{k\to\infty} \pi_{n+k}(\mathrm{Th}(\gamma_k)).$$

This implies

Theorem 18.35 (The Stable Pontrjagin–Thom Construction). *There is an isomorphism of abelian groups natural in* ξ_*

$$P \colon \Omega_n(\xi_*) \xrightarrow{\cong} \pi_n^s(\mathrm{Th}(\gamma_*)).$$

18.11   The Oriented Bordism Ring

Let $\Omega_n(X)$ be the bordism group of pairs (M, f) of oriented closed n-dimensional manifolds M together with reference maps $g \colon M \to X$. Let $\xi_k \colon E_k \to BSO(k)$ be the universal oriented k-dimensional vector bundle. Let $\overline{j_k} \colon \xi_k \oplus \mathbb{R} \to \xi_{k+1}$ be a bundle map covering a map $j_k \colon BSO(k) \to BSO(k+1)$. Up to homotopy of bundle maps this map is unique. The map j_k can be arranged to be an inclusion of CW-complexes. Denote by γ_k the bundle $X \times E_k \to X \times BSO(k)$ and by $(\overline{i_k}, i_k) \colon \gamma_k \oplus \mathbb{R} \to \gamma_{k+1}$ the bundle map $\mathrm{id}_X \times(\overline{j_k}, j_k)$. The bundle map $(\overline{i_k}, i_k)$ is unique up to homotopy of bundle maps and hence induces a well-defined map $\Omega_n(\overline{i_k}) \colon \Omega_n(\gamma_k) \to \Omega_n(\gamma_{k+1})$, which sends the class of $(M, i, f, \overline{f})$ to the class of the quadruple which comes from the embedding $j \colon M \overset{i}{\to} \mathbb{R}^{n+k} \subseteq \mathbb{R}^{n+k+1}$ and the canonical isomorphism $\nu(i) \oplus \mathbb{R} = \nu(j)$. Consider the homomorphism

$$V_k \colon \Omega_n(\gamma_k) \to \Omega_n(X)$$

which sends the class of $(M, i, f, \overline{f})$ to $(M, \mathrm{pr}_X \circ f)$, where we equip M with the orientation determined by $\overline{f}$. The system of the maps V_k induce a homomorphism

$$V \colon \Omega_n(\gamma_*) \overset{\cong}{\longrightarrow} \Omega_n(X). \tag{18.36}$$

This map is bijective because of the classifying property of γ_k and the facts that for $k > n+1$ any closed manifold M of dimension n can be embedded into $\mathbb{R}^{n+k}$ and two such embeddings are isotopic. This implies

Theorem 18.37 (Pontrjagin–Thom Construction and Oriented Bordism). *There is an isomorphism of abelian groups natural in X,*

$$P \colon \Omega_n(X) \overset{\cong}{\longrightarrow} \pi_n^s(\mathrm{Th}(\gamma_*)).$$

Remark 18.38 (The Thom spectrum of a stable vector space). Notice that this is the beginning of the theory of spectra and stable homotopy theory. Obviously the spaces $\mathrm{Th}(\xi_k)$ together with the identification (18.30) and the maps $\mathrm{Th}(\overline{i_k})$ form a spectrum $\mathbf{Th}(\xi_*)$, the *Thom spectrum* associated to the stable vector bundle ξ_*.

Remark 18.39 (Stable homotopy theory and bordism). Theorem 18.37 is a kind of mile stone in homotopy theory since it is the prototype of a result, where the computation of geometrically defined objects are translated into a computation of (stable) homotopy groups. It applies to all other kind of bordism groups, where one puts additional structures on the manifolds, for instance a Spin-structure or a framing. The bijection is always of the same type, but the sequence of bundles ξ_k depends on the additional structure. If we want to deal with the unoriented bordism ring we have to replace the bundle $\xi_k \to BSO(k)$ by the universal k-dimensional vector bundle over $BO(k)$.

18.12 Stable Homotopy

Given a space X, the trivial stable vector bundle $\underline{\mathbb{R}}^*$ over X is given by the trivial k-dimensional bundle $\underline{\mathbb{R}}^k$ as k-th bundle and the identity maps as structure maps. The stable homotopy groups $\pi_n^s(X)$ are the abelian groups $\pi_n^s(\underline{\mathbb{R}}^*)$. This is the same as the colimit of the system of suspension homomorphisms

$$\pi_n(X_+) \to \pi_{n+1}(S^1 \wedge X_+) \to \pi_{n+2}(S^2 \wedge X_+).$$

They can also be defined for pairs, and one obtains a generalized homology theory π_*^s on the category of pairs of CW-complexes which satisfies the disjoint union axiom. The Hurewicz homomorphism induces natural maps

$$h_n(X,A) \colon \pi_n^s(X,A) \quad \to \quad H_n(X,A). \tag{18.40}$$

It turns out that one obtains a transformation of homology theories

$$h_* \colon \pi_*^s \quad \to \quad H_*. \tag{18.41}$$

The next result is due to Serre [217] (see also [131]).

Theorem 18.42 (Higher stable stems are finite). *The groups $\pi_n^s(\{\bullet\})$ are finite for $n \geq 1$.*

It is a basic and still unsolved problem in algebraic topology to compute the groups $\pi_n^s(\{\bullet\})$. The first stable stems are given by

n	0	1	2	3	4	5	6	7	8	9
π_n^s	$\mathbb{Z}$	$\mathbb{Z}/2$	$\mathbb{Z}/2$	$\mathbb{Z}/24$	0	0	$\mathbb{Z}/2$	$\mathbb{Z}/240$	$\mathbb{Z}/2 \oplus \mathbb{Z}/2$	$\mathbb{Z}/2 \oplus \mathbb{Z}/2 \oplus \mathbb{Z}/2$

Since $\mathbb{Q}$ is flat as $\mathbb{Z}$-module and $- \otimes_{\mathbb{Z}} \mathbb{Q}$ is compatible with direct sums over arbitrary index sets, $\pi_*^s \otimes \mathbb{Q}$ defines a homology theory satisfying the disjoint union axiom and we obtain a natural transformation of homology theories satisfying the disjoint union axiom

$$h_* \otimes_{\mathbb{Z}} \mathbb{Q} \colon \pi_*^s \otimes_{\mathbb{Z}} \mathbb{Q} \quad \to \quad H_*(-;\mathbb{Q}). \tag{18.43}$$

By Serre's Theorem 18.42 the map $h_n(\{\bullet\}) \otimes_{\mathbb{Z}} \mathbb{Q} \colon \pi_*^s(\{\bullet\}) \otimes_{\mathbb{Z}} \mathbb{Q} \to H_*(\{\bullet\};\mathbb{Q})$ is bijective for all $n \in \mathbb{Z}$. Any transformation of homology theories satisfying the disjoint union axiom is an equivalence of homology theories if and only if it induces isomorphisms for $\{\bullet\}$ for all $n \in \mathbb{Z}$ (see for instance [235, 7.55 on page 123]). We conclude

Theorem 18.44 (Rationalized stable homotopy agrees with rational homology). *For every CW-pair (X,A) and $n \in \mathbb{Z}$ the homomorphism*

$$h_n(X,A) \otimes_{\mathbb{Z}} \mathbb{Q} \colon \pi_*^s(X,A) \otimes_{\mathbb{Z}} \mathbb{Q} \xrightarrow{\cong} H_*(X,A;\mathbb{Q})$$

is bijective.

18.13   The Thom Isomorphism

We recall the Thom isomorphism (see for instance [171, Chapter 10], [235, 15.51 on page 367]).

Theorem 18.45 (Thom Isomorphism). *Let ξ given by $p\colon E \to X$ be an oriented k-dimensional vector bundle over a connected CW-complex X. Then there exists a so-called* Thom class $t_\xi \in H^k(DE, E)$ *such that the composites*

$$H_q(DE, SE) \xrightarrow{t_\xi \cap ?} H_{q-k}(DE) \xrightarrow{H_{q-k}(p)} H_{q-k}(X), \qquad (18.46)$$

$$H_q(DE, SE) \xrightarrow{t_\xi \cap ?} H_{q-k}(DE) \xrightarrow{H_{q-k}(p)} H_{q-k}(X), \qquad (18.47)$$

$$H^q(X) \xrightarrow{H^p(q)} H^q(DE) \xrightarrow{? \cup t_\xi} H^{q+k}(DE, SE), \qquad (18.48)$$

$$H^q(X) \xrightarrow{H^q(p)} H^q(DE) \xrightarrow{? \cup t_\xi} H^{q+k}(DE, SE), \qquad (18.49)$$

are bijective. These maps are called Thom isomorphisms.

Notice that $H_q(DE, SE) \cong H_q(\mathrm{Th}(\xi), \infty)$ and $H^q(DE, SE) \cong H^q(\mathrm{Th}(\xi), \infty)$ holds by excision.

18.14   The Rationalized Oriented Bordism Ring

Given an oriented closed oriented manifold M, we can choose $k \geq 0$ and an embedding $i_k\colon M \to \mathbb{R}^{n+k}$. Let $\overline{c_k}\colon \nu(i_k) \to \gamma_k$ be a bundle map covering a map $c_k\colon M \to BSO(k)$ for γ_k the universal k-dimensional bundle over $BSO(k)$. Define the space

$$BSO \;=\; \mathrm{colim}_{k\to\infty}\, BSO(k) \qquad (18.50)$$

with respect to the system of the maps $j_k\colon BSO(k) \to BSO(k+1)$ which can be arranged to be inclusions of CW-complexes. Let $c_M\colon M \to BSO$ be the composite of c_k with the canonical map $BSO(k) \to BO$. The homotopy class of c_M depends only on M and is independent of all other choices.

If we combine Theorem 18.37, Theorem 18.44 and Theorem 18.45, we obtain

Theorem 18.51 (The rationalized oriented bordism ring). *There is for every CW-complex X and $n \in \mathbb{Z}$ a natural isomorphism*

$$\nu_n(X)\colon \Omega_n(X) \xrightarrow{\;\cong\;} H_n(X \times BSO), \quad [f\colon M \to X] \mapsto H_n(f \times c_M)([M]).$$

In particular the n-th oriented bordism group $\Omega_n = \Omega_n(\{\bullet\})$ is isomorphic to $H_n(BSO)$ by the map which sends the class of an oriented n-dimensional manifold M to the image of its fundamental class $[M]$ under the map $H_n(c_M)\colon H_n(M) \to H_n(BSO)$ induced by the classifying map $c_M\colon M \to BSO$.

A standard inductive argument gives a computation of $H^*(BSO; \mathbb{Q})$:

$$H^*(BSO; \mathbb{Q}) \; \cong \; \mathbb{Q}[p_1, p_2, p_3, \ldots], \qquad (18.52)$$

where p_i are given by the Pontrjagin classes of the universal classifying bundles γ_k for large k. Since $(c_M)^*(p_i) = p_i(\nu(M))$ is the i-th Pontrjagin class of the normal bundle of M, we conclude that the Pontrjagin numbers of the normal bundle determine the rational bordism class of a manifold M. Since the Pontrjagin classes of the normal bundle determine those of the tangent bundle (and vice versa) we conclude the following result due to Thom.

Recall that a *partition* I of an integer $k \geq 0$ is an unordered sequence i_1, i_2, $\ldots$, i_r of positive integers with $\sum_{l=1}^r i_l = k$. Let $p(k)$ be the number of partitions of k. The first values are

$$
\begin{array}{c||c|c|c|c|c|c|c|c|c|c|c}
k & 0 & 1 & 2 & 3 & 4 & 5 & 6 & 7 & 8 & 9 & 10 \\
\hline
p(k) & 1 & 1 & 2 & 3 & 5 & 7 & 11 & 15 & 22 & 30 & 42
\end{array}
\qquad (18.53)
$$

The I-th *Pontrjagin number* is defined by

$$p_I[M] \; := \; \langle p_{i_1}(TM) \cup \ldots \cup p_{i_s}(TM), [M] \rangle. \qquad (18.54)$$

Theorem 18.55 (The rationalized oriented bordism ring and Pontrjagin classes). *We obtain for $k \in \mathbb{Z}$ natural isomorphisms*

$$\Omega_{4k} \otimes \mathbb{Q} \xrightarrow{\cong} \mathbb{Q}^{p(k)}, \quad [M] \mapsto (p_I(M))_I,$$

for $m \neq 0 \mod 4$,

$$\Omega_m \otimes \mathbb{Q} = 0.$$

One can give an explicit basis of $\Omega_{4k} \otimes \mathbb{Q}$. One "only" has to find for each k and each partition J of k manifolds M_J such that the matrix with entries

$$(p_I[M_J])_{I,J}$$

has non-trivial determinant. We refer to for the proof of the next result for instance to [171, Chapter 17].

Theorem 18.56 (Generators for the rational bordism ring). *The products of complex projective spaces*

$$\mathbb{CP}^{2i_1} \times \mathbb{CP}^{2i_2} \times \cdots \times \mathbb{CP}^{2i_r},$$

where $I = (i_1, i_2, \ldots, i_r)$ runs through all partitions of k are a $\mathbb{Q}$-basis of $\Omega_{4k} \otimes \mathbb{Q}$.

18.15 The Integral Oriented Bordism Ring

The first oriented bordism groups are given by

$$
\begin{array}{c||c|c|c|c|c|c|c|c|c|c}
n & 0 & 1 & 2 & 3 & 4 & 5 & 6 & 7 & 8 & 9 \\
\hline
\Omega_n & \mathbb{Z} & 0 & 0 & 0 & \mathbb{Z} & \mathbb{Z}/2 & 0 & 0 & \mathbb{Z} \oplus \mathbb{Z} & \mathbb{Z}/2 \oplus \mathbb{Z}/2
\end{array}
\qquad (18.57)
$$

The infinite cyclic group Ω_4 is generated by $\mathbb{CP}^2$ and an explicit isomorphism $\Omega_4 \xrightarrow{\cong} \mathbb{Z}$ is given by the signature which has been introduced in Chapter 3.

A complete computation of the oriented bordism ring has been carried out by Wall [245].

Chapter 19

Classifying Spaces of Families

In this section we introduce the classifying space of a family of subgroups.

To read the remaining chapters is suffices to understand Definition 19.1 and Section 19.2.

19.1 Basics about G-CW-Complexes

Definition 19.1 (G-CW-complex). *A G-CW-complex X is a G-space together with a G-invariant filtration*

$$\emptyset = X_{-1} \subseteq X_0 \subseteq X_1 \subseteq \ldots \subseteq X_n \subseteq \ldots \subseteq \bigcup_{n \geq 0} X_n = X$$

such that X carries the colimit topology with respect to this filtration (i.e., a set $C \subseteq X$ is closed if and only if $C \cap X_n$ is closed in X_n for all $n \geq 0$) and X_n is obtained from X_{n-1} for each $n \geq 0$ by attaching equivariant n-dimensional cells, i.e., there exists a G-pushout

$$
\begin{array}{ccc}
\coprod_{i \in I_n} G/H_i \times S^{n-1} & \xrightarrow{\coprod_{i \in I_n} q_i^n} & X_{n-1} \\
\downarrow & & \downarrow \\
\coprod_{i \in I_n} G/H_i \times D^n & \xrightarrow[\coprod_{i \in I_n} Q_i^n]{} & X_n
\end{array}
$$

The space X_n is called the *n-skeleton* of X. Notice that only the filtration by skeletons belongs to the G-CW-structure but not the G-pushouts, only their existence is required. An *equivariant open n-dimensional cell* is a G-component of $X_n - X_{n-1}$, i.e., the preimage of a path component of $G\backslash(X_n - X_{n-1})$. The closure of an equivariant open n-dimensional cell is called an *equivariant closed n-dimensional cell*. If one has chosen the G-pushouts in Definition 19.1, then the equivariant open n-dimensional cells are the G-subspaces $Q_i(G/H_i \times (D^n - S^{n-1}))$ and the equivariant closed n-dimensional cells are the G-subspaces $Q_i(G/H_i \times D^n)$.

Remark 19.2 (Proper G-CW-Complexes). A G-space X is called *proper* if for each pair of points x and y in X there are open neighborhoods V_x of x and W_y of y in X such that the closure of the subset $\{g \in G \mid gV_x \cap W_y \neq \emptyset\}$ of G is finite. A G-CW-complex X is proper if and only if all its isotropy groups are finite [152, Theorem 1.23]. In particular a free G-CW-complex is always proper. However, not every free G-space is proper.

Remark 19.3 (Cell-preserving G-actions). Let X be a G-space with G-invariant filtration

$$\emptyset = X_{-1} \subseteq X_0 \subseteq X_1 \subseteq \ldots \subseteq X_n \subseteq \ldots \subseteq \bigcup_{n \geq 0} X_n = X.$$

Then the following assertions are equivalent. i) The filtration above yields a G-CW-structure on X. ii) The filtration above yields a (non-equivariant) CW-structure on X such that for each open cell $e \subseteq X$ and each $g \in G$ with $ge \cap e \neq \emptyset$ we have $gx = x$ for all $x \in e$.

Notice that Definition 19.1 of a G-CW-complexes makes sense also for arbitrary topological groups, but then the characterization above is not valid in general.

Example 19.4 (Lie groups acting properly and smoothly on manifolds). Let M be a (smooth) proper G-manifold, then an equivariant smooth triangulation induces a G-CW-structure on M. For the proof and for equivariant smooth triangulations we refer to [117].

Example 19.5 (Simplicial Actions). Let X be a simplicial complex on which the group G acts by simplicial automorphisms. Then G acts also on the barycentric subdivision X' by simplicial automorphisms. The filtration of the barycentric subdivision X' by the simplicial n-skeleton yields the structure of a G-CW-complex which is not necessarily true for X.

A G-space is called *cocompact* if $G \backslash X$ is compact. A G-CW-complex X is *finite* if X has only finitely many equivariant cells. A G-CW-complex is finite if and only if it is cocompact. A G-CW-complex X is *of finite type* if each n-skeleton is finite. It is called *of dimension* $\leq n$ if $X = X_n$ and *finite dimensional* if it is of dimension $\leq n$ for some integer n. A free G-CW-complex X is the same as a G-principal bundle $X \to Y$ over a CW-complex Y.

Theorem 19.6 (Whitehead Theorem for Families). *Let $f: Y \to Z$ be a G-map of G-spaces. Let $\mathcal{F}$ be a set of (closed) subgroups of G which is closed under conjugation. Then the following assertions are equivalent:*

(1) *For any G-CW-complex X, whose isotropy groups belong to $\mathcal{F}$, the map induced by f*

$$f_*: [X, Y]^G \to [X, Z]^G, \quad [g] \mapsto [g \circ f]$$

between the set of G-homotopy classes of G-maps is bijective.

(2) *For any $H \in \mathcal{F}$ the map $f^H : Y^H \to Z^H$ is a weak homotopy equivalence, i.e.,
    for any base point $y \in Y^H$ and $n \in \mathbb{Z}$, $n \geq 0$, the map $\pi_n(f^H, y) : \pi_n(Y^H, y) \to
    \pi_n(Z^H, f^H(y))$ is bijective.*

Proof. (1) $\Rightarrow$ (2) Evaluation at $1H$ induces for any CW-complex A (equipped
with the trivial G-action) a bijection $[G/H \times A, Y]^G \xrightarrow{\cong} [A, Y^H]$. Hence for any
CW-complex A the map f^H induces a bijection

$$(f^H)_* : [A, Y^H] \to [A, Z^H], \quad [g] \mapsto [g \circ f^H].$$

This is equivalent to f^H being a weak homotopy equivalence by the classical non-
equivariant Whitehead Theorem [255, Theorem 7.17 in Chapter IV.7 on page 182].

(2) $\Rightarrow$ (1) We only give the proof in the case, where Z is G/G since this is the
most important case for us and the basic idea becomes already clear. The general
case is treated for instance in [237, Proposition II.2.6 on page 107]. We have to
show for any G-CW-complex X that two G-maps $f_0, f_1 : X \to Y$ are G-homotopic
provided that for any isotropy group H of X the H-fixed point set Y^H is *weakly
contractible*, i.e., $\pi_n(Y^H, y)$ consists of one element for all base points $y \in Y^H$.
Since X is $\operatorname{colim}_{n \to \infty} X_n$ it suffices to construct inductively over n G-homotopies
$h[n] : X_n \times [0, 1] \to Z$ such that $h[n]_i = f_i$ holds for $i = 0, 1$ and $h[n]|_{X_{n-1} \times [0,1]} =
h[n-1]$. The start of the induction, $n = -1$, is trivial because of $X_{-1} = \emptyset$, the
induction step from $n-1$ to $n \geq 0$ is done as follows. Fix a G-pushout

$$
\begin{array}{ccc}
\coprod_{i \in I_n} G/H_i \times S^{n-1} & \xrightarrow{\coprod_{i \in I_n} q_i^n} & X_{n-1} \\
\downarrow & & \downarrow \\
\coprod_{i \in I_n} G/H_i \times D^n & \xrightarrow{\coprod_{i \in I_n} Q_i^n} & X_n
\end{array}
$$

One easily checks that the desired G-homotopy $h[n]$ exists if and only if for each
$i \in I$ we can find an extension of the G-map

$$f_0 \circ Q_i^n \cup f_1 \circ Q_i^n \cup h[n-1] \circ (q_i^n \times \mathrm{id}_{[0,1]}):$$
$$G/H_i \times D^n \times \{0\} \cup G/H_i \times D^n \times \{1\} \cup G/H_i \times S^{n-1} \times [0,1] \;\to\; Y$$

to a G-map $G/H_i \times D^n \times [0,1] \to Y$. This is the same problem as extending the
(non-equivariant) map $D^n \times \{0\} \cup D^n \times \{1\} \cup S^{n-1} \times [0,1] \to Y^{H_i}$, which is given by
restricting the G-map above to $1H_i$, to a (non-equivariant) map $D^n \times [0,1] \to Y^{H_i}$.
Such an extension exists since Y^{H_i} is weakly contractible. This finishes the proof
of Theorem 19.6. $\qquad\square$

A G-map $f : X \to Y$ of G-CW-complexes is a G-homotopy equivalence if
and only if for any subgroup $H \subseteq G$ which occurs as isotropy group of X or Y the

induced map $f^H \colon X^H \to Y^H$ is a weak homotopy equivalence. This follows from the Whitehead Theorem for families 19.6 above.

A G-map of G-CW-complexes $f \colon X \to Y$ is *cellular* if $f(X_n) \subseteq Y_n$ holds for all $n \geq 0$. There is an equivariant version of the *Cellular Approximation Theorem*, namely, each G-map of G-CW-complexes is G-homotopic to a cellular one and each G-homotopy between cellular G-maps can be replaced by a cellular G-homotopy [237, Theorem II.2.1 on page 104].

19.2   The Classifying Space for a Family

Definition 19.7 (Family of subgroups). A *family* $\mathcal{F}$ *of subgroups* of G is a set of (closed) subgroups of G which is closed under conjugation and finite intersections.

Examples for $\mathcal{F}$ are

$\mathcal{TR}$	$=$	{trivial subgroup};
$\mathcal{FIN}$	$=$	{finite subgroups};
$\mathcal{CYC}$	$=$	{cyclic subgroups};
$\mathcal{FCYC}$	$=$	{finite cyclic subgroups};
$\mathcal{VCYC}$	$=$	{virtually cyclic subgroups};
$\mathcal{ALL}$	$=$	{all subgroups}.

Definition 19.8 (Classifying G-CW-complex for a family of subgroups). *Let $\mathcal{F}$ be a family of subgroups of G. A* model $E_{\mathcal{F}}(G)$ *for the classifying G-CW-complex for the family $\mathcal{F}$ of subgroups is a G-CW-complex $E_{\mathcal{F}}(G)$ which has the following properties:*

   i) *All isotropy groups of $E_{\mathcal{F}}(G)$ belong to $\mathcal{F}$.*

   ii) *For any G-CW-complex Y, whose isotropy groups belong to $\mathcal{F}$, there is up to G-homotopy precisely one G-map $Y \to E_{\mathcal{F}}(G)$.*

We abbreviate $\underline{E}G := E_{\mathcal{FIN}}(G)$ *and call it the* universal G-CW-complex for proper G-actions.

Remark 19.9 (Change of Families). In other words, $E_{\mathcal{F}}(G)$ is a terminal object in the G-homotopy category of G-CW-complexes, whose isotropy groups belong to $\mathcal{F}$. In particular two models for $E_{\mathcal{F}}(G)$ are G-homotopy equivalent and for two families $\mathcal{F}_0 \subseteq \mathcal{F}_1$ there is up to G-homotopy precisely one G-map $E_{\mathcal{F}_0}(G) \to E_{\mathcal{F}_1}(G)$.

Theorem 19.10 (Homotopy Characterization of $E_{\mathcal{F}}(G)$). *Let $\mathcal{F}$ be a family of subgroups.*

   (1) *There exists a model for $E_{\mathcal{F}}(G)$ for any family $\mathcal{F}$;*

   (2) *A G-CW-complex X is a model for $E_{\mathcal{F}}(G)$ if and only if all its isotropy groups belong to $\mathcal{F}$ and for each $H \in \mathcal{F}$ the H-fixed point set X^H is weakly contractible.*

Proof. (1) A model can be obtained by attaching equivariant cells $G/H \times D^n$ for all $H \in \mathcal{F}$ to make the H-fixed point sets weakly contractible. See for instance [152, Proposition 2.3 on page 35]. A functorial construction can be found in [64, page 230 and Lemma 7.6].

(2) This follows from the Whitehead Theorem for families 19.6 applied to $f \colon X \to G/G$. $\qquad\qquad\square$

19.3 Special Models

In this section we present some interesting geometric models for the space $E_{\mathcal{F}}(G)$ focusing on $E_{\mathcal{FIN}}(G) = \underline{E}G$.

19.3.1 The Family of All Subgroups and the Trivial Family

One extreme case is, when we take $\mathcal{F}$ to be the family $\mathcal{ALL}$ of all subgroups. Then a model for $E_{\mathcal{ALL}}(G)$ is G/G. The other extreme case is the family $\mathcal{TR}$ consisting of the trivial subgroup. Then $E_{\mathcal{TR}}(G)$ is the same as EG which is the total space of the universal G-principal bundle $G \to EG \to BG$.

19.3.2 Operator Theoretic Model

Let G be discrete. A model for $\underline{E}G$ is the space

$$X_G = \{f \colon G \to [0,1] \mid f \text{ has finite support}, \sum_{g \in G} f(g) = 1\}$$

with the topology coming from the supremum norm [22, page 248].

Let $P_\infty(G)$ be the geometric realization of the simplicial set whose k-simplices consist of $(k+1)$-tuples $(g_0, g_1, \ldots, g_k)$ of elements g_i in G. This also a model for $\underline{E}G$ [1, Example 2.6]. The spaces X_G and $P_\infty(G)$ have the same underlying sets but in general they have different topologies. The identity map induces a (continuous) G-map $P_\infty(G) \to X_G$ which is a G-homotopy equivalence, but in general not a G-homeomorphism (see also [240, A.2]).

19.3.3 Discrete Subgroups of Almost Connected Lie Groups

The next result is a special case of a result due to Abels [1, Corollary 4.14].

Theorem 19.11 (Discrete Subgroups of Almost Connected Lie Groups). *Let L be an almost connected Lie group, i.e., a Lie group with finitely many path components. Let $K \subseteq L$ be a maximal compact subgroup. Let $G \subseteq L$ be a discrete subgroup of L.*

Then L/K with the obvious left G-action is a finite-dimensional G-CW-model for $\underline{E}G$.

19.3.4   Simply Connected Non-Positively Curved Manifolds

The next theorem is due to Abels [1, Theorem 4.15].

Theorem 19.12 (Actions on Simply Connected Non-Positively Curved Manifolds). *Suppose that G acts properly and isometrically on the simply-connected complete Riemannian manifold M with non-positive sectional curvature. Then M is a model for $\underline{E}G$.*

19.3.5   CAT(0)-Spaces

Theorem 19.13 (Actions on CAT(0)-spaces). *Let G be a (locally compact Hausdorff) topological group. Let X be a proper G-CW-complex. Suppose that X has the structure of a complete $\mathrm{CAT}(0)$-space for which G acts by isometries. Then X is a model for $\underline{E}G$.*

Proof. By [30, Corollary II.2.8 on page 179] the K-fixed point set of X is non-empty convex subset of X and hence contractible for any compact subgroup $K \subseteq G$. $\square$

19.3.6   Trees with Finite Isotropy Groups

A *tree* is a 1-dimensional CW-complex which is contractible.

Theorem 19.14 (Actions on Trees). *Suppose that G acts continuously on a tree T such that for each element $g \in G$ and each open cell e with $g \cdot e \cap e \neq \emptyset$ we have $gx = x$ for any $x \in e$. Assume that the isotropy group of each $x \in T$ is finite. Then T is a 1-dimensional model for $\underline{E}G$.*

Proof. Let $H \subseteq G$ be finite. If e_0 is a zero-cell in T, then $H \cdot e_0$ is finite. Let T' be the union of all geodesics with extremities in $H \cdot e$. This is a H-invariant subtree of T of finite diameter. One shows now inductively over the diameter of T' that T' has a vertex which is fixed under the H-action (see [220, page 20] or [66, Proposition 4.7 on page 17]). Hence T^H is non-empty. If e and f are vertices in T^H, the geodesic in T from e to f must be H-invariant. Hence T^H is a connected CW-subcomplex of the tree T and hence is itself a tree. This shows that T^H is contractible. $\square$

19.3.7   Amalgamated Products and HNN-Extensions

Consider groups H, K_{-1} and K_1 together with injective group homomorphisms $\varphi_i \colon H \to K_i$ for $i \in \{-1, 1\}$. Let G be the amalgamated product $K_1 *_H K_1$ with respect to φ_{-1} and φ_1, i.e., the pushout of groups

$$
\begin{array}{ccc}
H & \xrightarrow{\ \varphi_{-1}\ } & K_{-1} \\
{\scriptstyle \varphi_1}\big\downarrow & & \big\downarrow \\
K_1 & \longrightarrow & G
\end{array}
$$

Choose φ_i-equivariant maps $f_i \colon \underline{E}H \to \underline{E}K_i$. They induce G-maps

$$F_i \colon G \times_H \underline{E}H \ \to \ G \times_{K_i} \underline{E}K_i, \quad (g,x) \mapsto (g, f_i(x)).$$

We get a model for $\underline{E}G$ as the G-pushout

$$
\begin{array}{ccc}
G \times_H \underline{E}H \times \{-1,1\} & \xrightarrow{F_{-1} \amalg F_1} & G \times_{K_{-1}} \underline{E}K_{-1} \amalg G \times_{K_1} \underline{E}K_1 \\
\downarrow & & \downarrow \\
G \times_H \underline{E}H \times [-1,1] & \longrightarrow & \underline{E}G
\end{array}
$$

Consider two groups H and K and two injective group homomorphisms $\varphi_i \colon H \to K$ for $i \in \{-1,1\}$. Let G be the HNN-extension associated to the data $\varphi_i \colon H \to K$ for $i \in \{-1,1\}$, i.e., the group generated by the elements of K and a letter t whose relations are those of K and the relations $t^{-1}\varphi_{-1}(h)t = \varphi_1(h)$ for all $h \in H$. The natural map $K \to G$ is injective and we will identify K with its image in G. Choose φ_i-equivariant maps $f_i \colon \underline{E}H \to \underline{E}K$. Let $F_i \colon G \times_{\varphi_{-1}} \underline{E}H \to G \times_K \underline{E}K$ be the G-map which sends (g,x) to $gf_{-1}(x)$ for $i = -1$ and to $gtf_1(x)$ for $i = 1$. Then a model for $\underline{E}G$ is given by the G-pushout

$$
\begin{array}{ccc}
G \times_{\varphi_{-1}} \underline{E}H \times \{-1,1\} & \xrightarrow{F_{-1} \amalg F_1} & G \times_K \underline{E}K \\
\downarrow & & \downarrow \\
G \times_{\varphi_{-1}} \underline{E}H \times [-1,1] & \longrightarrow & \underline{E}G
\end{array}
$$

Consider the special case, where $H = K$, $\varphi_{-1} = \mathrm{id}$ and φ_1 is an automorphism. Then G is the semidirect product $K \rtimes_{\varphi_1} \mathbb{Z}$. Choose a φ_1-equivariant map $f_1 \colon \underline{E}K \to \underline{E}K$. Then a model for $\underline{E}G$ is given by the to both side infinite mapping telescope of f_1 with the $K \rtimes_{\varphi_1} \mathbb{Z}$ action, for which $\mathbb{Z}$ acts by shifting to the right and $k \in K$ acts on the part belonging to $n \in \mathbb{Z}$ by multiplication with $\varphi_1^n(k)$. If we additionally assume that $\varphi_1 = \mathrm{id}$, then $G = K \times \mathbb{Z}$ and we get $\underline{E}K \times \mathbb{R}$ as model for $\underline{E}G$.

All these constructions also yield models for $EG = E_{\mathcal{TR}}(G)$ if one replaces everywhere the spaces $\underline{E}H$ and $\underline{E}K$ by the spaces EH and EK.

19.3.8 Arithmetic Groups

Arithmetic groups in a semisimple connected linear $\mathbb{Q}$-algebraic group possess finite models for $\underline{E}G$. Namely, let $G(\mathbb{R})$ be the $\mathbb{R}$-points of a semisimple $\mathbb{Q}$-group $G(\mathbb{Q})$ and let $K \subseteq G(\mathbb{R})$ a maximal compact subgroup. If $A \subseteq G(\mathbb{Q})$ is an arithmetic group, then $G(\mathbb{R})/K$ with the left A-action is a model for $E_{\mathcal{FIN}}(A)$ as already explained in Theorem 19.11. The A-space $G(\mathbb{R})/K$ is not necessarily cocompact. The Borel–Serre completion of $G(\mathbb{R})/K$ (see [29], [219]) is a finite A-CW-model for $E_{\mathcal{FIN}}(A)$ as pointed out in [4, Remark 5.8], where a private communication with Borel and Prasad is mentioned.

19.3.9   Outer Automorphism Groups of Free groups

Let F_n be the free group of rank n. Denote by $\mathrm{Out}(F_n)$ the group of outer automorphisms of F_n, i.e., the quotient of the group of all automorphisms of F_n by
the normal subgroup of inner automorphisms. Culler and Vogtmann [61], [242]
have constructed a space X_n called *outer space* on which $\mathrm{Out}(F_n)$ acts with finite
isotropy groups. It is analogous to the Teichmüller space of a surface with the action of the mapping class group of the surface. Fix a graph R_n with one vertex v
and n edges and identify F_n with $\pi_1(R_n, v)$. A *marked metric graph* (g, Γ) consists
of a graph Γ with all vertices of valence at least three, a homotopy equivalence
$g \colon R_n \to \Gamma$ called marking and to each edge of Γ there is assigned a positive length
which makes Γ into a metric space by the path metric. We call two marked metric
graphs (g, Γ) and (g', Γ') equivalent of there is a homothety $h \colon \Gamma \to \Gamma'$ such that
$g \circ h$ and h' are homotopic. Homothety means that there is a constant $\lambda > 0$ with
$d(h(x), h(y)) = \lambda \cdot d(x, y)$ for all x, y. Elements in outer space X_n are equivalence
classes of marked graphs. The main result in [61] is that X is contractible. Actually,
for each finite subgroup $H \subseteq \mathrm{Out}(F_n)$ the H-fixed point set X_n^H is contractible
[137, Proposition 3.3 and Theorem 8.1], [254, Theorem 5.1].

The space X_n contains a *spine* K_n which is an $\mathrm{Out}(F_n)$-equivariant deformation retraction. This space K_n is a simplicial complex of dimension $(2n - 3)$ on
which the $\mathrm{Out}(F_n)$-action is by simplicial automorphisms and cocompact. Actually
the group of simplicial automorphisms of K_n is $\mathrm{Out}(F_n)$ [31]. Hence the second
barycentric subdivision K_n'' is a finite $(2n - 3)$-dimensional model of $\underline{E}\,\mathrm{Out}(F_n)$.

19.3.10   Mapping Class groups

Let $\Gamma_{g,r}^s$ be the *mapping class group* of an orientable compact surface F of genus
g with s punctures and r boundary components. This is the group of isotopy
classes of orientation preserving self-diffeomorphisms $F_g \to F_g$, which preserve the
punctures individually and restrict to the identity on the boundary. We require
that the isotopies leave the boundary pointwise fixed. We will always assume that
$2g + s + r > 2$, or, equivalently, that the Euler characteristic of the punctured
surface F is negative. It is well-known that the associated *Teichmüller space* $\mathcal{T}_{g,r}^s$
is a contractible space on which $\Gamma_{g,r}^s$ acts properly. Actually $\mathcal{T}_{g,r}^s$ is a model for
$E_{\mathcal{FIN}}(\Gamma_{g,r}^s)$ by the results of Kerckhoff [127].

19.3.11   One-Relator Groups

Let G be a one-relator group. Let $G = \langle (q_i)_{i \in I} \mid r \rangle$ be a presentation with one
relation. There is up to conjugacy one maximal finite subgroup C which is cyclic.
There exists a 2-dimensional G-CW-model for $\underline{E}G$ such that $\underline{E}G$ is obtained
from G/C for a maximal finite cyclic subgroup $C \subseteq G$ by attaching free cells
of dimensions ≤ 2 and the CW-complex structure on the quotient $G \backslash \underline{E}G$ has
precisely one 0-cell, precisely one 2-cell and as many 1-cells as there are elements

in I. All these claims follow from [36, Exercise 2 (c) II. 5 on page 44].

If G is torsionfree, the 2-dimensional complex associated to a presentation with one relation is a model for BG (see also [164, Chapter III §§9–11]).

19.3.12 Special Linear Groups of (2,2)-Matrices

In order to illustrate some of the general statements above we consider the special example $SL_2(\mathbb{R})$ and $SL_2(\mathbb{Z})$.

Let $\mathbb{H}^2$ be the 2-dimensional hyperbolic space. We will use either the upper half-plane model or the Poincaré disk model. The group $SL_2(\mathbb{R})$ acts by isometric diffeomorphisms on the upper half-plane by Moebius transformations, i.e., a matrix $\begin{pmatrix} a & b \\ c & d \end{pmatrix}$ acts by sending a complex number z with positive imaginary part to $\frac{az+b}{cz+d}$. This action is proper and transitive. The isotropy group of $z = i$ is $SO(2)$. Since $\mathbb{H}^2$ is a simply-connected Riemannian manifold, whose sectional curvature is constant -1, the $SL_2(\mathbb{R})$-space $\mathbb{H}^2$ is a model for $\underline{E}SL_2(\mathbb{R})$ by Theorem 19.12.

One easily checks that $SL_2(\mathbb{R})$ is a connected Lie group and $SO(2) \subseteq SL_2(\mathbb{R})$ is a maximal compact subgroup. Hence $SL_2(\mathbb{R})/SO(2)$ is a model for $\underline{E}SL_2(\mathbb{R})$ by Theorem 19.11. Since the $SL_2(\mathbb{R})$-action on $\mathbb{H}^2$ is transitive and $SO(2)$ is the isotropy group at $i \in \mathbb{H}^2$, we see that the $SL_2(\mathbb{R})$-manifolds $SL_2(\mathbb{R})/SO(2)$ and $\mathbb{H}^2$ are $SL_2(\mathbb{R})$-diffeomorphic.

Since $SL_2(\mathbb{Z})$ is a discrete subgroup of $SL_2(\mathbb{R})$, the space $\mathbb{H}^2$ with the $SL_2(\mathbb{Z})$-action is a model for $\underline{E}SL_2(\mathbb{Z})$ (see Theorem 19.11).

The group $SL_2(\mathbb{Z})$ is isomorphic to the amalgamated product $\mathbb{Z}/4 *_{\mathbb{Z}/2} \mathbb{Z}/6$. From Section 19.3.7 we conclude that a model for $\underline{E}SL_2(\mathbb{Z})$ is given by the $SL_2(\mathbb{Z})$-pushout

$$\begin{array}{ccc} SL_2(\mathbb{Z})/(\mathbb{Z}/2) \times \{-1,1\} & \xrightarrow{F_{-1} \amalg F_1} & SL_2(\mathbb{Z})/(\mathbb{Z}/4) \amalg SL_2(\mathbb{Z})/(\mathbb{Z}/6) \\ \downarrow & & \downarrow \\ SL_2(\mathbb{Z})/(\mathbb{Z}/2) \times [-1,1] & \longrightarrow & \underline{E}SL_2(\mathbb{Z}) \end{array}$$

where F_{-1} and F_1 are the obvious projections. This model for $\underline{E}SL_2(\mathbb{Z})$ is a tree, which has alternately two and three edges emanating from each vertex. The other model $\mathbb{H}^2$ is a manifold. These two models must be $SL_2(\mathbb{Z})$-homotopy equivalent. They can explicitly be related by the following construction.

Divide the Poincaré disk into fundamental domains for the $SL_2(\mathbb{Z})$-action. Each fundamental domain is a geodesic triangle with one vertex at infinity, i.e., a vertex on the boundary sphere, and two vertices in the interior. Then the union of the edges, whose end points lie in the interior of the Poincaré disk, is a tree T with $SL_2(\mathbb{Z})$-action. This is the tree model above. The tree is a $SL_2(\mathbb{Z})$-equivariant deformation retraction of the Poincaré disk. A retraction is given by moving a point p in the Poincaré disk along a geodesic starting at the vertex at infinity, which belongs to the triangle containing p, through p to the first intersection point of this geodesic with T.

   More information about classifying spaces for families of subgroups can be found for instance in [158].

Chapter 20

Equivariant Homology Theories and the Meta-Conjecture

In this section we formulate a Meta-Conjecture for a group G which depends on a choice of an equivariant homology theory and a family of subgroups. If we insert for them certain values, we obtain the Farrell–Jones and the Baum–Connes Conjectures. We will explain the notion of an equivariant homology theory and how it can be constructed from covariant functors GROUPOIDS $\to$ SPECTRA.

20.1   The Meta-Conjecture

Suppose we are given

- a discrete group G;
- a family $\mathcal{F}$ of subgroups of G;
- a G-homology theory $\mathcal{H}_*^G(-)$.

Then one can formulate the following Meta-Conjecture.

Conjecture 20.1 (Meta-Conjecture). *The assembly map*

$$A_{\mathcal{F}} \colon \mathcal{H}_n^G(E_{\mathcal{F}}(G)) \to \mathcal{H}_n^G(\{\bullet\})$$

which is the map induced by the projection $E_{\mathcal{F}}(G) \to \{\bullet\}$ is an isomorphism for $n \in \mathbb{Z}$.

Remark 20.2 (Discussion of the Meta-Conjecture). Of course the conjecture above is not true for arbitrary G, $\mathcal{F}$ and $\mathcal{H}_*^G(-)$, but the Farrell–Jones and Baum–Connes Conjectures state that for specific G-homology theories there is a natural choice of a family $\mathcal{F} = \mathcal{F}(G)$ of subgroups for every group G such that $A_{\mathcal{F}(G)}$ becomes an isomorphism for all groups G. The point of this Meta-Conjecture 20.1 is that one wants to compute the target $\mathcal{H}_n^G(\{\bullet\})$ and that the source $\mathcal{H}_n^G(E_{\mathcal{F}}(G))$ is easier to understand since it only involves the subgroups appearing in $\mathcal{F}$. Given a G-homology theory $\mathcal{H}_*^G(-)$, the point will be to choose $\mathcal{F}$ as small as possible. If one takes $\mathcal{F}$ to be the family $\mathcal{ALL}$ of all subgroups, then $E_{\mathcal{F}}(G) = \{\bullet\}$ and the Meta-Conjecture 20.1 is obviously true but gives no information.

20.2   Formulation of the Farrell–Jones and the Baum–Connes Conjecture

Let R be a ring (with involution). We will describe in Example 20.15 the construction of G-homology theories which will be denoted

$$H_n^G(-;\mathbf{K}_R), \quad H_n^G(-;\mathbf{L}_R^{\langle-\infty\rangle}) \quad \text{and} \quad H_n^G(-;\mathbf{K}^{\text{top}}).$$

The main feature of these homology theories is that evaluated on the one point space $\{\bullet\}$ (considered as a trivial G-space) we obtain the K- and L-theory of the group ring RG, respectively the topological K-theory of the reduced C^*-algebra

$$
\begin{aligned}
K_n(RG) &\cong H_n^G(\{\bullet\};\mathbf{K}_R), \\
L_n^{\langle-\infty\rangle}(RG) &\cong H_n^G(\{\bullet\};\mathbf{L}_R^{\langle-\infty\rangle}) \quad \text{and} \\
K_n(C_r^*(G)) &\cong H_n^G(\{\bullet\};\mathbf{K}^{\text{top}}).
\end{aligned}
$$

Let $\mathcal{FIN}$ be the family of finite subgroups and let $\mathcal{VCYC}$ be the family of virtually cyclic subgroups.

Conjecture 20.3 (Farrell–Jones Conjecture for K- and L-theory). *Let R be a ring (with involution) and let G be a group. Then for all $n \in \mathbb{Z}$ the maps*

$$
\begin{aligned}
A_{\mathcal{VCYC}} \colon H_n^G(E_{\mathcal{VCYC}}(G);\mathbf{K}_R) &\rightarrow H_n^G(\{\bullet\};\mathbf{K}_R) \cong K_n(RG), \\
A_{\mathcal{VCYC}} \colon H_n^G(E_{\mathcal{VCYC}}(G);\mathbf{L}_R^{\langle-\infty\rangle}) &\rightarrow H_n^G(\{\bullet\};\mathbf{L}_R^{\langle-\infty\rangle}) \cong L_n^{\langle-\infty\rangle}(RG),
\end{aligned}
$$

which are induced by the projection $E_{\mathcal{VCYC}}(G) \rightarrow \{\bullet\}$, are isomorphisms.

Conjecture 20.4 (Baum–Connes Conjecture). *Let G be a group. Then for all $n \in \mathbb{Z}$ the map*

$$A_{\mathcal{FIN}} \colon H_n^G(E_{\mathcal{FIN}}(G);\mathbf{K}^{\text{top}}) \rightarrow H_n^G(\{\bullet\};\mathbf{K}^{\text{top}}) \cong K_n(C_r^*(G))$$

induced by the projection $E_{\mathcal{FIN}}(G) \rightarrow \{\bullet\}$ is an isomorphism.

Roughly speaking, these conjectures yield a way to compute the various $K-$ and L-groups $K_n(RG)$, $L_n^{\langle-\infty\rangle}(RG)$ and $K_n(C_r^*(G))$ from the knowledge of their values $K_n(RH)$, $L_n^{\langle-\infty\rangle}(RH)$ and $K_n(C_r^*(H))$, where H runs through all virtually cyclic or finite subgroups of G.

20.3 Equivariant Homology Theories

In order to understand the Meta-Conjecture 20.1 we have to explain the notion of a G-homology theory. It will be the obvious extension of the notion of a homology theory for CW-complexes to G-CW-complexes.

Fix a group G and an associative commutative ring Λ with unit. A *G-homology theory $\mathcal{H}_*^G$ with values in Λ-modules* is a collection of covariant functors $\mathcal{H}_n^G$ from the category of G-CW-pairs to the category of Λ-modules indexed by $n \in \mathbb{Z}$ together with natural transformations

$$\partial_n^G(X, A)\colon \mathcal{H}_n^G(X, A) \to \mathcal{H}_{n-1}^G(A) := \mathcal{H}_{n-1}^G(A, \emptyset)$$

for $n \in \mathbb{Z}$ such that the following axioms are satisfied:

(1) *G-homotopy invariance*

If f_0 and f_1 are G-homotopic maps $(X, A) \to (Y, B)$ of G-CW-pairs, then $\mathcal{H}_n^G(f_0) = \mathcal{H}_n^G(f_1)$ for $n \in \mathbb{Z}$.

(2) *Long exact sequence of a pair*

Given a pair (X, A) of G-CW-complexes, there is a long exact sequence

$$\ldots \xrightarrow{\mathcal{H}_{n+1}^G(j)} \mathcal{H}_{n+1}^G(X, A) \xrightarrow{\partial_{n+1}^G} \mathcal{H}_n^G(A) \xrightarrow{\mathcal{H}_n^G(i)} \mathcal{H}_n^G(X)$$
$$\xrightarrow{\mathcal{H}_n^G(j)} \mathcal{H}_n^G(X, A) \xrightarrow{\partial_n^G} \mathcal{H}_{n-1}^G(A) \xrightarrow{\mathcal{H}_{n-1}^G(i)} \ldots,$$

where $i\colon A \to X$ and $j\colon X \to (X, A)$ are the inclusions.

(3) *Excision*

Let (X, A) be a G-CW-pair and let $f\colon A \to B$ be a cellular G-map of G-CW-complexes. Equip $(X \cup_f B, B)$ with the induced structure of a G-CW-pair. Then the canonical map $(F, f)\colon (X, A) \to (X \cup_f B, B)$ induces for each $n \in \mathbb{Z}$ an isomorphism

$$\mathcal{H}_n^G(F, f)\colon \mathcal{H}_n^G(X, A) \xrightarrow{\cong} \mathcal{H}_n^G(X \cup_f B, B).$$

(4) *Disjoint union axiom*

Let $\{X_i \mid i \in I\}$ be a family of G-CW-complexes. Denote by $j_i\colon X_i \to \coprod_{i \in I} X_i$ the canonical inclusion. Then the map

$$\bigoplus_{i \in I} \mathcal{H}_n^G(j_i)\colon \bigoplus_{i \in I} \mathcal{H}_n^G(X_i) \xrightarrow{\cong} \mathcal{H}_n^G\left(\coprod_{i \in I} X_i\right)$$

is bijective for each $n \in \mathbb{Z}$.

Of course a G-homology theory for the trivial group $G = \{1\}$ is a homology theory (satisfying the disjoint union axiom) in the classical non-equivariant sense.

If $\mathcal{H}_*^G$ is defined or considered only for proper G-CW-pairs (X, A), we call it a *proper G-homology theory $\mathcal{H}_*^G$ with values in Λ-modules*.

The disjoint union axiom ensures that we can pass from finite G-CW-complexes to arbitrary ones using the following lemma.

Lemma 20.5. *Let $\mathcal{H}_*^G$ be a G-homology theory. Let X be a G-CW-complex and $\{X_i \mid i \in I\}$ be a directed system of G-CW-subcomplexes directed by inclusion such that $X = \cup_{i \in I} X_i$. Then for all $n \in \mathbb{Z}$ the natural map*

$$\operatorname{colim}_{i \in I} \mathcal{H}_n^G(X_i) \xrightarrow{\cong} \mathcal{H}_n^G(X)$$

is bijective.

Proof. Compare for example with [235, Proposition 7.53 on page 121], where the non-equivariant case for $I = \mathbb{N}$ is treated.   □

In all cases it will turn out and be important that we get actually for each group G a homology theory $\mathcal{H}_*^G$ and that the various $\mathcal{H}_*^G$ are linked by a so-called induction structure. Let us axiomatize the situation.

Let $\alpha \colon H \to G$ be a group homomorphism. Given an H-space X, define the *induction of X with α* to be the G-space $\operatorname{ind}_\alpha X$ which is the quotient of $G \times X$ by the right H-action $(g, x) \cdot h := (g\alpha(h), h^{-1}x)$ for $h \in H$ and $(g, x) \in G \times X$. If $\alpha \colon H \to G$ is an inclusion, we also write $\operatorname{ind}_H^G X$ instead of $\operatorname{ind}_\alpha X$.

Definition 20.6 (Equivariant homology theory). A *(proper) equivariant homology theory $\mathcal{H}_*^?$ with values in Λ-modules* consists of a (proper) G-homology theory $\mathcal{H}_*^G$ with values in Λ-modules for each group G together with the following so-called *induction structure*: given a group homomorphism $\alpha \colon H \to G$ and a H-CW-pair (X, A) such that $\ker(\alpha)$ acts freely on X, there are for each $n \in \mathbb{Z}$ natural isomorphisms

$$\operatorname{ind}_\alpha \colon \mathcal{H}_n^H(X, A) \xrightarrow{\cong} \mathcal{H}_n^G(\operatorname{ind}_\alpha(X, A)) \tag{20.7}$$

satisfying

(1) Compatibility with the boundary homomorphisms

$$\partial_n^G \circ \operatorname{ind}_\alpha = \operatorname{ind}_\alpha \circ \partial_n^H .$$

(2) Functoriality

Let $\beta \colon G \to K$ be another group homomorphism such that $\ker(\beta \circ \alpha)$ acts freely on X. Then we have for $n \in \mathbb{Z}$

$$\operatorname{ind}_{\beta \circ \alpha} = \mathcal{H}_n^K(f_1) \circ \operatorname{ind}_\beta \circ \operatorname{ind}_\alpha : \mathcal{H}_n^H(X, A) \to \mathcal{H}_n^K(\operatorname{ind}_{\beta \circ \alpha}(X, A)),$$

where $f_1 \colon \operatorname{ind}_\beta \operatorname{ind}_\alpha(X, A) \xrightarrow{\cong} \operatorname{ind}_{\beta \circ \alpha}(X, A)$, $(k, g, x) \mapsto (k\beta(g), x)$ is the natural K-homeomorphism.

(3) Compatibility with conjugation

For $n \in \mathbb{Z}$, $g \in G$ and a (proper) G-CW-pair (X, A) the homomorphism $\mathrm{ind}_{c(g)\colon G \to G}\colon \mathcal{H}_n^G(X, A) \to \mathcal{H}_n^G(\mathrm{ind}_{c(g)\colon G \to G}(X, A))$ agrees with $\mathcal{H}_n^G(f_2)$ for the G-homeomorphism $f_2\colon (X, A) \to \mathrm{ind}_{c(g)\colon G \to G}(X, A)$ which sends x to $(1, g^{-1}x)$ in $G \times_{c(g)} (X, A)$.

If $\mathcal{H}_*^?$ is defined or considered only for proper G-CW-pairs (X, A), we call it a *proper equivariant homology theory* $\mathcal{H}_*^?$ *with values in* Λ-*modules*.

Example 20.8 (The Borel construction). Let $\mathcal{K}_*$ be a homology theory for (non-equivariant) CW-pairs with values in Λ-modules. Examples are singular homology, oriented bordism theory or topological K-homology. Then we obtain two equivariant homology theories with values in Λ-modules by the following constructions

$$
\begin{aligned}
\mathcal{H}_n^G(X, A) &= \mathcal{K}_n(G \backslash X, G \backslash A), \\
\mathcal{H}_n^G(X, A) &= \mathcal{K}_n(EG \times_G (X, A)).
\end{aligned}
$$

The second one is called the *equivariant Borel homology associated to* $\mathcal{K}$. In both cases $\mathcal{H}_*^G$ inherits the structure of a G-homology theory from the homology structure on $\mathcal{K}_*$. Let $a\colon H \backslash X \xrightarrow{\cong} G \backslash (G \times_\alpha X)$ be the homeomorphism sending Hx to $G(1, x)$. Define $b\colon EH \times_H X \to EG \times_G G \times_\alpha X$ by sending (e, x) to $(E\alpha(e), 1, x)$ for $e \in EH$, $x \in X$ and $E\alpha\colon EH \to EG$ the α-equivariant map induced by α. Induction for a group homomorphism $\alpha\colon H \to G$ is induced by these maps a and b. If the kernel $\ker(\alpha)$ acts freely on X, the map b is a homotopy equivalence and hence in both cases ind_α is bijective.

Example 20.9 (Equivariant bordism). Given a proper G-CW-pair (X, A), one can define the G-bordism group $\Omega_n^G(X, A)$ as the abelian group of G-bordism classes of G-maps $f\colon (M, \partial M) \to (X, A)$ whose sources are oriented smooth manifolds with orientation preserving cocompact proper smooth G-actions. The definition is analogous to the one in the non-equivariant case. This is also true for the proof that this defines a proper G-homology theory. There is an obvious induction structure coming from induction of equivariant spaces. It is well-defined because of the following fact. Let $\alpha\colon H \to G$ be a group homomorphism. Let M be an oriented smooth H-manifold with orientation preserving proper smooth H-action such that $H \backslash M$ is compact and $\ker(\alpha)$ acts freely. Then $\mathrm{ind}_\alpha M$ is an oriented smooth G-manifold with orientation preserving proper smooth G-action such that $G \backslash M$ is compact. The boundary of $\mathrm{ind}_\alpha M$ is $\mathrm{ind}_\alpha \partial M$.

In case of equivariant bordism one can see geometrically what the role of the classifying spaces for families and the idea of the Meta Conjecture 20.1 is.

Namely, let $\mathcal{F} \subseteq \mathcal{FIN}$ be a family. Then $\Omega_n^G(E_\mathcal{F}(G))$ is given by the G-bordism classes of n-dimensional oriented smooth G-manifolds with proper cocompact orientation preserving smooth G-action whose isotropy groups belong to $\mathcal{F}$. In the case $\mathcal{FIN} = \mathcal{TR}$, one can identify $E_{\mathcal{TR}}(G) = EG$ and taking the quotient space yields an isomorphism

$$\Omega_n^G(E_{\mathcal{TR}}(G)) \xrightarrow{\cong} \Omega_n(BG)$$

to the non-equivariant bordism groups of BG. Now suppose that G is finite. Then for a family $\mathcal{F}$ the assembly map

$$\Omega_n^G(E_{\mathcal{F}}(G)) \to \Omega_n^G(\{\bullet\})$$

is given by forgetting the condition that all isotropy groups of M belong to $\mathcal{F}$. It follows from the equivariant Chern character constructed in [154] that for the family of cyclic subgroups $\mathcal{CYC}$ of the finite group G the assembly map yields rationally an isomorphism

$$\mathbb{Q} \otimes_{\mathbb{Z}} \Omega_n^G(E_{\mathcal{CYC}}(G)) \xrightarrow{\cong} \mathbb{Q} \otimes_{\mathbb{Z}} \Omega_n^G(\{\bullet\}).$$

20.4   The Construction of Equivariant Homology Theories from Spectra

Recall from Lemma 18.14 that a (non-equivariant) spectrum yields an associated (non-equivariant) homology theory. In this section we explain how a covariant functor GROUPOIDS $\to$ SPECTRA defines an equivariant homology theory.

In the sequel $\mathcal{C}$ is a small category. Our main example is the *orbit category* $\mathrm{Or}(G)$ *of a group* G whose objects are homogeneous G-spaces G/H and whose morphisms are G-maps.

Definition 20.10. *A* covariant (contravariant) $\mathcal{C}$-space X *is a covariant (contravariant) functor*

$$X \colon \mathcal{C} \to \text{SPACES}.$$

A map between $\mathcal{C}$-spaces is a natural transformation of such functors. Analogously a pointed $\mathcal{C}$-space *is a functor from $\mathcal{C}$ to* SPACES$^+$ *and a $\mathcal{C}$-spectrum a functor to* SPECTRA.

Example 20.11. Let Y be a left G-space. Define the associated *contravariant* $\mathrm{Or}(G)$-*space* $\mathrm{map}_G(-,Y)$ by

$$\mathrm{map}_G(-,Y) \colon \mathrm{Or}(G) \to \text{SPACES}, \qquad G/H \mapsto \mathrm{map}_G(G/H, Y) = Y^H.$$

If Y is pointed then $\mathrm{map}_G(-,Y)$ takes values in pointed spaces.

Let X be a contravariant and Y be a covariant $\mathcal{C}$-space. Define their *balanced product* to be the space

$$X \times_{\mathcal{C}} Y := \coprod_{c \in \mathrm{ob}(\mathcal{C})} X(c) \times Y(c) / \sim$$

where $\sim$ is the equivalence relation generated by $(x\varphi, y) \sim (x, \varphi y)$ for all morphisms $\varphi\colon c \to d$ in $\mathcal{C}$ and points $x \in X(d)$ and $y \in Y(c)$. Here $x\varphi$ stands for $X(\varphi)(x)$ and φy for $Y(\varphi)(y)$. If X and Y are pointed, then one defines analogously their *balanced smash product* to be the pointed space

$$X \wedge_{\mathcal{C}} Y = \bigvee_{c \in \mathrm{ob}(\mathcal{C})} X(c) \wedge Y(c)/\sim .$$

In [64] the notation $X \otimes_{\mathcal{C}} Y$ was used for this space. Doing the same construction level-wise one defines the *balanced smash product* $X \wedge_{\mathcal{C}} \mathbf{E}$ of a contravariant pointed $\mathcal{C}$-space and a covariant $\mathcal{C}$-spectrum $\mathbf{E}$.

The proof of the next result is analogous to the non-equivariant case. Details can be found in [64, Lemma 4.4], where also cohomology theories are treated.

Lemma 20.12 (Constructing G-Homology Theories). *Let $\mathbf{E}$ be a covariant $\mathrm{Or}(G)$-spectrum. It defines a G-homology theory $H_*^G(-; \mathbf{E})$ by*

$$H_n^G(X, A; \mathbf{E}) = \pi_n\left(\mathrm{map}_G\left(-, (X_+ \cup_{A_+} \mathrm{cone}(A_+))\right) \wedge_{\mathrm{Or}(G)} \mathbf{E}\right).$$

In particular we have

$$H_n^G(G/H; \mathbf{E}) = \pi_n(\mathbf{E}(G/H)).$$

Recall that we seek an equivariant homology theory and not only a G-homology theory. If the $\mathrm{Or}(G)$-spectrum in Lemma 20.12 is obtained from a GROUPOIDS-spectrum in a way we will now describe, then automatically we obtain the desired induction structure.

For a G-set S we denote by $\mathcal{G}^G(S)$ its associated *transport groupoid*. Its objects are the elements of S. The set of morphisms from s_0 to s_1 consists of those elements $g \in G$ which satisfy $gs_0 = s_1$. Composition in $\mathcal{G}^G(S)$ comes from the multiplication in G. Thus we obtain for a group G a covariant functor

$$\mathcal{G}^G\colon \mathrm{Or}(G) \to \mathrm{GROUPOIDS}^{\mathrm{inj}}, \quad G/H \mapsto \mathcal{G}^G(G/H). \tag{20.13}$$

A functor of small categories $F\colon \mathcal{C} \to \mathcal{D}$ is called an *equivalence* if there exists a functor $G\colon \mathcal{D} \to \mathcal{C}$ such that both $F \circ G$ and $G \circ F$ are naturally equivalent to the identity functor. This is equivalent to the condition that F induces a bijection on the set of isomorphisms classes of objects and for any objects $x, y \in \mathcal{C}$ the map $\mathrm{mor}_{\mathcal{C}}(x, y) \to \mathrm{mor}_{\mathcal{D}}(F(x), F(y))$ induced by F is bijective.

Lemma 20.14 (Constructing Equivariant Homology Theories). *Consider a covariant GROUPOIDS^{inj}-spectrum*

$$\mathbf{E}\colon \mathrm{GROUPOIDS}^{\mathrm{inj}} \to \mathrm{SPECTRA}.$$

Suppose that $\mathbf{E}$ respects equivalences, i.e., it sends an equivalence of groupoids to a weak equivalence of spectra. Then $\mathbf{E}$ defines an equivariant homology theory

$H^?_*(-;\mathbf{E})$, whose underlying G-homology theory for a group G is the G-homology theory associated to the covariant $\mathrm{Or}(G)$-spectrum $\mathbf{E} \circ \mathcal{G}^G \colon \mathrm{Or}(G) \to \mathrm{SPECTRA}$ in the previous Lemma 20.12, i.e.,

$$H^G_*(X, A; \mathbf{E}) = H^G_*(X, A; \mathbf{E} \circ \mathcal{G}^G).$$

In particular we have

$$H^G_n(G/H; \mathbf{E}) \cong H^H_n(\{\bullet\}; \mathbf{E}) \cong \pi_n(\mathbf{E}(I(H))),$$

where $I(H)$ denotes H considered as a groupoid with one object. The whole construction is natural in $\mathbf{E}$.

Proof. We have to specify the induction structure for a homomorphism $\alpha \colon H \to G$. We only sketch the construction in the special case where α is injective and $A = \emptyset$. The details of the full proof can be found in [212, Theorem 2.10 on page 21].

The functor induced by α on the orbit categories is denoted in the same way

$$\alpha \colon \mathrm{Or}(H) \to \mathrm{Or}(G), \quad H/L \mapsto \mathrm{ind}_\alpha(H/L) = G/\alpha(L).$$

There is an obvious natural equivalence of functors $\mathrm{Or}(H) \to \mathrm{GROUPOIDS}^{\mathrm{inj}}$

$$T \colon \mathcal{G}^H \to \mathcal{G}^G \circ \alpha.$$

Its evaluation at H/L is the equivalence of groupoids $\mathcal{G}^H(H/L) \to \mathcal{G}^G(G/\alpha(L))$ which sends an object hL to the object $\alpha(h)\alpha(L)$ and a morphism given by $h \in H$ to the morphism $\alpha(h) \in G$. The desired isomorphism

$$\mathrm{ind}_\alpha \colon H^H_n(X; \mathbf{E} \circ \mathcal{G}^H) \to H^G_n(\mathrm{ind}_\alpha X; \mathbf{E} \circ \mathcal{G}^G)$$

is induced by the following map of spectra

$$\mathrm{map}_H(-, X_+) \wedge_{\mathrm{Or}(H)} \mathbf{E} \circ \mathcal{G}^H \xrightarrow{\mathrm{id} \wedge \mathbf{E}(T)} \mathrm{map}_H(-, X_+) \wedge_{\mathrm{Or}(H)} \mathbf{E} \circ \mathcal{G}^G \circ \alpha$$

$$\xleftarrow{\cong} (\alpha_* \, \mathrm{map}_H(-, X_+)) \wedge_{\mathrm{Or}(G)} \mathbf{E} \circ \mathcal{G}^G \xleftarrow{\cong} \mathrm{map}_G(-, \mathrm{ind}_\alpha X_+) \wedge_{\mathrm{Or}(G)} \mathbf{E} \circ \mathcal{G}^G.$$

Here $\alpha_* \, \mathrm{map}_H(-, X_+)$ is the pointed $\mathrm{Or}(G)$-space which is obtained from the pointed $\mathrm{Or}(H)$-space $\mathrm{map}_H(-, X_+)$ by induction, i.e., by taking the balanced product over $\mathrm{Or}(H)$ with the $\mathrm{Or}(H)$-$\mathrm{Or}(G)$ bimodule $\mathrm{mor}_{\mathrm{Or}(G)}(??, \alpha(?))$ [64, Definition 1.8]. Notice that $\mathbf{E} \circ \mathcal{G}^G \circ \alpha$ is the same as the restriction of the $\mathrm{Or}(G)$-spectrum $\mathbf{E} \circ \mathcal{G}^G$ along α which is often denoted by $\alpha^*(\mathbf{E} \circ \mathcal{G}^G)$ in the literature [64, Definition 1.8]. The second map is given by the adjunction homeomorphism of induction α_* and restriction α^* (see [64, Lemma 1.9]). The third map is the homeomorphism of $\mathrm{Or}(G)$-spaces which is the adjoint of the obvious map of $\mathrm{Or}(H)$-spaces $\mathrm{map}_H(-, X_+) \to \alpha^* \mathrm{map}_G(-, \mathrm{ind}_\alpha X_+)$ whose evaluation at H/L is given by ind_α. $\square$

Example 20.15 (The equivariant homology theories associated to K-theory and L-theory). We have constructed in Theorem 18.24 spectra $\mathbf{K}_R$, $\mathbf{L}_R^{\langle j \rangle}$ and $\mathbf{K}^{\mathrm{top}}$. Because of Lemma 20.14 they define equivariant homology theories $H_*^?(-;\mathbf{K}_R)$, $H_*^?(-;\mathbf{L}_R^{\langle j \rangle})$ and $H_*^?(-;\mathbf{K}^{\mathrm{top}})$. These are the ones which we have promised in Subsection 20.2 and appear in the formulations of the Farrell–Jones Conjecture 20.3 and the Baum–Connes Conjecture 20.4.

Chapter 21

The Farrell–Jones Conjecture

We have already stated the Farrell–Jones Conjecture 20.3 which says that the assembly maps

$$A_{\mathcal{VC}yc}: H_n^G(E_{\mathcal{VC}yc}(G); \mathbf{K}_R) \ \rightarrow \ H_n^G(\{\bullet\}; \mathbf{K}_R) \cong K_n(RG),$$
$$A_{\mathcal{VC}yc}: H_n^G(E_{\mathcal{VC}yc}(G); \mathbf{L}_R^{\langle-\infty\rangle}) \ \rightarrow \ H_n^G(\{\bullet\}; \mathbf{L}_R^{\langle-\infty\rangle}) \cong L_n^{\langle-\infty\rangle}(RG),$$

are isomorphisms for $n \in \mathbb{Z}$. In this section we discuss its meaning, give some motivation and evidence for it and discuss some special cases like torsionfree groups and $R = \mathbb{Z}$ when the statement becomes much simpler. We also explain why it implies the Borel Conjecture 1.10. Its connection to the Novikov Conjecture 1.2 will be explained in Section 23.1.

21.1 The Bass–Heller–Swan Decomposition in Arbitrary Dimensions

We have already discussed the case $n = 1$ of the following result in Section 5.4.

Theorem 21.1 (Bass–Heller–Swan Decomposition). *The so-called* Bass–Heller–Swan decomposition, *also known as the* Fundamental Theorem of algebraic K-theory, *computes the algebraic K-groups of $R[\mathbb{Z}]$ in terms of the algebraic K-groups and Nil-groups of R for all $n \in \mathbb{Z}$:*

$$K_n(R[\mathbb{Z}]) \ \cong \ K_{n-1}(R) \oplus K_n(R) \oplus NK_n(R) \oplus NK_n(R).$$

The group $NK_n(R)$ is defined as the cokernel of the split injection $K_n(R) \to K_n(R[t])$. It can be identified with the cokernel of the split injection $K_{n-1}(R) \to K_{n-1}(\mathcal{N}il(R))$. Here $K_n(\mathcal{N}il(R))$ denotes the K-theory of the exact category of nilpotent endomorphisms of finitely generated projective R-modules. For negative

n it is defined with the help of Bass' contracting functor [19] (see also [48]). The groups are known as *Nil-groups* and often denoted $\text{Nil}_{n-1}(R)$.

For proofs of these facts and more information the reader should consult [19, Chapter XII], [20], [100, Theorem on page 236], [192, Corollary in §6 on page 38], [208, Theorems 3.3.3 and 5.3.30], [225, Theorem 9.8] and [234, Theorem 10.1].

The Nil-terms $NK_n(R)$ seem to be hard to compute. For instance $NK_1(R)$ either vanishes or is infinitely generated as an abelian group [74]. For more information about Nil-groups see for instance [57], [56], [109], [251] and [252].

Remark 21.2 (Negative K-groups must appear in the Farrell–Jones Conjecture). The Bass–Heller–Swan decomposition (see Theorem 21.1) shows that it is necessary to formulate the Farrell–Jones Conjecture 20.3 with the non-connective K-theory spectrum. Namely, $K_n(RG)$ can be affected by K_m-groups for arbitrary $m \leq n$.

21.2   Decorations in L-Theory and the Shaneson Splitting

L-groups are designed as obstruction groups for surgery problems. They come with so-called decorations which reflect what kind of surgery problem one is interested in, up to simple homotopy equivalence, up homotopy equivalence or a non-compact version. We will deal with the quadratic algebraic L-groups and denote them by $L_n^{\langle j \rangle}(R)$. Here $n \in \mathbb{Z}$ and we call $j \in \{-\infty\} \amalg \{j \in \mathbb{Z} \mid j \leq 2\}$ the *decoration*. The decorations $j = 0, 1$ correspond to the decorations p, h appearing in the literature (see also Section 17.2). The decoration $j = 2$ corresponds to the decoration s provided $R = \mathbb{Z}$ and one uses the subgroup given by the trivial units $\{\pm g \mid g \in G\} \subseteq K_1(\mathbb{Z}G)$ in the definition of the corresponding L-group. The L-groups $L_n^{\langle j \rangle}(R)$ are 4-periodic, i.e., $L_n^{\langle j \rangle}(R) \cong L_{n+4}^{\langle j \rangle}(R)$ for $n \in \mathbb{Z}$.

There are forgetful maps $L_n^{\langle j+1 \rangle}(R) \to L_n^{\langle j \rangle}(R)$. The group $L_n^{\langle -\infty \rangle}(R)$ is defined as the colimit over these maps. For details the reader should consult [195], [201].

For $j \leq 1$ there is the so-called *Rothenberg sequence* [198, Proposition 1.10.1 on page 104], [201, 17.2].

$$\ldots \to L_n^{\langle j+1 \rangle}(R) \to L_n^{\langle j \rangle}(R) \to \widehat{H}^n(\mathbb{Z}/2; \widetilde{K}_j(R))$$
$$\to L_{n-1}^{\langle j+1 \rangle}(R) \to L_{n-1}^{\langle j \rangle}(R) \to \ldots . \quad (21.3)$$

Here $\widehat{H}^n(\mathbb{Z}/2; \widetilde{K}_j(R))$ is the Tate cohomology of the group $\mathbb{Z}/2$ with coefficients in the $\mathbb{Z}[\mathbb{Z}/2]$-module $\widetilde{K}_j(R)$. The involution on $\widetilde{K}_j(R)$ comes from the involution on R. Note that Tate-cohomology groups of the group $\mathbb{Z}/2$ are always annihilated by multiplication with 2. In particular $L_n^{\langle j \rangle}(R)[\frac{1}{2}] = L_n^{\langle j \rangle}(R) \otimes_{\mathbb{Z}} \mathbb{Z}[\frac{1}{2}]$ is always independent of j. To get the passage from s to h in the special case $R = \mathbb{Z}$ one must use for the Tate-cohomology term $\widehat{H}^n(\mathbb{Z}/2; \mathrm{Wh}(G))$.

The Bass–Heller–Swan decomposition (see Theorem 21.1) has the following analogue for the algebraic L-groups.

Theorem 21.4 (Shaneson splitting). *There is an explicit isomorphism, called Shaneson splitting [221]*

$$L_n^{\langle j \rangle}(R[\mathbb{Z}]) \cong L_{n-1}^{\langle j-1 \rangle}(R) \oplus L_n^{\langle j \rangle}(R).$$

Here for the decoration $j = -\infty$ one has to interpret $j - 1$ as $-\infty$.

Remark 21.5 (The decoration $\langle -\infty \rangle$ must appear in the Farrell–Jones Conjecture). The Shaneson splitting explains why in the formulation of the L-theoretic Farrell–Jones Conjecture 20.3 we use the decoration $j = -\infty$. Namely, the Shaneson splitting does not mix two different decorations only in the case $j = -\infty$. In fact, for the decorations p, h and s there are counterexamples even for $R = \mathbb{Z}$ (see [88]).

Remark 21.6 (UNil-Terms). Even though in the above Shaneson splitting (see Theorem 21.4) there are no terms analogous to the Nil-terms in the Bass–Heller–Swan decomposition (see Theorem 21.1), Nil-phenomena do also occur in L-theory, as soon as one considers amalgamated free products. The corresponding groups are the UNil-groups. They vanish if one inverts 2 [43]. For more information about the UNil-groups we refer to [13] [40], [41], [57], [60], [75], [202].

Remark 21.7. (The Shaneson splitting and the Novikov Conjecture for $G = \mathbb{Z}^n$). Rationally all the decorations of the L-groups do not matter because of the Rothenberg sequence (21.3). Thus rationally the Shaneson splitting gives an isomorphism

$$L_m(G \times \mathbb{Z}) \cong L_m(G) \oplus L_{m-1}(G)$$

if we write $\mathbb{Z}[G \times \mathbb{Z}]$ as RG for $R = \mathbb{Z}[G]$ and use the notation $L_m(G) = L_m^h(\mathbb{Z}[G])$.

Since $h_*(X)$ introduced in (9.2) is a homology theory, there is for all $m \in \mathbb{Z}$ a splitting

$$h_m(X \times S^1) \cong h_m(X) \oplus h_{m-1}(X).$$

These two splittings are compatible with the assembly maps defined in (9.4), i.e., we get for all $m \in \mathbb{Z}$

$$A_m^{G \times \mathbb{Z}} = A_m^G \oplus A_{m-1}^G.$$

This implies that $A_m^{G \times \mathbb{Z}}$ is bijective for all $m \in \mathbb{Z}$ if and only if A_m^G is bijective for all $m \in \mathbb{Z}$. It is well-known that A_m^G is bijective for all $m \in \mathbb{Z}$ if G is the trivial group. We conclude that A_m^G is bijective for all $m \in \mathbb{Z}$ if $G = \mathbb{Z}^n$ for some $n \geq 0$. This implies the Novikov Conjecture 1.2 for $G = \mathbb{Z}^n$ for all $n \geq 0$ by Proposition 15.4.

21.3   Changing the Family

We next try to explain the role of the family of subgroups.

Theorem 21.8 (Transitivity Principle). *Let $\mathcal{H}_*^?(-)$ be an equivariant homology theory in the sense of Definition 20.6. Suppose $\mathcal{F} \subseteq \mathcal{F}'$ are two families of subgroups of G. Suppose that $K \cap H \in \mathcal{F}$ for each $K \in \mathcal{F}$ and $H \in \mathcal{F}'$ (this is automatic if $\mathcal{F}$ is closed under taking subgroups). Let N be an integer. If for every $H \in \mathcal{F}'$ and every $n \leq N$ the assembly map*

$$A_{\mathcal{F} \cap H \to \mathcal{ALL}} \colon \mathcal{H}_n^H(E_{\mathcal{F} \cap H}(H)) \; \to \; \mathcal{H}_n^H(\{\bullet\})$$

is an isomorphism, then for every $n \leq N$ the relative assembly map

$$A_{\mathcal{F} \to \mathcal{F}'} \colon \mathcal{H}_n^G(E_{\mathcal{F}}(G)) \; \to \; \mathcal{H}_n^G(E_{\mathcal{F}'}(G))$$

is an isomorphism.

Proof. If we equip $E_{\mathcal{F}}(G) \times E_{\mathcal{F}'}(G)$ with the diagonal G-action, it is a model for $E_{\mathcal{F}}(G)$. Now the claim follows from the more general Lemma 21.9 below applied to the special case $Z = E_{\mathcal{F}'}(G)$. $\square$

Lemma 21.9. *Let $\mathcal{H}_*^?$ be an equivariant homology theory with values in Λ-modules. Let G be a group and let $\mathcal{F}$ a family of subgroups of G. Let Z be a G-CW-complex. Consider $N \in \mathbb{Z} \cup \{\infty\}$. For $H \subseteq G$ let $\mathcal{F} \cap H$ be the family of subgroups of H given by $\{K \cap H \mid K \in \mathcal{F}\}$. Suppose for each $H \subseteq G$, which occurs as isotropy group in Z, that the map induced by the projection $\mathrm{pr} \colon E_{\mathcal{F} \cap H}(H) \to \{\bullet\}$,*

$$\mathcal{H}_n^H(\mathrm{pr}) \colon \mathcal{H}_n^H(E_{\mathcal{F} \cap H}(H)) \to \mathcal{H}_n^H(\{\bullet\}),$$

is bijective for all $n \in \mathbb{Z}, n \leq N$.
   Then the map induced by the projection $\mathrm{pr}_2 \colon E_{\mathcal{F}}(G) \times Z \to Z$,

$$\mathcal{H}_n^G(\mathrm{pr}_2) \colon \mathcal{H}_n^G(E_{\mathcal{F}}(G) \times Z) \; \to \; \mathcal{H}_n^G(Z),$$

is bijective for $n \in \mathbb{Z}, n \leq N$.

Proof. We first prove the claim for finite-dimensional G-CW-complexes by induction over $d = \dim(Z)$. The induction beginning $\dim(Z) = -1$, i.e., $Z = \emptyset$, is trivial. In the induction step from $(d-1)$ to d we choose a G-pushout

$$
\begin{array}{ccc}
\coprod_{i \in I_d} G/H_i \times S^{d-1} & \longrightarrow & Z_{d-1} \\
\downarrow & & \downarrow \\
\coprod_{i \in I_d} G/H_i \times D^d & \longrightarrow & Z_d
\end{array}
$$

If we cross it with $E_{\mathcal{F}}(G)$, we obtain another G-pushout of G-CW-complexes. The various projections induce a map from the Mayer–Vietoris sequence of the latter G-pushout to the Mayer–Vietoris sequence of the first G-pushout. By the 5-Lemma it suffices to prove that the following maps

$$\mathcal{H}_n^G(\mathrm{pr}_2)\colon \mathcal{H}_n^G\left(E_{\mathcal{F}}(G)\times\coprod_{i\in I_d}G/H_i\times S^{d-1}\right)\ \to\ \mathcal{H}_n^G\left(\coprod_{i\in I_d}G/H_i\times S^{d-1}\right),$$

$$\mathcal{H}_n^G(\mathrm{pr}_2)\colon \mathcal{H}_n^G(E_{\mathcal{F}}(G)\times Z_{d-1})\ \to\ \mathcal{H}_n^G(Z_{d-1}),$$

$$\mathcal{H}_n^G(\mathrm{pr}_2)\colon \mathcal{H}_n^G\left(E_{\mathcal{F}}(G)\times\coprod_{i\in I_d}G/H_i\times D^{d}\right)\ \to\ \mathcal{H}_n^G\left(\coprod_{i\in I_d}G/H_i\times D^{d}\right)$$

are bijective for $n\in\mathbb{Z}, n\le N$. This follows from the induction hypothesis for the first two maps. Because of the disjoint union axiom and G-homotopy invariance of $\mathcal{H}_*^?$ the claim follows for the third map if we can show for any $H\subseteq G$ which occurs as isotropy group in Z that the map

$$\mathcal{H}_n^G(\mathrm{pr}_2)\colon \mathcal{H}_n^G(E_{\mathcal{F}}(G)\times G/H)\ \to\ \mathcal{H}^G(G/H) \tag{21.10}$$

is bijective for $n\in\mathbb{Z}, n\le N$. The G-map

$$G\times_H \mathrm{res}_G^H E_{\mathcal{F}}(G)\to G/H\times E_{\mathcal{F}}(G)\quad (g,x)\mapsto (gH,gx)$$

is a G-homeomorphism where res_G^H denotes the restriction of the G-action to an H-action. Obviously $\mathrm{res}_G^H E_{\mathcal{F}}(G)$ is a model for $E_{\mathcal{F}\cap H}(H)$. We conclude from the induction structure that the map (21.10) can be identified with the map

$$\mathcal{H}_n^G(\mathrm{pr})\colon \mathcal{H}_n^H(E_{\mathcal{F}\cap H}(H))\ \to\ \mathcal{H}^H(\{\bullet\})$$

which is bijective for all $n\in\mathbb{Z}, n\le N$ by assumption. This finishes the proof in the case that Z is finite-dimensional. The general case follows by a colimit argument using Lemma 20.5. $\qquad\square$

21.4 The Farrell–Jones Conjecture for Torsionfree Groups

Recall that R is *Noetherian* if any submodule of a finitely generated R-module is again finitely generated. It is called *regular* if it is Noetherian and any R-module has a finite-dimensional projective resolution. Any principal ideal domain such as $\mathbb{Z}$ or a field is regular.

The Farrell–Jones Conjecture for algebraic K-theory reduces for a torsionfree group to the following conjecture provided that R is regular.

Conjecture 21.11 (Farrell–Jones Conjecture for Torsionfree Groups). *Let G be a torsionfree group.*

(1) *Let R be a regular ring. Then the assembly map for the trivial family $\mathcal{TR}$*

$$H_n(BG; \mathbf{K}(R)) = H_n^G(E_{\mathcal{TR}}(G); \mathbf{K}_R) \xrightarrow{A_{\mathcal{TR}}} K_n(RG)$$

is an isomorphism for $n \in \mathbb{Z}$. In particular $K_n(RG) = 0$ for $n \leq -1$.

(2) *The assembly map for the trivial family $\mathcal{TR}$*

$$H_n(BG; \mathbf{L}^{\langle -\infty \rangle}(R)) = H_n^G(E_{\mathcal{TR}}(G); \mathbf{L}_R^{\langle -\infty \rangle}(R)) \xrightarrow{A_{\mathcal{TR}}} L_n^{\langle -\infty \rangle}(RG)$$

is an isomorphism for $n \in \mathbb{Z}$.

This follows from the Transitivity Principle 21.8, the Bass–Heller–Swan decomposition (see Theorem 21.1), the Shaneson splitting 21.4 and the facts that for a regular ring R we have

$$\begin{aligned} K_n(R) &= 0 \quad \text{for } n \leq -1, \\ \mathrm{Nil}(R) &= 0 \quad \text{for all } n \in \mathbb{Z}, \end{aligned}$$

and we get natural isomorphisms

$$\begin{aligned} H_n^G(B\mathbb{Z}; \mathbf{K}(R)) &\cong K_n(R) \oplus K_{n-1}(R), \\ H_n^G(B\mathbb{Z}; \mathbf{L}^{\langle -\infty \rangle}(R)) &\cong L_n^{\langle -\infty \rangle}(R) \oplus L_{n-1}^{\langle -\infty \rangle}(R). \end{aligned}$$

Remark 21.12 ($K_n(\mathbb{Z}G) \otimes_{\mathbb{Z}} \mathbb{Q}$ for torsionfree groups). Note that the Farrell–Jones Conjecture for Torsionfree Groups Conjecture 21.11 can only help us to explicitly compute the K-groups of RG in cases where we know enough about the K-groups of R. We obtain no new information about the K-theory of R itself. However, already for very simple rings the computation of their algebraic K-groups is an extremely hard problem.

It is known that the groups $K_n(\mathbb{Z})$ are finitely generated abelian groups [191]. Due to Borel [28] we know that

$$K_n(\mathbb{Z}) \otimes_{\mathbb{Z}} \mathbb{Q} \cong \begin{cases} \mathbb{Q} & \text{if } n = 0; \\ \mathbb{Q} & \text{if } n = 4k + 1 \text{ with } k \geq 1; \\ 0 & \text{otherwise.} \end{cases}$$

Since $\mathbb{Z}$ is regular we know that $K_n(\mathbb{Z})$ vanishes for $n \leq -1$. Moreover, $K_0(\mathbb{Z}) \cong \mathbb{Z}$ and $K_1(\mathbb{Z}) \cong \{\pm 1\}$, where the isomorphisms are given by the rank and the determinant. One also knows that $K_2(\mathbb{Z}) \cong \mathbb{Z}/2$, $K_3(\mathbb{Z}) \cong \mathbb{Z}/48$ [147] and $K_4(\mathbb{Z}) \cong 0$ [204]. Finite fields belong to the few rings where one has a complete and explicit knowledge of all K-groups [190]. We refer the reader for example to [132], [177], [205], and Soulé's article in [150] for more information about the algebraic K-theory of the integers or more generally of rings of integers in number fields.

Because of Borel's calculation and Theorem 18.28 the rationalization of the left hand side described in the Farrell–Jones Conjecture for Torsionfree Groups and K-theory 21.11 (1) specializes for $R = \mathbb{Z}$ to

$$H_n(BG; \mathbf{K}(\mathbb{Z})) \otimes_{\mathbb{Z}} \mathbb{Q} \;\cong\; H_n(BG; \mathbb{Q}) \oplus \bigoplus_{k=1}^{\infty} H_{n-(4k+1)}(BG; \mathbb{Q}), \qquad (21.13)$$

and it is predicted that this $\mathbb{Q}$-module is isomorphic to $K_n(\mathbb{Z}G) \otimes_{\mathbb{Z}} \mathbb{Q}$ for a torsionfree group G.

Remark 21.14 ($L_n(\mathbb{Z}G) \otimes_{\mathbb{Z}} \mathbb{Q}$ for torsionfree groups). The corresponding calculation in L-theory is much simpler. If we rationalize, the decoration j does not matter and one knows that $L_n^{\langle j \rangle}(\mathbb{Z}) \otimes_{\mathbb{Z}} \mathbb{Q}$ is $\mathbb{Q}$ for $n \equiv 0 \mod 4$ and vanishes otherwise. By Theorem 18.28 the rationalization of the left hand side described in the Farrell–Jones Conjecture for Torsionfree Groups and L-theory 21.11 (2) specializes for $R = \mathbb{Z}$ to

$$H_n(BG; \mathbf{L}^{\langle -\infty \rangle}(\mathbb{Z})) \otimes_{\mathbb{Z}} \mathbb{Q} \;\cong\; \bigoplus_{k \in \mathbb{Z}} H_{n-4k}(BG; \mathbb{Q}), \qquad (21.15)$$

and it is predicted that this $\mathbb{Q}$-module is isomorphic to $L_n^{\langle -\infty \rangle}(\mathbb{Z}G) \otimes_{\mathbb{Z}} \mathbb{Q}$ for a torsionfree group G.

An easy spectral sequence argument shows that for $R = \mathbb{Z}$ the Farrell–Jones Conjecture for Torsionfree Groups and K-Theory 21.11 (1) reduces for $n \le 1$ to

Conjecture 21.16 (Vanishing of low dimensional K-theory for torsionfree groups and integral coefficients). *For every torsionfree group G we have*

$$\begin{aligned} K_n(\mathbb{Z}G) &= 0 \quad \text{for } n \le -1 \\ \widetilde{K}_0(\mathbb{Z}G) &= 0; \\ \mathrm{Wh}(G) &= 0. \end{aligned}$$

Remark 21.17 (Finiteness Obstructions). Let X be a CW-complex. It is called *finite* if it consists of finitely many cells. It is called *finitely dominated* if there is a finite CW-complex Y together with maps $i\colon X \to Y$ and $r\colon Y \to X$ such that $r \circ i$ is homotopic to the identity on X. The fundamental group of a finitely dominated CW-complex is always finitely presented.

Wall's finiteness obstruction of a connected finitely dominated CW-complex X is a certain element $\widetilde{o}(X) \in \widetilde{K}_0(\mathbb{Z}\pi_1(X))$. A connected finitely dominated CW-complex X is homotopy equivalent to a finite CW-complex if and only if $\widetilde{o}(X) = 0 \in \widetilde{K}_0(\mathbb{Z}\pi_1(X))$. Every element in $\widetilde{K}_0(\mathbb{Z}G)$ can be realized as the finiteness obstruction $\widetilde{o}(X)$ of a connected finitely dominated CW-complex X with $G = \pi_1(X)$, provided that G is finitely presented. This implies that for a finitely presented group G the vanishing of $\widetilde{K}_0(\mathbb{Z}G)$ (as predicted in Conjecture 21.16 for

torsionfree groups) is equivalent to the statement that every connected finitely
dominated CW-complex X with $G \cong \pi_1(X)$ is homotopy equivalent to a finite
CW-complex. For more information about the finiteness obstruction we refer for
instance to [89], [90], [151], [175], [199], [206], [241], [246] and [247].

Remark 21.18 (The Farrell–Jones Conjecture and the s-Cobordism Theorem).
The s-Cobordism Theorem 7.1 tells us that the vanishing of the Whitehead group
(as predicted in Conjecture 21.16 for torsionfree groups) has the following geometric interpretation. For a finitely presented group G the vanishing of the Whitehead group $\mathrm{Wh}(G)$ is equivalent to the statement that each h-cobordism over a
closed connected manifold M of dimension $\dim(M) \geq 5$ with fundamental group
$\pi_1(M) \cong G$ is trivial.

21.5   The Farrell–Jones Conjecture and the Borel Conjecture

The Borel Conjecture 1.10 can be reformulated in the language of surgery theory
to the statement that the topological structure set $\mathcal{S}^{\mathrm{top}}(M)$ of an aspherical closed
topological manifold M consists of a single point. This set is the set of equivalence
classes of homotopy equivalences $f \colon M' \to M$ with a topological closed manifold
as source and M as target under the equivalence relation, for which $f_0 \colon M_0 \to M$
and $f_1 \colon M_1 \to M$ are equivalent if there is a homeomorphism $g \colon M_0 \to M_1$ such
that $f_1 \circ g$ and f_0 are homotopic.

  The *exact surgery sequence* of a closed orientable topological manifold M of
dimension $n \geq 5$ is the exact sequence

$$\ldots \to \mathcal{N}_{n+1}(M \times [0,1], M \times \{0,1\}) \xrightarrow{\sigma} L_{n+1}^s(\mathbb{Z}\pi_1(M)) \xrightarrow{\partial} \mathcal{S}^{\mathrm{top}}(M)$$
$$\xrightarrow{\eta} \mathcal{N}_n(M) \xrightarrow{\sigma} L_n^s(\mathbb{Z}\pi_1(M)),$$

which extends infinitely to the left. It is the basic tool for the classification of
topological manifolds. (There is also a smooth version of it.) The map σ appearing in the sequence sends a normal map of degree one to its surgery obstruction. This map can be identified with the version of the L-theory assem-
bly map where one works with the 1-connected cover $\mathbf{L}^s(\mathbb{Z})\langle 1 \rangle$ of $\mathbf{L}^s(\mathbb{Z})$. The
map $H_k(M; \mathbf{L}^s(\mathbb{Z})\langle 1 \rangle) \to H_k(M; \mathbf{L}^s(\mathbb{Z}))$ is injective for $k = n$ and an isomorphism for $k > n$. Because of the K-theoretic assumptions and the Rothenberg
sequence (21.3) we can replace the s-decoration with the $\langle -\infty \rangle$-decoration. Therefore the Farrell–Jones Conjecture 21.11 implies that the maps $\sigma \colon \mathcal{N}_n(M) \to$
$L_n^s(\mathbb{Z}\pi_1(M))$ and $\mathcal{N}_{n+1}(M \times [0,1], M \times \{0,1\}) \xrightarrow{\sigma} L_{n+1}^s(\mathbb{Z}\pi_1(M))$ are injective
respectively bijective and thus by the surgery sequence that $\mathcal{S}^{\mathrm{top}}(M)$ is a point
and hence the Borel Conjecture 1.10 holds for M. More details can be found e.g.
in [91, pages 17,18,28], [200, Chapter 18]. For more information about surgery

theory we refer for instance to [33], [38], [39], [86], [87], [121], [133], [153], [194], [228], [227], and [249].

21.6 The Passage from $\mathcal{FIN}$ to $\mathcal{VCYC}$

The following information about virtually cyclic groups is useful. Its elementary proof can be found in [81].

Lemma 21.19. *If G is an infinite virtually cyclic group then we have the following dichotomy.*

(I) *Either G admits a surjection with finite kernel onto the infinite cyclic group $\mathbb{Z}$, or*

(II) *G admits a surjection with finite kernel onto the infinite dihedral group $\mathbb{Z}/2 * \mathbb{Z}/2$.*

The next result is due to Bartels [14].

Theorem 21.20 (Passage from $\mathcal{FIN}$ to $\mathcal{VCYC}$). (1) *For every group G, every ring R and every $n \in \mathbb{Z}$ the relative assembly map*

$$A_{\mathcal{FIN} \to \mathcal{VCYC}} \colon H_n^G(E_{\mathcal{FIN}}(G); \mathbf{K}_R) \to H_n^G(E_{\mathcal{VCYC}}(G); \mathbf{K}_R)$$

is split-injective.

(2) *Suppose R is such that $K_{-i}(RV) = 0$ for all virtually cyclic subgroups V of G and for sufficiently large i (for example $R = \mathbb{Z}$ will do). Then the relative assembly map*

$$A_{\mathcal{FIN} \to \mathcal{VCYC}} \colon H_n^G(E_{\mathcal{FIN}}(G); \mathbf{L}_R^{\langle -\infty \rangle}) \to H_n^G(E_{\mathcal{VCYC}}(G); \mathbf{L}_R^{\langle -\infty \rangle})$$

is split-injective.

Combined with the Farrell–Jones Conjectures we obtain that the homology group $H_n^G(E_{\mathcal{FIN}}(G); \mathbf{K}_R)$ is a direct summand in $K_n(RG)$. It is much better understood than the remaining summand $H_n^G(E_{\mathcal{VCYC}}(G), E_{\mathcal{FIN}}(G); \mathbf{K}_R)$. This remaining summand is the one which plays the role of the Nil-terms for a general group. It is known that for $R = \mathbb{Z}$ the negative dimensional Nil-groups which occur for virtually cyclic groups vanish [81]. For $R = \mathbb{Z}$ they vanish rationally, in dimension 0 by [59] and in higher dimensions by [139]. For more information see also [58]. This implies

Lemma 21.21. *We have*

$$\begin{aligned} H_n^G(E_{\mathcal{VCYC}}(G), E_{\mathcal{FIN}}(G); \mathbf{K}_{\mathbb{Z}}) &= 0 & \text{for } n < 0; \\ H_n^G(E_{\mathcal{VCYC}}(G), E_{\mathcal{FIN}}(G); \mathbf{K}_{\mathbb{Z}}) \otimes_{\mathbb{Z}} \mathbb{Q} &= 0 & \text{for all } n \in \mathbb{Z}. \end{aligned}$$

Lemma 21.22. *For every group G, every ring R with involution, every decoration j and all $n \in \mathbb{Z}$ the relative assembly map*

$$A_{\mathcal{FIN} \to \mathcal{VCYC}} \colon H_n^G(E_{\mathcal{FIN}}(G); \mathbf{L}_R^{\langle j \rangle})[\tfrac{1}{2}] \to H_n^G(E_{\mathcal{VCYC}}(G); \mathbf{L}_R^{\langle j \rangle})[\tfrac{1}{2}]$$

is an isomorphism.

Proof. According to the Transitivity Principle it suffices to prove the claim for a virtually cyclic group. Now proceed as in the proof of Lemma 21.24 using the exact sequences in [42] and the fact that the UNil-terms appearing there vanish after inverting two [42]. □

Remark 21.23 (Rationally $\mathcal{FIN}$ suffices for the Farrell–Jones Conjecture). Hence the Farrell–Jones Conjecture 20.3 predicts that the map

$$A_{\mathcal{FIN}} \otimes_{\mathbb{Z}} \mathrm{id}_{\mathbb{Q}} \colon H_n^G(E_{\mathcal{FIN}}(G); \mathbf{K}_{\mathbb{Z}}) \otimes_{\mathbb{Z}} \mathbb{Q} \to K_n(\mathbb{Z}G) \otimes_{\mathbb{Z}} \mathbb{Q}$$

is always an isomorphism and that for every decoration j the assembly map

$$A_{\mathcal{FIN}}[\tfrac{1}{2}] \colon H_n^G(E_{\mathcal{FIN}}(G); \mathbf{L}_R^{\langle j \rangle})[\tfrac{1}{2}] \to L_n^{\langle j \rangle}(RG)[\tfrac{1}{2}]$$

is an isomorphism.

Lemma 21.24. *Let R be a regular ring with $\mathbb{Q} \subseteq R$, for instance a field of characteristic zero.*

(1) *Then for each group G the relative assembly map*

$$A_{\mathcal{FIN} \subseteq \mathcal{VCYC}} \colon H_n^G(E_{\mathcal{FIN}}(G); \mathbf{K}_R) \ \to \ H_n^G(E_{\mathcal{VCYC}}(G); \mathbf{K}_R)$$

   is bijective for all $n \in \mathbb{Z}$.

(2) *If the Farrell–Jones Conjecture 20.3 is true for G and R, then the assembly map*

$$A_{\mathcal{FIN}} \colon H_n^G(E_{\mathcal{FIN}}(G); \mathbf{K}_R) \ \to \ K_n(RG)$$

   is bijective for all $n \in \mathbb{Z}$.

Proof. (1) We first show that RH is regular for a finite group H. Since R is Noetherian and H is finite, RH is Noetherian. It remains to show that every RH-module M has a finite dimensional projective resolution. By assumption M considered as an R-module has a finite dimensional projective resolution. If one applies $RH \otimes_R -$ this yields a finite dimensional RH-resolution of $RH \otimes_R \operatorname{res} M$. Since $|H|$ is invertible, the RH-module M is a direct summand of $RH \otimes_R \operatorname{res} M$ and hence has a finite dimensional projective resolution.

Because of the Transitivity Principle 21.8, it suffices to prove for any virtually finite cyclic group V that

$$A_{\mathcal{FIN}} \colon H_n^V(E_{\mathcal{FIN}}(V); \mathbf{K}_R) \ \to \ K_n(RV)$$

is bijective. Because of Lemma 21.19 we can assume that either $V \cong K_1 *_H K_2$ or $V \cong H \rtimes \mathbb{Z}$ with finite groups H, K_1 and K_2. From [244] we obtain in both cases long exact sequences involving the algebraic K-theory of the group rings of H, K_1, K_2 and V and also additional Nil-terms. However, in both cases the Nil-terms vanish if RH is a regular ring (compare Theorem 4 on page 138 and the Remark on page 216 in [244]). Thus we get long exact sequences

$$\ldots \to K_n(RH) \to K_n(RH) \to K_n(RV) \to K_{n-1}(RH) \to K_{n-1}(RH) \to \ldots$$

and

$$\ldots \to K_n(RH) \to K_n(RK_1) \oplus K_n(RK_2) \to K_n(RV)$$
$$\to K_{n-1}(RH) \to K_{n-1}(RK_1) \oplus K_{n-1}(RK_2) \to \ldots$$

One obtains analogous exact sequences for the sources of the various assembly maps from the fact that the sources are equivariant homology theories and one can find specific models for $E_{\mathcal{FIN}}(V)$. For instance in the case $V \cong K_1 *_H K_2$ a specific model is given by the V-pushout

$$
\begin{array}{ccc}
V/H \times \{0,1\} & \longrightarrow & V/K_1 \amalg V/K_2 \\
\downarrow & & \downarrow \\
V/H \times [0,1] & \longrightarrow & E_{\mathcal{FIN}}(V)
\end{array}
$$

and in the case $V \cong H \rtimes \mathbb{Z}$ a model for $E_{\mathcal{FIN}}(V)$ is given by $E\mathbb{Z}$ considered as V space by the obvious epimorphism $V \to \mathbb{Z}$. These sequences are compatible with the assembly maps. The assembly maps for the finite groups H, K_1 and K_2 are bijective. Now a 5-Lemma argument shows that also the one for V is bijective.

(2) This follows from (1). □

We obtain a covariant functor $K_q(R?)$ from the orbit category $\mathrm{Or}(G)$ to the category of abelian groups as follows. It sends an object G/H to $K_q(RH)$. A morphisms $G/H \to G/K$ can be written as $gH \mapsto gg_0K$ for some $g_0 \in G$ with $g_0^{-1}Hg_0 \subseteq K$. Let $K_0(RH) \to K_0(RK)$ be the homomorphisms induced by the group homomorphism $H \to K$, $h \mapsto g_0^{-1}hg_0$. This is independent of the choice of g_0 since inner automorphisms of K induce the identity on $K_q(RK)$. Let $\mathrm{Or}(G; \mathcal{FIN})$ be the full subcategory of $\mathrm{Or}(G)$ consisting of objects G/H with finite H.

Lemma 21.25. (1) *Let R be a regular ring with $\mathbb{Q} \subseteq R$. Suppose that the Farrell–Jones Conjecture 20.3 for algebraic K-theory holds for the group G and the coefficient ring R. Then the canonical map*

$$a \colon \mathrm{colim}_{G/H \in \mathrm{Or}(G;\mathcal{FIN})} K_0(RH) \to K_0(RG)$$

is bijective.

(2) *Suppose that the Farrell–Jones Conjecture 20.3 holds for algebraic K-theory for the group G and the coefficient ring R = $\mathbb{Z}$. Then*

$$K_{-n}(\mathbb{Z}G) = 0 \quad for \; n \geq 2,$$

and the map

$$a \colon \operatorname{colim}_{G/H \in \operatorname{Or}(G;\mathcal{FIN})} K_{-1}(\mathbb{Z}H) \; \to \; K_{-1}(\mathbb{Z}G)$$

is an isomorphism.

Proof. (1) By Lemma 21.24 the assembly map

$$A_{\mathcal{FIN}} \colon H_0^G(E_{\mathcal{FIN}}(G); \mathbf{K}_R) \; \to \; K_0(RG)$$

is bijective. There is an equivariant version of the Atiyah–Hirzebruch spectral sequence converging to $H_{p+q}^G(E_{\mathcal{FIN}}(G); \mathbf{K}_R)$ whose $E_{p,q}^2$-term is given by the p-th Bredon homology $H_p^{\operatorname{Or}(G)}(E_{\mathcal{FIN}}(G); K_q(R?))$ with coefficients in the covariant functor $K_q(R?)$. Since for a finite group RH is regular, $K_q(RH) = 0$ for $q \leq -1$. Hence the edge homomorphism in the equivariant Atiyah–Hirzebruch spectral sequence yields an isomorphism

$$H_0^{\operatorname{Or}(G)}(E_{\mathcal{FIN}}(G); K_0(R?)) \; \xrightarrow{\cong} \; H_0^G(E_{\mathcal{FIN}}(G); \mathbf{K}_R).$$

Its composition with the assembly map $H_0^G(E_{\mathcal{FIN}}(G); \mathbf{K}_R) \to K_0(RG)$ can be identified with the map $a \colon \operatorname{colim}_{G/H \in \operatorname{Or}(G;\mathcal{FIN})} K_0(RH) \to K_0(RG)$.

(2) This is proven analogous to assertion (1) using Lemma 21.21 and the fact that $K_n(\mathbb{Z}) = 0$ for $n \leq -1$ holds since $\mathbb{Z}$ is regular. $\square$

A systematically study how small one can choose the family appearing in the assembly maps of the Farrell–Jones Conjecture 20.3 and the Baum–Connes Conjecture 20.4 is presented in [16].

Chapter 22

The Baum–Connes Conjecture

We have already stated the Baum–Connes Conjecture 20.4. In this section we discuss its meaning, give some motivation and evidence for it.

22.1 Index Theoretic Interpretation of the Baum–Connes Assembly Map

We have already explained in Section 21.5 that the Farrell–Jones assembly map for L-theory has a surgery theoretic interpretation which allows to deduce the Borel Conjecture 1.10 from the Farrell–Jones Conjecture 20.3. We want to explain briefly that the Baum–Connes assembly map

$$A_{\mathcal{FIN}} \colon H_n^G(E_{\mathcal{F}}(G); \mathbf{K}^{\mathrm{top}}) \quad \to \quad K_n(C_r^*(G)) \tag{22.1}$$

has an index theoretic interpretation. Recall that the Baum–Connes Conjecture 20.4 predicts that it is bijective for all $n \in \mathbb{Z}$.

We begin with a discussion of the target of the Baum–Connes assembly map. Let $\mathcal{B}(l^2(G))$ denote the bounded linear operators on the Hilbert space $l^2(G)$ whose orthonormal basis is G. The *reduced complex group C^*-algebra* $C_r^*(G)$ is the closure in the norm topology of the image of the regular representation $\mathbb{C}G \to \mathcal{B}(l^2(G))$, which sends an element $u \in \mathbb{C}G$ to the (left) G-equivariant bounded operator $l^2(G) \to l^2(G)$ given by right multiplication with u. In particular one has natural inclusions

$$\mathbb{C}G \subseteq C_r^*(G) \subseteq \mathcal{B}(l^2(G))^G \subseteq \mathcal{B}(l^2(G)).$$

It is essential to use the reduced group C^*-algebra in the Baum–Connes Conjecture, there are counterexamples for the version with the maximal group C^*-algebra. The *topological K-groups* $K_n(A)$ of a C^*-algebra A are 2-periodic. Whereas $K_0(A)$ coincides with the algebraically defined K_0-group, the other groups $K_n(A)$ take the topology of the C^*-algebra A into account, for instance

$K_n(A) = \pi_{n-1}(GL(A))$ for $n \geq 1$. For information about C^*-algebras and their topological K-theory we refer for instance to [26], [55], [62], [113], [143], [178], [215] and [250].

The target of the Baum–Connes assembly map $H_n^G(E_{\mathcal{F}}(G); \mathbf{K}^{\mathrm{top}})$ can be identified with the equivariant K-homology $K_n^G(E_{\mathcal{FIN}}(G))$ which is defined in terms of equivariant Kasparov KK-theory (see [124]). An element in dimension $n = 0$ can be interpreted as a pair $[M, P^*]$ which consists of a proper smooth G-manifold M with G-invariant Riemannian metric together with an elliptic G-complex P^* of bounded linear G-equivariant differential operators of order 1 [23]. To such a pair (M, P^*) one can assign an index $\mathrm{ind}_{C_r^*(G)}(M, P^*)$ in $K_0(C_r^*(G))$ [174]. This is the image of the element in $K_0^G(E_{\mathcal{FIN}}(G))$ represented by (M, P^*) under the original Baum–Connes assembly map. The original Baum–Connes assembly map has been identified with the one defined in these notes in [107] using the universal characterization of the assembly map of [64].

22.2   The Baum–Connes Conjecture for Torsionfree Groups

We denote by $K_*(Y)$ the complex K-homology of a topological space Y and by $K_*(C_r^*(G))$ the (topological) K-theory of the reduced group C^*-algebra. More explanations will follow below. If G is torsionfree, the families $\mathcal{TR}$ and $\mathcal{FIN}$ coincide. Hence the Baum–Connes Conjecture 20.4 reduces for a torsionfree group to the following statement.

Conjecture 22.2 (Baum–Connes Conjecture for Torsionfree Groups). *Let G be a torsionfree group. Then the Baum–Connes assembly map for the trivial family $\mathcal{TR}$*

$$K_n(BG) = H_n(BG; \mathbf{K}^{\mathrm{top}}(\mathbb{C})) = H_n^G(E_{\mathcal{TR}}(G); \mathbf{K}^{\mathrm{top}}) \to K_n(C_r^*(G))$$

is bijective for all $n \in \mathbb{Z}$.

Complex K-homology $K_*(Y)$ is the homology theory associated to the topological (complex) K-theory spectrum $\mathbf{K}^{\mathrm{top}}$, which is often denoted $\mathbf{BU}$ and has been introduced in Example 18.16 and can also be written as $K_*(Y) = H_*(Y; \mathbf{K}^{\mathrm{top}})$.

Remark 22.3 $(K_n(C_r^*(G)) \otimes_{\mathbb{Z}} \mathbb{Q}$ **for torsionfree groups).** One knows that $K_n(C_r^*(G)) \otimes_{\mathbb{Z}} \mathbb{Q}$ is $\mathbb{Q}$ for even n and vanishes for n odd. By Theorem 18.28 the rationalization of the left hand side described in the Baum–Connes Conjecture for Torsionfree Groups 22.2 is

$$H_n(BG; \mathbf{L}^{\mathrm{top}}(\mathbb{C})) \otimes_{\mathbb{Z}} \mathbb{Q} \quad \cong \quad \bigoplus_{k \in \mathbb{Z}} H_{n-2k}(BG; \mathbb{Q}), \tag{22.4}$$

and it is predicted that this $\mathbb{Q}$-module is isomorphic to $K_n(C_r^*(G))$ for a torsionfree group G.

22.3   The Trace Conjecture in the Torsionfree Case

The *standard trace*

$$\text{tr}_{C_r^*(G)} \colon C_r^*(G) \;\rightarrow\; \mathbb{C} \tag{22.5}$$

sends an element $f \in C_r^*(G) \subseteq \mathcal{B}(l^2(G))$ to $\langle f(1), 1\rangle_{l^2(G)}$. Applying the trace to idempotent matrices yields a homomorphism

$$\text{tr}_{C_r^*(G)} \colon K_0(C_r^*(G)) \to \mathbb{R}.$$

The following conjecture is taken from [21, page 21].

Conjecture 22.6 (Trace Conjecture for Torsionfree Groups). *For a torsionfree group G the image of*

$$\text{tr}_{C_r^*(G)} \colon K_0(C_r^*(G)) \to \mathbb{R}$$

consists of the integers.

Lemma 22.7. *Let G be a torsionfree group. The surjectivity of the Baum–Connes assembly map*

$$K_0(BG) \cong H_n^G(E_{TR}(G), \mathbf{K}^{\text{top}}) = H_n^G(E_{\mathcal{FIN}}(G), \mathbf{K}^{\text{top}}) \;\rightarrow\; K_0(C_r^*(G))$$

implies the Trace Conjecture for Torsionfree Groups 22.6.

Proof. Let $\text{pr} \colon BG \to \{\bullet\}$ be the projection. For a every group G the following diagram commutes:

$$\begin{array}{ccccc}
K_0(BG) & \xrightarrow{\;A_{TR}\;} & K_0(C_r^*(G)) & \xrightarrow{\;\text{tr}_{C_r^*(G)}\;} & \mathbb{R} \\
{\scriptstyle K_0(\text{pr})}\downarrow & & & & \uparrow{\scriptstyle i} \\
K_0(\{\bullet\}) & \xrightarrow[\cong]{} & K_0(\mathbb{C}) & \xrightarrow[\text{tr}_{\mathbb{C}}]{\cong} & \mathbb{Z}
\end{array} \tag{22.8}$$

Here $i \colon \mathbb{Z} \to \mathbb{R}$ is the inclusion and A_{TR} is the assembly map associated to the family $\mathcal{TR}$. The commutativity follows from Atiyah's L^2-index theorem [9]. Atiyah's theorem says that the L^2-index $\text{tr}_{C_r^*(G)} \circ A(\eta)$ of an element $\eta \in K_0(BG)$ which is represented by a pair (M, P^*) agrees with the ordinary index of $(G\backslash M; G\backslash P^*)$, which is $\text{tr}_{\mathbb{C}} \circ K_0(\text{pr})(\eta) \in \mathbb{Z}$. $\qquad\square$

For a discussion of the Trace Conjecture for arbitrary groups see [156].

22.4   The Kadison Conjecture

Conjecture 22.9 (Kadison Conjecture). *If G is a torsionfree group, then the only idempotent elements in $C_r^*(G)$ are 0 and 1.*

Lemma 22.10. *The Trace Conjecture for Torsionfree Groups 22.6 implies the Kadison Conjecture 22.9.*

Proof. Assume that $p \in C_r^*(G)$ is an idempotent different from 0 or 1. From p one can construct a non-trivial projection $q \in C_r^*(G)$, i.e., $q^2 = q$, $q^* = q$, with $\text{im}(p) = \text{im}(q)$ and hence with $0 < q < 1$. Since the standard trace $\text{tr}_{C_r^*(G)}$ is faithful, we conclude $\text{tr}_{C_r^*(G)}(q) \in \mathbb{R}$ with $0 < \text{tr}_{C_r^*(G)}(q) < 1$. Since $\text{tr}_{C_r^*(G)}(q)$ is by definition the image of the element $[\text{im}(q)] \in K_0(C_r^*(G))$ under $\text{tr}_{C_r^*(G)} \colon K_0(C_r^*(G)) \to \mathbb{R}$, we get a contradiction to the assumption $\text{im}(\text{tr}_{C_r^*(G)}) \subseteq \mathbb{Z}$.   □

Recall that a ring R is called an *integral domain* if it has no non-trivial zero-divisors, i.e., if $r, s \in R$ satisfy $rs = 0$, then r or s is 0. Obviously the Kadison Conjecture 22.9 implies for $R \subseteq \mathbb{C}$ the following.

Conjecture 22.11 (Idempotent Conjecture). *Let R be an integral domain and let G be a torsionfree group. Then the only idempotents in RG are 0 and 1.*

The statement in the conjecture above is a purely algebraic statement. If $R \subseteq \mathbb{C}$, it is by the arguments above related to questions about operator algebras, and thus methods from operator algebras can be used to attack it.

22.5   The Stable Gromov–Lawson–Rosenberg Conjecture

The Stable Gromov–Lawson–Rosenberg Conjecture is a typical conjecture relating Riemannian geometry to topology. It is concerned with the question when a given closed manifold admits a metric of positive scalar curvature. To discuss its relation with the Baum–Connes Conjecture we will need the real version of the Baum–Connes Conjecture.

Let $\Omega_n^{\text{Spin}}(BG)$ be the bordism group of closed Spin-manifolds M of dimension n with a reference map to BG. Let $C_r^*(G; \mathbb{R})$ be the real reduced group C^*-algebra and let $KO_n(C_r^*(G; \mathbb{R})) = K_n(C_r^*(G; \mathbb{R}))$ be its topological K-theory. We use KO instead of K as a reminder that we here use the real reduced group C^*-algebra. Given an element $[u \colon M \to BG] \in \Omega_n^{\text{Spin}}(BG)$, we can take the $C_r^*(G; \mathbb{R})$-valued index of the equivariant Dirac operator associated to the G-covering $\overline{M} \to M$ determined by u. Thus we get a homomorphism

$$\text{ind}_{C_r^*(G;\mathbb{R})} \colon \Omega_n^{\text{Spin}}(BG) \quad \to \quad KO_n(C_r^*(G; \mathbb{R})). \tag{22.12}$$

A *Bott manifold* is any simply connected closed Spin-manifold B of dimension 8 whose $\widehat{A}$-genus $\widehat{A}(B)$ is 8. We fix such a choice, the particular choice does not matter for the sequel. Notice that $\text{ind}_{C_r^*(\{1\};\mathbb{R})}(B) \in KO_8(\mathbb{R}) \cong \mathbb{Z}$ is a generator and the product with this element induces the Bott periodicity isomorphisms $KO_n(C_r^*(G; \mathbb{R})) \xrightarrow{\cong} KO_{n+8}(C_r^*(G; \mathbb{R}))$. In particular

$$\text{ind}_{C_r^*(G;\mathbb{R})}(M) \quad = \quad \text{ind}_{C_r^*(G;\mathbb{R})}(M \times B), \tag{22.13}$$

if we identify $KO_n(C_r^*(G;\mathbb{R})) = KO_{n+8}(C_r^*(G;\mathbb{R}))$ via Bott periodicity.

Conjecture 22.14 (Stable Gromov–Lawson–Rosenberg Conjecture). *Let M be a closed connected* Spin*-manifold of dimension* $n \geq 5$. *Let* $u_M \colon M \to B\pi_1(M)$ *be the classifying map of its universal covering. Then* $M \times B^k$ *carries for some integer* $k \geq 0$ *a Riemannian metric with positive scalar curvature if and only if*

$$\mathrm{ind}_{C_r^*(\pi_1(M);\mathbb{R})}([M, u_M]) \;=\; 0 \quad \in KO_n(C_r^*(\pi_1(M);\mathbb{R})).$$

If M carries a Riemannian metric with positive scalar curvature, then the index of the Dirac operator must vanish by the Bochner–Lichnerowicz formula [207]. The converse statement that the vanishing of the index implies the existence of a Riemannian metric with positive scalar curvature is the hard part of the conjecture. The following result is due to Stolz. A sketch of the proof can be found in [230, Section 3], details are announced to appear in a different paper.

Theorem 22.15 (The Baum–Connes Conjecture implies the stable Gromov–Lawson–Rosenberg Conjecture). *If the assembly map for the real version of the Baum–Connes Conjecture*

$$KO_n(E_{\mathcal{FIN}}(G)) \to KO_n(C_r^*(G;\mathbb{R}))$$

is injective for the group G, *then the Stable Gromov–Lawson–Rosenberg Conjecture 22.14 is true for all closed* Spin*-manifolds of dimension* ≥ 5 *with* $\pi_1(M) \cong G$.

The following result appears in [24].

Lemma 22.16. *The Baum–Connes Conjecture 20.4 implies the real version of the Baum–Connes Conjecture.*

The requirement $\dim(M) \geq 5$ is essential in the Stable Gromov–Lawson–Rosenberg Conjecture, since in dimension four new obstructions, the Seiberg–Witten invariants, occur. The unstable version of this conjecture says that M carries a Riemannian metric with positive scalar curvature if and only if we have $\mathrm{ind}_{C_r^*(\pi_1(M);\mathbb{R})}([M, u_M]) = 0$. Schick [214] constructs counterexamples to the unstable version using minimal hypersurface methods due to Schoen and Yau (see also [72]). It is not known at the time of writing whether the unstable version is true for finite fundamental groups. Since the Baum–Connes Conjecture 20.4 is true for finite groups (for the trivial reason that $E_{\mathcal{FIN}}(G) = \{\bullet\}$ for finite groups G), Theorem 22.15 implies that the Stable Gromov–Lawson Conjecture 22.14 holds for finite fundamental groups (see also [210]).

22.6 The Choice of the Family $\mathcal{FIN}$ in the Baum–Connes Conjecture

In contrast to the Farrell–Jones Conjecture 20.3 one does not have to use the family of virtually cyclic subgroups in the Baum–Connes Conjecture 20.4, the family $\mathcal{FIN}$ of finite groups suffices. This is an important simplification and can be interpreted as the fact that there are no Nil-phenomenons for the topological K-theory of group C^*-algebras. Recall that the occurrence of Nil-groups complicates the computations of algebraic K and L-groups of integral group rings considerably. It is not hard to prove analogously to Lemma 21.24 using the so-called Pimsner–Voiculescu sequences [187] that for any virtually cyclic group V the assembly map

$$A_{\mathcal{FIN}}\colon H_n^V(E_{\mathcal{FIN}}(V); \mathbf{K}^{\mathrm{top}}) \to K_n(C_r^*(V))$$

is bijective for all $n \in \mathbb{Z}$. Now the Transitivity Principle 21.8 implies that the relative assembly map

$$A_{\mathcal{FIN}\to\mathcal{VCYC}}\colon H_n^G(E_{\mathcal{FIN}}(G); \mathbf{K}^{\mathrm{top}}) \to H_n^G(E_{\mathcal{VCYC}}(G); \mathbf{K}^{\mathrm{top}})$$

is bijective for all groups G. Hence it does not matter whether one uses the family $\mathcal{FIN}$ or $\mathcal{VCYC}$ in the Baum–Connes Conjecture 20.4.

Remark 22.17 (Pimsner–Voiculescu splitting). There is an analogue of the Bass–Heller–Swan decomposition in algebraic K-theory (see Theorem 21.1) or of the Shaneson splitting in L-theory (see Theorem 21.4) for topological K-theory. Namely we have

$$K_n(C_r^*(G \times \mathbb{Z})) \cong K_n(C_r^*(G)) \oplus K_{n-1}(C_r^*(G)),$$

see [187, Theorem 3.1 on page 151] or more generally [188, Theorem 18 on page 632]. This is consistent with the Baum–Connes Conjecture 20.4 since $E_{\mathcal{FIN}}(G) \times E\mathbb{Z}$ is a model for $E_{\mathcal{FIN}}(G \times \mathbb{Z})$ and we have obvious isomorphisms

$$H_n^{G \times \mathbb{Z}}(E_{\mathcal{FIN}}(G \times \mathbb{Z}); \mathbf{K}^{\mathrm{top}}) \cong H_n^{G \times \mathbb{Z}}(E_{\mathcal{FIN}}(G) \times E\mathbb{Z}; \mathbf{K}^{\mathrm{top}})$$
$$\cong H_n^G(E_{\mathcal{FIN}}(G) \times \mathbb{Z}\backslash E\mathbb{Z}; \mathbf{K}^{\mathrm{top}}) \cong H_n^G(E_{\mathcal{FIN}}(G) \times S^1; \mathbf{K}^{\mathrm{top}})$$
$$\cong H_n^G(E_{\mathcal{FIN}}(G); \mathbf{K}^{\mathrm{top}}) \oplus H_{n-1}^G(E_{\mathcal{FIN}}(G); \mathbf{K}^{\mathrm{top}}).$$

Chapter 23

Relating the Novikov, the Farrell–Jones and the Baum–Connes Conjectures

In this chapter we explain why the Novikov Conjecture 1.2 follows from the Farrell–Jones Conjecture 20.3 or from the Baum–Connes Conjecture 20.4.

23.1 The Farrell–Jones Conjecture and the Novikov Conjecture

Lemma 23.1. (1) *The relative assembly map*

$$H_n(BG; \mathbf{L}(R)) = H_n^G(E_{\mathcal{TR}}(G); \mathbf{L}_R) \xrightarrow{A_{\mathcal{TR} \to \mathcal{FIN}}} H_n^G(E_{\mathcal{FIN}}(G), \mathbf{L}_R)$$

is rationally injective.

(2) *If G satisfies the Farrell–Jones Conjecture for L-theory 20.3, then the assembly map for the trivial family $\mathcal{TR}$*

$$H_n(BG; \mathbf{L}(\mathbb{Z})) = H_n^G(E_{\mathcal{TR}}(G); \mathbf{L}_R) \xrightarrow{A_{\mathcal{TR}}} L_n^{\langle -\infty \rangle}(\mathbb{Z}G)$$

is rationally injective.

Proof. (1) The induction structure of an equivariant homology theory (see Definition 20.6) yields for any group homomorphism $\alpha \colon H \to G$ a map $\mathrm{ind}_\alpha \colon \mathcal{H}_n^H(X) \to \mathcal{H}_n^G(G \times_\alpha X)$, the condition that $\ker(\alpha)$ acts freely on X is only needed to ensure that this map is bijective. In particular we get a map

$$H_n^G(E_{\mathcal{FIN}}(G); \mathbf{L}_R) \to H_n(G \backslash E_{\mathcal{FIN}}(G); \mathbf{L}(R)).$$

Its composition with the relative assembly map $A_{\mathcal{TR}\to\mathcal{FIN}}$ is the same as the map

$$H_n(G\backslash f; \mathbf{L}(R)) \colon H_n(BG; \mathbf{L}(R)) \to H_n(G\backslash E_{\mathcal{FIN}}(G); \mathbf{L}(R)),$$

where $f \colon EG = E_{\mathcal{TR}}(G) \to E_{\mathcal{FIN}}(G)$ is the (up to G-homotopy unique) G-map. It is not hard to check that for any non-equivariant homology theory $\mathcal{H}_*$ (satisfying the disjoint union axiom) the map $\mathcal{H}_n(G\backslash f)$ is rationally bijective for any group G.

(2) This follows from Lemma 21.22 and assertion (1). $\square$

Lemma 23.2. *Let G be a group. Then:*

(1) *The Novikov Conjecture 1.2 follows from the rational injectivity of the assembly map for the family $\mathcal{TR}$*

$$H_n(BG; \mathbf{L}^{\langle -\infty\rangle}(\mathbb{Z})) = H_n^G(E_{\mathcal{TR}}(G); \mathbf{L}_{\mathbb{Z}}^{\langle -\infty\rangle}) \xrightarrow{A_{\mathcal{TR}}} L_n^{\langle -\infty\rangle}(\mathbb{Z}G).$$

(2) *If G satisfies the Farrell–Jones Conjecture for L-theory 20.3, or more generally, if the assembly map appearing in the Farrell–Jones Conjecture for L-theory 20.3 is rationally injective, then G satisfies the Novikov Conjecture 1.2.*

Proof. (1) In the sequel we will suppress any decorations since we will finally invert 2 and hence they do not matter. The default is the decoration h. For any n-dimensional closed manifold M together with a reference map $u \colon M \to BG$ we introduced in (17.34) its symmetric signature taking values in the symmetric L-groups

$$\sigma(M, u) \ \in \ L^n(\mathbb{Z}G).$$

It has the two important features that it is a homotopy invariant i.e., for two closed oriented smooth manifolds M and N with reference maps $u \colon M \to BG$ and $v \colon N \to BG$ we have

$$\sigma(M, u) = \sigma(N, v),$$

if there is an orientation preserving homotopy equivalence $f \colon M \to N$ such that $v \circ f$ and u are homotopic and that it is a bordism invariant. Hence it defines for each connected CW-complex X a map (see Theorem 17.35)

$$\sigma(X) \colon \Omega_n(X) \ \to \ L^n(\mathbb{Z}\pi_1(X)), \quad [M, u] \mapsto \sigma(M, u). \tag{23.3}$$

The assembly map for the trivial family has a universal property. Roughly speaking it says that the source of the assembly map is the best approximation of a given homotopy invariant functor from the left by a generalized homology theory (see [64], [253]). This implies that the map (23.3) above factorizes as a composition

$$\Omega_n(X) \xrightarrow{\sigma'(X)} H_n(X; \mathbf{L}^{\mathrm{sym}}(\mathbb{Z})) \xrightarrow{u_X} H_n(B\pi_1(X); \mathbf{L}^{\mathrm{sym}}(\mathbb{Z})) \xrightarrow{A_{\mathcal{TR}}} L^n(\mathbb{Z}\pi_1(X)),$$

where $\mathbf{L}^{\mathrm{sym}}(\mathbb{Z})$ denotes the symmetric version of the L-theory spectrum of $\mathbb{Z}$ and $u_X \colon X \to B\pi_1(X)$ is the classifying map of X. The symmetric L-theory spectrum is four-periodic, the periodicity isomorphism is geometrically given by crossing with $\mathbb{CP}^2$. Hence we can think of it in the sequel as graded by $\mathbb{Z}/4$. Let $\mathbb{Q}_*$ be the $\mathbb{Z}$-graded vector space, which is $\mathbb{Q}$ in dimensions divisible by four and zero otherwise. Let $\Omega_*(\{\bullet\}) \to \mathbb{Q}_*$ be the map of $\mathbb{Z}$-graded vector spaces which sends the bordism class of an oriented closed $4k$-dimensional manifold to its signature. Then we can form for each CW-complex X the $\mathbb{Z}/4$-graded rational vector space $\Omega_*(X) \otimes_{\Omega_*(\{\bullet\})} \mathbb{Q}_*$. This is in degree $\overline{n} \in \mathbb{Z}/4$ given by $\bigoplus_{p\in\mathbb{Z},\overline{p}=\overline{n}} \Omega_p(X)/ \sim$, where $\sim$ is given by $[M \times N, u \circ \mathrm{pr}_M] \sim \mathrm{sign}(N) \cdot [M, u]$ for $[M, u] \in \Omega_p(X)$ and $[N] \in \Omega_{4k}(\{\bullet\})$.

We obtain for every CW-complex X from $\sigma'(X)$ a map of $\mathbb{Z}/4$-graded rational vector spaces

$$\overline{\sigma}(X) \colon \Omega_*(X) \otimes_{\Omega_*(\{\bullet\})} \mathbb{Q}_* \quad \to \quad H_*(X; \mathbf{L}^{\mathrm{sym}}(\mathbb{Z})) \otimes_\mathbb{Z} \mathbb{Q}. \tag{23.4}$$

It is a transformation of homology theories satisfying the disjoint union axiom and induces an isomorphism for $X = \{\bullet\}$ (see [145, Example 3.4]). Hence $\overline{\sigma}(X)$ is bijective for all CW-complexes X (see for instance [235, 7.55 on page 123]). For each connected CW-complex X the following square commutes:

$$\begin{array}{ccc}
\Omega_*(X) \otimes_\mathbb{Z} \mathbb{Q} & \xrightarrow{\ \sigma(X)\otimes_\mathbb{Z}\mathrm{id}_\mathbb{Q}\ } & L^*(\mathbb{Z}\pi_1(X)) \otimes_\mathbb{Z} \mathbb{Q} \\
{\scriptstyle\mathrm{pr}}\downarrow & {\scriptstyle (A_{\mathcal{TR}}\circ H_n(u_X,\mathbf{L}^{\mathrm{sym}}(\mathbb{Z})))\otimes_\mathbb{Z}\mathrm{id}_\mathbb{Q}}\uparrow & \\
\Omega_*(X) \otimes_{\Omega_*(\{\bullet\})} \mathbb{Q}_* & \xrightarrow[\cong]{\ \overline{\sigma}(X)\ } & H_*(X; \mathbf{L}^{\mathrm{sym}}(\mathbb{Z})) \otimes_\mathbb{Z} \mathbb{Q}
\end{array} \tag{23.5}$$

There is a Chern character (see Theorem 18.28), which is an isomorphism of $\mathbb{Z}/4$-graded rational vector spaces

$$\mathrm{ch}_* \colon H_*(X; \mathbf{L}^{\mathrm{sym}}(\mathbb{Z})) \otimes_\mathbb{Z} \mathbb{Q} \xrightarrow{\cong} \bigoplus_{n\in\mathbb{Z}} H_{*+4n}(X; \mathbb{Q}). \tag{23.6}$$

It turns out that the composition

$$\Omega_*(BG) \xrightarrow{\ \mathrm{pr}\ } \Omega_*(BG) \otimes_{\Omega_*(\{\bullet\})} \mathbb{Q}_* \xrightarrow{\ \overline{\sigma}(BG)\otimes_\mathbb{Z}\mathrm{id}_\mathbb{Q}\ } H_*(BG; \mathbf{L}^{\mathrm{sym}}(\mathbb{Z})) \otimes_\mathbb{Z} \mathbb{Q}$$
$$\xrightarrow{\ \mathrm{ch}\ } \bigoplus_{n\in\mathbb{Z}} H_{*+4n}(BG; \mathbb{Q}) \quad (23.7)$$

sends the class of $[u \colon M \to BG]$ to $u_*(\mathcal{L}(M) \cap [M])$. We conclude from the commutative diagram (23.6) that the composition

$$\bigoplus_{n\in\mathbb{Z}} H_{*+4n}(BG; \mathbb{Q}) \xrightarrow{\ \mathrm{ch}^{-1}\ } H_*(BG; \mathbf{L}^{\mathrm{sym}}(\mathbb{Z})) \otimes_\mathbb{Z} \mathbb{Q} \xrightarrow{\ A_{\mathcal{TR}}\otimes_\mathbb{Z}\mathrm{id}_\mathbb{Q}\ } L^n(\mathbb{Z}G) \otimes_\mathbb{Z} \mathbb{Q}$$

sends $u_*(\mathcal{L}(M) \cap [M])$ to $\sigma(M,u) \otimes_{\mathbb{Z}} 1$. Since $\sigma(M,u)$ is a homotopy invariant, $u_*(\mathcal{L}(M) \cap [M])$ is a homotopy invariant and hence the Novikov Conjecture 1.2 follows for G (see Remark 1.8) if we can show that

$$A_{TR} \colon H_*(BG; \mathbf{L}^{\mathrm{sym}}(\mathbb{Z})) \to L^n(\mathbb{Z}G) \otimes_{\mathbb{Z}} \mathbb{Q}$$

is injective. There is a *symmetrization map*

$$s \colon L_n(\mathbb{Z}G) \to L^n(\mathbb{Z}G)$$

which is an isomorphism after inverting 2 and can also be defined on spectrum level. It yields a commutative diagram

$$
\begin{array}{ccc}
L_n(\mathbb{Z}G) \otimes_{\mathbb{Z}} \mathbb{Q} & \xrightarrow[\cong]{s_n \otimes_{\mathbb{Z}} \mathbb{Q}} & L^n(\mathbb{Z}G) \otimes_{\mathbb{Z}} \mathbb{Q} \\[2mm]
{\scriptstyle A_{TR} \otimes_{\mathbb{Z}} \mathrm{id}_{\mathbb{Q}}} \Big\uparrow & & {\scriptstyle A_{TR} \otimes_{\mathbb{Z}} \mathrm{id}_{\mathbb{Q}}} \Big\uparrow \\[2mm]
H_n(BG; \mathbf{L}(\mathbb{Z})) \otimes_{\mathbb{Z}} \mathbb{Q} & \xrightarrow[\cong]{H_n(\mathrm{id}_{BG};s) \otimes_{\mathbb{Z}} \mathbb{Q}} & H_n(BG; \mathbf{L}^{\mathrm{sym}}(\mathbb{Z})) \otimes_{\mathbb{Z}} \mathbb{Q}
\end{array}
$$

whose horizontal arrows are bijective. Hence the left vertical arrow is injective if and only if the right vertical arrow is. Hence the Novikov Conjecture 1.2 is true for G if the assembly map

$$H_n(BG; \mathbf{L}(\mathbb{Z})) = H_n^G(E_{TR}(G); \mathbf{L}_{\mathbb{Z}}) \xrightarrow{A_{TR}} L_n^{\langle -\infty \rangle}(\mathbb{Z}G)$$

is rationally injective.

(2) This follows from assertion (1) and Lemma 23.1 (2).   □

Remark 23.8. (Identifying various assembly maps). We mention that the assembly map of (9.4)

$$A_m^G \colon h_m(BG; \mathbb{Q}) \to L_m(G) \otimes \mathbb{Q}$$

agrees with the composition

$$h_m(BG; \mathbb{Q}) = \bigoplus_{n \in \mathbb{Z}} H_{*+4n}(BG; \mathbb{Q}) \xrightarrow{\mathrm{ch}^{-1}} H_*(BG; \mathbf{L}^{\mathrm{sym}}(\mathbb{Z})) \otimes_{\mathbb{Z}} \mathbb{Q}$$

$$\xrightarrow{A_{TR}} L_m^{\langle -\infty \rangle}(\mathbb{Z}G) \cong L_m(G) \otimes \mathbb{Q}.$$

up to a non-zero factor in $\mathbb{Q}$. The same is true for the map $\widehat{A}_m^G$ in its form presented in (9.11) and the composition given in (23.7).

Also Lemma 23.2 and Proposition 15.4 are closely related.

23.2 Relating Topological K-Theory and L-Theory

Theorem 23.9 (Relating L-theory and K-theory for C^*-algebras). *For every C^*-algebra A and $n \in \mathbb{Z}$ there is an isomorphism*

$$\operatorname{sign}_n^{(2)} \colon L_n^p(A) \xrightarrow{\cong} K_n(A).$$

For every real C^*-algebra A and $n \in \mathbb{Z}$ there is an isomorphism

$$\operatorname{sign}_n^{(2)} \colon L_n^p(A)[1/2] \xrightarrow{\cong} K_n(A)[1/2].$$

Proof. The details of the proof can be found in [209, Theorem 1.8 and Theorem 1.11]. At least we explain the map in dimension $n = 0$ in the complex case which is essentially given by a C^*-signature. Since $1/2$ is contained in A, the symmetrization map $L_n^p(A) \xrightarrow{\cong} L_p^n(A)$ is an isomorphism. There are maps

$$\operatorname{sign}^{(2)} \colon L_p^0(A) \quad \to \quad K_0(A), \tag{23.10}$$

$$\iota \colon K_0(A) \quad \to \quad L_p^0(A), \tag{23.11}$$

which turn out to be inverse to one another. The map ι of (23.11) above sends the class $[P] \in K_0(A)$ of a finitely generated projective A-module P to the class of $\overline{\mu} \colon P \to P^*$ coming from some inner product μ on P. Such an inner product exists and the class of $\overline{\mu} \colon P \to P^*$ in $L_p^0(A)$ is independent of the choice of the inner product.

Next we define $\operatorname{sign}^{(2)}([a])$ for the class $[a] \in L^0(A)$ represented by a nonsingular symmetric form $a \colon P \to P^*$. Choose a finitely generated projective A-module Q together with an isomorphism $u \colon A^m \to P \oplus Q$. Let $i \colon (A^m)^* \to A^m$ be the standard isomorphism. Let $\overline{a} \colon A^m \to A^m$ be the endomorphism $i \circ u^* \circ (a+0) \circ u$. This is a selfadjoint element in the C^*-algebra $M_m(A)$. We get by spectral theory projections $\chi_{(0,\infty)}(\overline{a}) \colon A^m \to A^m$ and $\chi_{(-\infty,0)}(\overline{a}) \colon A^m \to A^m$. Define P_+ and P_- to be image of $\chi_{(0,\infty)}(\overline{a})$ and $\chi_{(-\infty,0)}(\overline{a})$. Put $\operatorname{sign}^{(2)}([a]) = [P_+] - [P_-]$. We omit the proof that this yields a well defined homomorphism $\operatorname{sign}^{(2)}$. The non-singular symmetric form $a \colon P \to P^*$ is isomorphic to the orthogonal sum of $a_+ \colon P_+ \to P_+^*$ and $a_- \colon P_- \to P_-^*$, where a_+ and $-a_-$ come from inner products. This implies that $\operatorname{sign}^{(2)}$ and ι are inverse to one another. $\qquad\square$

Remark 23.12 (Relevance of C^*-algebras for the Novikov Conjecture). Theorem 23.9 explains why C^*-algebras enter when dealing with the Novikov Conjecture 1.2. If G is a finite group, one can define an isomorphism

$$\operatorname{sign}^{(2)} \colon L_0^p(\mathbb{R}G) \to K_0(\mathbb{R}G)$$

by sending a symmetric non-degenerate G-equivariant form $\alpha \colon P \to P^*$, which can be viewed as a symmetric non-degenerate bilinear pairing $\alpha \colon P \times P \to \mathbb{R}$ compatible with the G-action, to the difference of the class of the subspace P^+ given by the direct sum of the eigenspaces of positive eigenvalues minus the class of the subspace P^- given by the direct sum of the eigenspaces of negative eigenvalues. These spaces P^+ and P^- are G-invariant direct summands in P and therefore define elements in $K_0(\mathbb{R}G)$. This construction does not work for $\mathbb{R}G$ if G is infinite. The idea is to complete $\mathbb{R}G$ to the reduced C^*-algebra $C_r^*(G)$ so that the constructions above do work because we now have functional calculus available.

Lemma 23.13. *Let $\mathcal{F} \subseteq \mathcal{FIN}$ be a family of finite subgroups of G. If the assembly map for $\mathcal{F}$ and topological K-theory*

$$A_{\mathcal{F}} \colon H_n^G(E_{\mathcal{F}}(G); \mathbf{K}^{\mathrm{top}})[\tfrac{1}{2}] \to K_n(C_r^*(G))[\tfrac{1}{2}]$$

is injective, then for an arbitrary decoration j also the map

$$A_{\mathcal{F}} \colon H_n^G(E_{\mathcal{F}}(G); \mathbf{L}_{\mathbb{Z}}^{\langle j \rangle})[\tfrac{1}{2}] \to L_n^{\langle j \rangle}(\mathbb{Z}G)[\tfrac{1}{2}]$$

is injective.

Proof. Recall that after inverting 2 there is no difference between the different decorations and we can hence work with the p-decoration. One can construct for any subfamily $\mathcal{F} \subseteq \mathcal{FIN}$ the commutative diagram [153, Section 7.5]

$$
\begin{array}{ccc}
H_n^G(E_{\mathcal{F}}(G); \mathbf{L}_{\mathbb{Z}}^p[1/2]) & \xrightarrow{A_{\mathcal{F}}^1} & L_n^p(\mathbb{Z}G)[1/2] \\
{\scriptstyle i_1} \downarrow {\scriptstyle \cong} & & {\scriptstyle j_1} \downarrow {\scriptstyle \cong} \\
H_n^G(E_{\mathcal{F}}(G); \mathbf{L}_{\mathbb{Q}}^p[1/2]) & \xrightarrow{A_{\mathcal{F}}^2} & L_n^p(\mathbb{Q}G)[1/2] \\
{\scriptstyle i_2} \downarrow {\scriptstyle \cong} & & {\scriptstyle j_2} \downarrow \\
H_n^G(E_{\mathcal{F}}(G); \mathbf{L}_{\mathbb{R}}^p[1/2]) & \xrightarrow{A_{\mathcal{F}}^3} & L_n^p(\mathbb{R}G)[1/2] \\
{\scriptstyle i_3} \downarrow {\scriptstyle \cong} & & {\scriptstyle j_3} \downarrow \\
H_n^G(E_{\mathcal{F}}(G); \mathbf{L}_{C_r^*(?;\mathbb{R})}^p[1/2]) & \xrightarrow{A_{\mathcal{F}}^4} & L_n^p(C_r^*(G;\mathbb{R}))[1/2] \\
{\scriptstyle i_4} \downarrow {\scriptstyle \cong} & & {\scriptstyle j_4} \downarrow {\scriptstyle \cong} \\
H_n^G(E_{\mathcal{F}}(G); \mathbf{K}_{\mathbb{R}}^{\mathrm{top}}[1/2]) & \xrightarrow{A_{\mathcal{F}}^5} & K_n(C_r^*(G;\mathbb{R}))[1/2] \\
{\scriptstyle i_5} \downarrow & & {\scriptstyle j_5} \downarrow \\
H_n^G(E_{\mathcal{F}}(G); \mathbf{K}_{\mathbb{C}}^{\mathrm{top}}[1/2]) & \xrightarrow{A_{\mathcal{F}}^6} & K_n(C_r^*(G))[1/2]
\end{array}
$$

Here

$$\mathbf{L}_{\mathbb{Z}}^p[1/2], \quad \mathbf{L}_{\mathbb{Q}}^p[1/2], \quad \mathbf{L}_{\mathbb{R}}^p[1/2], \quad \mathbf{L}_{C_r^*(?;\mathbb{R})}[1/2],$$
$$\mathbf{K}_{\mathbb{R}}^{\mathrm{top}}[1/2] \quad \text{and} \quad \mathbf{K}_{\mathbb{C}}^{\mathrm{top}}[1/2]$$

are covariant GROUPOIDS^{inj}-spectra and yield by Lemma 20.14 equivariant homology theories such that the n-th homotopy group of their evaluations at G/H are given by

$$L_n^p(\mathbb{Z}H)[1/2], \quad L_n^p(\mathbb{Q}H)[1/2], \quad L_n^p(\mathbb{R}H)[1/2], \quad L_n^p(C_r^*(H;\mathbb{R}))[1/2],$$
$$K_n(C_r^*(H;\mathbb{R}))[1/2] \quad \text{respectively} \quad K_n(C_r^*(H)[1/2].$$

All horizontal maps are assembly maps induced by the projection pr: $E_{\mathcal{F}}(G) \to \{\bullet\}$. The maps i_k and j_k for $k = 1, 2, 3$ are induced from a change of rings. The isomorphisms i_4 and j_4 come from the general isomorphism for any real C^*-algebra A

$$L_n^p(A)[1/2] \xrightarrow{\cong} K_n(A)[1/2]$$

of Theorem 23.9 and its spectrum version. The maps i_1, j_1, i_2 are isomorphisms by [198, page 376] and [200, Proposition 22.34 on page 252]. The map i_3 is bijective since for a finite group H we have $\mathbb{R}H = C_r^*(H;\mathbb{R})$. The maps i_5 and j_5 are given by extending the scalars from $\mathbb{R}$ to $\mathbb{C}$ by induction. For every real C^*-algebra A the composition

$$K_n(A)[1/2] \to K_n(A \otimes_{\mathbb{R}} \mathbb{C})[1/2] \to K_n(M_2(A))[1/2]$$

is an isomorphism and hence j_5 is split injective. An $\text{Or}(G)$-spectrum version of this argument yields that also i_5 is split injective. $\qquad\square$

Remark 23.14 (Relation between the L-theoretic Farrell–Jones Conjecture and the Baum–Connes Conjecture over $\mathbb{R}$ after inverting 2). One may conjecture that the right vertical maps j_2 and j_3 are isomorphisms and try to prove this directly. Then if we invert 2 everywhere the Baum–Connes Conjecture 20.4 for the real reduced group C^*-algebra, would be equivalent to the Farrell–Jones Isomorphism Conjecture for $L_*(\mathbb{Z}G)[1/2]$.

23.3 The Baum–Connes Conjecture and the Novikov Conjecture

If we combine Lemma 23.2 and Lemma 23.13, we conclude

Corollary 23.15. *The Baum–Connes Conjecture 20.4, or more generally, the rational injectivity of the assembly map appearing in the Baum–Connes Conjecture 20.4, implies the Novikov Conjecture 1.2.*

It is worthwhile to sketch the C^*-theoretic analogue of the proof of Lemma 23.2. The key ingredient is the following commutative diagram of $\mathbb{Z}/4$-graded rational vector spaces

$$
\left(\Omega_*(BG)\otimes_{\Omega_*(*)}\mathbb{Q}\right)_n \xrightarrow[\cong]{\overline{D}} KO_n(BG)\otimes_{\mathbb{Z}}\mathbb{Q} \xrightarrow{j_1} K_n(BG)\otimes_{\mathbb{Z}}\mathbb{Q}
$$

$$
\sigma\downarrow \qquad\qquad A^{\mathbb{R}}_{TR}\otimes_{\mathbb{Z}}\mathrm{id}_{\mathbb{Q}}\downarrow \qquad\qquad A_{TR}\otimes_{\mathbb{Z}}\mathrm{id}_{\mathbb{Q}}\downarrow
$$

$$
L^n(C^*_r(G;\mathbb{R}))\otimes_{\mathbb{Z}}\mathbb{Q} \xrightarrow[\cong]{\mathrm{sign}} KO_n(C^*_r(G;\mathbb{R}))\otimes_{\mathbb{Z}}\mathbb{Q} \xrightarrow{j_2} K_n(C^*_r(G))\otimes_{\mathbb{Z}}\mathbb{Q}
$$

Here are some explanations (see also [148]). The map $\overline{D}$ is induced by the $\mathbb{Z}$-graded homomorphism

$$
D\colon \Omega_n(BG)\to KO_n(BG),
$$

which sends $[r\colon M\to BG]$ to the K-homology class of the signature operator of the covering $\overline{M}\to M$ associated to r. The homological Chern character is an isomorphism of $\mathbb{Z}/4$-graded rational vector spaces

$$
\mathrm{ch}\colon KO_n(BG)\otimes_{\mathbb{Z}}\mathbb{Q} \xrightarrow{\cong} \bigoplus_{k\in\mathbb{Z}} H_{4k+n}(BG;\mathbb{Q}). \tag{23.16}
$$

By the Atiyah–Hirzebruch index theorem the image $\mathrm{ch}\circ D([M,\mathrm{id}\colon M\to M])$ of the K-homology class of the signature operator of M in $K_{\dim(M)}(M)$ under the homological Chern character ch is $\mathcal{L}(M)\cap[M]$. Hence the composition $\mathrm{ch}\circ\overline{D}\colon \left(\Omega_*(BG)\otimes_{\Omega_*(*)}\mathbb{Q}\right)_n\to\bigoplus_{k\in\mathbb{Z}}H_{4k+n}(BG;\mathbb{Q})$ sends $[r\colon M\to BG]\otimes 1$ to the image under $H_*(r)\colon H_*(M;\mathbb{Q})\to H_*(BG;\mathbb{Q})$ of the Poincaré dual $\mathcal{L}(M)\cap[M]\in\bigoplus_{i\geq 0}H_{4i-\dim(M)}(M;\mathbb{Q})$ of the L-class $\mathcal{L}(M)$.

The map $\overline{D}$ is an isomorphism since it is a transformation of homology theories satisfying the disjoint union axiom [145, Example 3.4] and induces an isomorphism for the space consisting of one point. The map σ assigns to $[M,r]$ the class of the associated symmetric Poincaré $C^*_r(G;\mathbb{R})$-chain complex $C_*(\overline{M})\otimes_{\mathbb{Z}G}C^*_r(G;\mathbb{R})$. The map $A^{\mathbb{R}}_{TR}$ resp. A_{TR} are assembly maps for the trivial family TR for the real and the complex case. The map $\mathrm{sign}^{(2)}$ is in dimension $n=0$ mod 4 given by taking the signature of a non-degenerate symmetric bilinear form and has been explained in Theorem 23.9. The maps j_1 and j_2 are given by induction with the inclusion $\mathbb{R}\to\mathbb{C}$ and are injective. Obviously the right square commutes. In order to show that the diagram commutes it suffices to prove this for the outer square. Here the claim follows from the commutative diagram in [122, page 81].

The Novikov Conjecture is equivalent to the statement that $\mathcal{L}(M)\cap[M]\in\bigoplus_{i\geq 0}H_{4i-\dim(M)}(M;\mathbb{Q})$ is homotopy invariant (see Remark 1.8). Since the Chern character ch of (23.16) is bijective, this is equivalent to the homotopy invariance of $\overline{D}([M,r])$. Since $\sigma([M,r])$ is homotopy invariant, the commutative square above shows that the image of $\overline{D}([M,r])$ under the assembly map

$$
A^{\mathbb{R}}_{TR}\otimes_{\mathbb{Z}}\mathrm{id}_{\mathbb{Q}}\colon KO_n(BG)\otimes_{\mathbb{Z}}\mathbb{Q}\to K_n(C^*_r(G))\otimes_{\mathbb{Z}}\mathbb{Q}
$$

is homotopy invariant. Hence the Novikov Conjecture 1.2 holds for G, if $A^{\mathbb{R}}_{TR}\otimes_{\mathbb{Z}}\mathrm{id}_{\mathbb{Q}}$ is injective. But this is true if $A_{TR}\otimes_{\mathbb{Z}}\mathrm{id}_{\mathbb{Q}}$ is injective.

The connection between the Novikov Conjecture and operator theory was initialized by Lusztig [163].

Chapter 24

Miscellaneous

In this chapter we give a survey about the status of the Novikov Conjecture 1.2, Baum–Connes Conjecture 20.4 and the Farrell–Jones Conjecture 20.3, very briefly discuss methods of proof and explain how these conjectures can be used to give rational or integral computations of $K_n(RG)$, $L_n^{\langle j \rangle}(RG)$ and $K_n(C_r^*(G))$.

24.1 Status of the Conjectures

In the following table we list prominent classes of groups and state whether they are known to satisfy the Baum–Connes Conjecture 20.4 or the Farrell–Jones Conjecture 20.3. Some of the classes are redundant. A question mark means that the authors do not know about a corresponding result. The reader should keep in mind that there may exist results of which the authors are not aware. More information and explanations about the status of this conjectures and generalizations of them can be found in [159], where the tables below are taken from.

Remark 24.1 (Groups satisfying the Novikov Conjecture). *All groups appearing in the list below satisfy the Novikov Conjecture 1.2 by Lemma 23.2 (2) and Corollary 23.15.*

type of group	Baum–Connes Conjecture 20.4	Farrell–Jones Conjecture 20.3 for K-theory for $R = \mathbb{Z}$	Farrell–Jones Conjecture 20.3 for L-theory for $R = \mathbb{Z}$
a-T-menable groups	true [111, Theorem 1.1]	?	injectivity is true after inverting 2 (see Lemma 23.13)
amenable groups	true	?	injectivity is true after inverting 2
elementary amenable groups	true	?	true after inverting 2 [83, Theorem 5.2]
virtually poly-cyclic	true	true rationally	true
torsionfree virtually solvable subgroups of $GL(n, \mathbb{C})$	true	true in the range ≤ 1 [83, Theorem 1.1]	true after inverting 2 [83, Corollary 5.3]
discrete subgroups of Lie groups with finitely many path components	injectivity true [224, Theorem 6.1] and [112, Section 4]	?	injectivity is true after inverting 2 (see Lemma 23.13)
subgroups of groups which are discrete cocompact subgroups of Lie groups with finitely many path components	injectivity true	true in the range $n \leq 1$ and true rationally [80, Theorem 2.1 on page 263], [159, Proposition 4.14]	(probably) true [80, Remark 2.1.3 on page 263] Injectivity is true after inverting 2 (see Lemma 23.13)
countable groups admitting a uniform embedding into Hilbert space	injectivity is true [224, Theorem 6.1]		Injectivity is true after inverting 2 (see Lemma 23.13)
G admits an amenable action on some compact space	injectivity is true [110, Theorem 1.1]		Injectivity is true after inverting 2 (see Lemma 23.13)
linear groups	injectivity is true [105])	?	injectivity is true after inverting 2 (see Lemma 23.13)

type of group	Baum–Connes Conjecture 20.4	Farrell–Jones Conjecture 20.3 for K-theory for $R = \mathbb{Z}$	Farrell–Jones Conjecture 20.3 for L-theory for $R = \mathbb{Z}$
torsionfree discrete subgroups of $GL(n, \mathbb{R})$	injectivity is true	true in the range $n \leq 1$ [82]	true [82]
Groups with finite BG and finite asymptotic dimension [257]	injectivity is true [15]	injectivity is true for arbitrary coefficients R [15]	injectivity is true for regular R as coefficients [15]
G acts properly and isometrically on a complete Riemannian manifold M with non-positive sectional curvature	rational injectivity is true [125]	?	rational injectivity is true (see Lemma 23.13)
$\pi_1(M)$ for a complete Riemannian manifold M with non-positive sectional curvature	rational injectivity is true [125]	?	injectivity true [92, Corollary 2.3]
$\pi_1(M)$ for a complete Riemannian manifold M with non-positive sectional curvature which is A-regular	rational injectivity is true	true in the range $n \leq 1$, rationally surjective [82], [120]	true [82]
$\pi_1(M)$ for a complete Riemannian manifold M with pinched negative sectional curvature	rational injectivity is true	true in the range $n \leq 1$, rationally surjective	true
$\pi_1(M)$ for a closed Riemannian manifold M with non-positive sectional curvature	rational injectivity is true	true in the range $n \leq 1$, true rationally [79]	true
$\pi_1(M)$ for a closed Riemannian manifold M with negative sectional curvature	true for all subgroups [172, Theorem 20]	true for all coefficients R [18]	true

type of group	Baum–Connes Conjecture 20.4	Farrell–Jones Conjecture 20.3 for K-theory for $R = \mathbb{Z}$	Farrell–Jones Conjecture 20.3 for L-theory for $R = \mathbb{Z}$
word hyperbolic groups	true for all subgroups [172, Theorem 20]	?	injectivity is true after inverting 2 (see Lemma 23.13)
one-relator groups	true [176, Theorem 5.18]	injectivity is true rationally [17]	true after inverting 2 [17]
torsionfree one-relator groups	true	true for R regular [244, Theorem 19.4 on page 249 and Theorem 19.5 on page 250]	true after inverting 2 [41, Corollary 8]
Haken 3-manifold groups (in particular knot groups)	true [176, Theorem 5.18]	true in the range $n \leq 1$ for R regular [244, Theorem 19.4 on page 249 and Theorem 19.5 on page 250]	true after inverting 2 [41, Corollary 8], [44]
$SL(n, \mathbb{Z})$, $n \geq 3$	injectivity is true	?	injectivity is true after inverting 2 (see Lemma 23.13)
Artin's braid group B_n	true [176, Theorem 5.25], [213]	true in the range $n \leq 1$, true rationally [84]	injectivity is true after inverting 2 (see Lemma 23.13)
pure braid group C_n	true	true in the range $n \leq 1$ [7]	injectivity is true after inverting 2 (see Lemma 23.13)
Thompson's group F	true [73]	?	injectivity is true after inverting 2 (see Lemma 23.13)

Here are some explanations about the classes of groups appearing in the list above.

A group G is *a-T-menable*, or, equivalently, has the *Haagerup property* if G admits a metrically proper isometric action on some affine Hilbert space. *Metrically proper* means that for any bounded subset B the set $\{g \in G \mid gB \cap B \neq \emptyset\}$ is finite. An extensive treatment of such groups is presented in [52]. Any a-T-menable group is countable. The class of a-T-menable groups is closed under taking subgroups, under extensions with finite quotients and under finite products. It is not closed under semi-direct products. Examples of a-T-menable groups are countable amenable groups, countable free groups, discrete subgroups of $SO(n,1)$ and $SU(n,1)$, Coxeter groups, countable groups acting properly on trees, products of trees, or simply connected CAT(0) cubical complexes. A group G has Kazhdan's *property (T)* if, whenever it acts isometrically on some affine Hilbert space, it has a fixed point. An infinite a-T-menable group does not have property (T). Since $SL(n, \mathbb{Z})$ for $n \geq 3$ has property (T), it cannot be a-T-menable.

The *asymptotic dimension* of a proper metric space X is the infimum over all integers n such that for any $R > 0$ there exists a cover $\mathcal{U}$ of X with the property that the diameter of the members of $\mathcal{U}$ is uniformly bounded and every ball of radius R intersects at most $(n + 1)$ elements of $\mathcal{U}$ (see [101, page 28]).

A complete Riemannian manifold M is called *A-regular* if there exists a sequence of positive real numbers $A_0, A_1, A_2, \ldots$ such that $\|\nabla^n K\| \leq A_n$, where $\|\nabla^n K\|$ is the supremum-norm of the n-th covariant derivative of the curvature tensor K. Every locally symmetric space is A-regular since ∇K is identically zero. Obviously every closed Riemannian manifold is A-regular. If M is a pinched negatively curved complete Riemannian manifold, then there is another Riemannian metric for which M is negatively curved complete and A-regular. This fact is mentioned in [82, page 216] and attributed there to Abresch [2] and Shi [222].

A metric space (X, d) admits a *uniform embedding into Hilbert space* if there exist a separable Hilbert space H, a map $f\colon X \to H$ and non-decreasing functions ρ_1 and ρ_2 from $[0, \infty) \to \mathbb{R}$ such that $\rho_1(d(x, y)) \leq \|f(x) - f(y)\| \leq \rho_2(d(x, y))$ for $x, y \in X$ and $\lim_{r \to \infty} \rho_i(r) = \infty$ for $i = 1, 2$. A metric is *proper* if for each $r > 0$ and $x \in X$ the closed ball of radius r centered at x is compact. The question whether a discrete group G equipped with a proper left G-invariant metric d admits a uniform embedding into Hilbert space is independent of the choice of d, since the induced coarse structure does not depend on d [224, page 808]. For more information about groups admitting a uniform embedding into Hilbert space we refer to [71], [105].

The class of finitely generated groups, which for some proper left G-invariant metric embed uniformly into Hilbert space, contains a subclass A, which contains all word hyperbolic groups, finitely generated discrete subgroups of connected Lie groups and finitely generated amenable groups and is closed under semi-direct products [258, Definition 2.1, Theorem 2.2 and Proposition 2.6]. Gromov [103], [104] has announced examples of finitely generated groups which do not admit a uniform embedding into Hilbert space. Details of the construction are described in Ghys [98].

A continuous action of a discrete group G on a compact space X is called *amenable* if there exists a sequence

$$p_n \colon X \to M^1(G) = \{f \colon G \to [0,1] \mid \sum_{g \in G} f(g) = 1\}$$

of weak-$*$-continuous maps such that for each $g \in G$ one has

$$\lim_{n \to \infty} \sup_{x \in X} \|g * (p_n(x) - p_n(g \cdot x))\|_1 = 0.$$

Note that a group G is amenable if and only if its action on the one-point-space is amenable. More information about this notion can be found for instance in [5], [6].

Remark 24.2. The authors have found no information about the status of these conjectures for mapping class groups of higher genus or the group of outer automorphisms of free groups in the literature.

24.2   Methods of Proof

A survey on the methods of proofs of these conjectures can be found in [159, Chapter 7], where also more references are given. We briefly mention that for the Baum–Connes Conjecture 20.4 the key tool is the Dirac-Dual Dirac method [123], [126], [239] and bivariant equivariant KK-theory [124] and that a further break through is based on studying Banach-KK-theory [140], [141], [142]. For the Farrell–Jones Conjecture 20.3 the interpretation of the assembly map as a forget control map allows to use methods from controlled topology [18], [47], [76], [77], [85], [119]. One can also compare K-theory using cyclotomic traces with topological cyclic homology and prove injectivity results for K-theory [27], [160].

24.3   Computations for Finite Groups

In this section we briefly state the computations of $K_n(\mathbb{Z}G)$, $L_n^{\langle -\infty \rangle}(\mathbb{Z}G)$, and $K_n(C_*^r(G))$ for finite groups which will be basic for the computations for arbitrary groups.

24.3.1   Topological K-Theory for Finite Groups

Let G be a finite group. By $r_F(G)$, we denote the number of isomorphism classes of irreducible representations of G over the field F. By $r_\mathbb{R}(G; \mathbb{R})$, $r_\mathbb{R}(G; \mathbb{C})$, respectively $r_\mathbb{R}(G; \mathbb{H})$ we denote the number of isomorphism classes of irreducible real G-representations V, which are of real, of complex respectively of quaternionic type, i.e., $\mathrm{Aut}_{\mathbb{R}G}(V)$ is isomorphic to the field of real numbers $\mathbb{R}$, complex numbers

$\mathbb{C}$ or quaternions $\mathbb{H}$. We mention that $r_{\mathbb{C}}(G)$ is the number of conjugacy classes of elements in G, $r_{\mathbb{R}}(G)$ is the number of $\mathbb{R}$-conjugacy classes of elements in G, where g_1 and g_2 are called $\mathbb{R}$-conjugate if g_1 and g_2 or g_1^{-1} and g_2 are conjugate, and $r_{\mathbb{Q}}(G)$ is the number of conjugacy classes of cyclic subgroups of G, Let $RO(G)$ respectively $R(G)$ be the *real* respectively the *complex representation ring*.

Theorem 24.3 (Topological K-theory of $C_r^*(G)$ for finite groups). *Let G be a finite group.*

(1) *We have*

$$K_n(C_r^*(G)) \cong \begin{cases} R(G) \cong \mathbb{Z}^{r_{\mathbb{C}}(G)} & \text{for } n \text{ even;} \\ 0 & \text{for } n \text{ odd.} \end{cases}$$

(2) *There is an isomorphism of topological K-groups*

$$K_n(C_r^*(G;\mathbb{R})) \cong K_n(\mathbb{R})^{r_{\mathbb{R}}(G;\mathbb{R})} \times K_n(\mathbb{C})^{r_{\mathbb{R}}(G;\mathbb{C})} \times K_n(\mathbb{H})^{r_{\mathbb{R}}(G;\mathbb{H})}.$$

Moreover $K_n(\mathbb{C})$ is 2-periodic with values $\mathbb{Z}$, 0 for $n = 0, 1$, $K_n(\mathbb{R})$ is 8-periodic with values $\mathbb{Z}$, $\mathbb{Z}/2$, $\mathbb{Z}/2$, 0, $\mathbb{Z}$, 0, 0, 0 for $n = 0, 1, \ldots, 7$ and $K_n(\mathbb{H}) = K_{n+4}(\mathbb{R})$ for $n \in \mathbb{Z}$.

Proof. One gets isomorphisms of C^*-algebras

$$C_r^*(G) \cong \prod_{j=1}^{r_{\mathbb{C}}(G)} M_{n_i}(\mathbb{C})$$

and

$$C_r^*(G;\mathbb{R}) \cong \prod_{i=1}^{r_{\mathbb{R}}(G;\mathbb{R})} M_{m_i}(\mathbb{R}) \times \prod_{i=1}^{r_{\mathbb{R}}(G;\mathbb{C})} M_{n_i}(\mathbb{C}) \times \prod_{i=1}^{r_{\mathbb{R}}(G;\mathbb{H})} M_{p_i}(\mathbb{H})$$

from [218, Theorem 7 on page 19, Corollary 2 on page 96, page 102, page 106]. Now the claim follows from Morita invariance and the well-known values for $K_n(\mathbb{R})$, $K_n(\mathbb{C})$ and $K_n(\mathbb{H})$ (see for instance [235, page 216]). $\qquad\square$

24.3.2 Algebraic K-Theory for Finite Groups

Theorem 24.4 (Algebraic K-theory of $\mathbb{Z}G$ for finite groups G). *Let G be a finite group.*

(1) $K_n(\mathbb{Z}G) = 0$ *for $n \leq -2$.*

(2) *We have*

$$K_{-1}(\mathbb{Z}G) \cong \mathbb{Z}^r \oplus (\mathbb{Z}/2)^s,$$

where

$$r = 1 - r_{\mathbb{Q}}(G) + \sum_{p \,|\, |G|} r_{\mathbb{Q}_p}(G) - r_{\mathbb{F}_p}(G)$$

and the sum runs over all primes dividing the order of G.

There is an explicit description of the integer s in terms of global and local Schur indices [49]. *If G contains a normal abelian subgroup of odd index, then $s = 0$;*

(3) *The group $\widetilde{K}_0(\mathbb{Z}G)$ is finite.*

(4) *The group $\mathrm{Wh}(G)$ is a finitely generated abelian group and its rank is $r_\mathbb{R}(G) - r_\mathbb{Q}(G)$. If G is cyclic, $\mathrm{Wh}(G)$ is torsionfree.*

(5) *The groups $K_n(\mathbb{Z}G)$ are finitely generated for all $n \in \mathbb{Z}$.*

Proof. (1) and (2) are proved in [49].

(3) is proved in [233, Proposition 9.1 on page 573].

(4) This is proved for instance in [184].

(5) See [138], [191]. □

24.3.3 Algebraic L-Theory for Finite Groups

Theorem 24.5 (Algebraic L-theory of $\mathbb{Z}G$ for finite groups). *Let G be a finite group. Then*

(1) *For each $j \le 1$ the groups $L_n^{\langle j \rangle}(\mathbb{Z}G)$ are finitely generated as abelian groups and contain no p-torsion for odd primes p. Moreover, they are finite for odd n.*

(2) *For every decoration $\langle j \rangle$ we have*

$$L_n^{\langle j \rangle}(\mathbb{Z}G)[1/2] \;\cong\; L_n^{\langle j \rangle}(\mathbb{R}G)[1/2] \;\cong\; \begin{cases} \mathbb{Z}[1/2]^{r_\mathbb{R}(G)} & n \equiv 0 \ (4); \\ \mathbb{Z}[1/2]^{r_\mathbb{C}(G)} & n \equiv 2 \ (4); \\ 0 & n \equiv 1,3 \ (4). \end{cases}$$

(3) *If G has odd order and n is odd, then $L_n^\varepsilon(\mathbb{Z}G) = 0$ for $\varepsilon = p, h, s$.*

Proof. (1) See for instance [108].

(2) See [200, Proposition 22.34 on page 253].

(3) See [11] or [108]. □

24.4 Rational Computations

In Remarks 21.12, 21.14 and 22.3 we have given explicit formulas for $K_n(\mathbb{Z}G) \otimes_\mathbb{Z} \mathbb{Q}$, $L_n^{\langle -\infty \rangle}(\mathbb{Z}G) \otimes_\mathbb{Z} \mathbb{Q}$, and $K_n(C_r^*(G)) \otimes_\mathbb{Z} \mathbb{Q}$ for a torsionfree group G provided that the Farrell–Jones Conjecture for torsionfree groups 21.11 and the Baum–Connes Conjecture for torsionfree groups 22.2 are true. They were based on the existence of a Chern character for (non-equivariant) homology theories (see Theorem 18.28).

Next we want to state what one gets in the general case provided the Farrell–Jones Conjecture 20.3 and the Baum–Connes Conjecture 20.4 are true. The key ingredient is the *equivariant Chern character* [154].

In the sequel let $(\mathcal{FCYC})$ be the set of conjugacy classes (C) for finite cyclic subgroups $C \subseteq G$. For $H \subseteq G$ let $N_G H = \{g \in G \mid gHg^{-1} = H\}$ be its *normalizer*, let $Z_G H = \{g \in G \mid ghg^{-1} = h \text{ for } h \in H\}$ be its *centralizer*, and put

$$W_G H := N_G H/(H \cdot Z_G H),$$

where $H \cdot Z_G H$ is the normal subgroup of $N_G H$ consisting of elements of the form hu for $h \in H$ and $u \in Z_G H$. Notice that $W_G H$ is finite if H is finite.

Recall that the *Burnside ring* $A(G)$ of a finite group is the Grothendieck group associated to the abelian monoid of isomorphism classes of finite G-sets with respect to the disjoint union. The ring multiplication comes from the cartesian product. The zero element is represented by the empty set, the unit is represented by $G/G = \{\bullet\}$. For a finite group G the abelian groups $K_q(C_r^*(G))$, $K_q(RG)$ and $L^{\langle -\infty \rangle}(RG)$ become modules over $A(G)$ because these functors come with a Mackey structure and $[G/H]$ acts by $\mathrm{ind}_H^G \circ \mathrm{res}_G^H$.

We obtain a ring homomorphism

$$\chi^G \colon A(G) \to \prod_{(H) \in \mathcal{FIN}} \mathbb{Z}, \quad [S] \mapsto (|S^H|)_{(H) \in \mathcal{FIN}}$$

which sends the class of a finite G-set S to the element given by the cardinalities of the H-fixed point sets. This is an injection with finite cokernel. We obtain an isomorphism of $\mathbb{Q}$-algebras

$$\chi_{\mathbb{Q}}^G := \chi^G \otimes_{\mathbb{Z}} \mathrm{id}_{\mathbb{Q}} \colon A(G) \otimes_{\mathbb{Z}} \mathbb{Q} \xrightarrow{\cong} \prod_{(H) \in (\mathcal{FIN})} \mathbb{Q}. \tag{24.6}$$

For a finite cyclic group C let

$$\theta_C \in A(C) \otimes_{\mathbb{Z}} \mathbb{Z}[1/|C|] \tag{24.7}$$

be the element which is sent under the isomorphism $\chi_{\mathbb{Q}}^C \colon A(C) \otimes_{\mathbb{Z}} \mathbb{Q} \xrightarrow{\cong} \prod_{(H) \in \mathcal{FIN}} \mathbb{Q}$ of (24.6) to the element, whose entry is one if $(H) = (C)$ and is zero if $(H) \neq (C)$. Notice that θ_C is an idempotent. In particular we get a direct summand $\theta_C \cdot K_q(C_r^*(C)) \otimes_{\mathbb{Z}} \mathbb{Q}$ in $K_q(C_r^*(C)) \otimes_{\mathbb{Z}} \mathbb{Q}$ and analogously for $K_q(RC) \otimes_{\mathbb{Z}} \mathbb{Q}$ and $L^{\langle -\infty \rangle}(RG) \otimes_{\mathbb{Z}} \mathbb{Q}$.

24.4.1 Rationalized Topological K-Theory for Infinite Groups

The next result is taken from [156, Theorem 0.4 and page 127]. Let Λ^G be the ring $\mathbb{Z} \subseteq \Lambda^G \subseteq \mathbb{Q}$ which is obtained from $\mathbb{Z}$ by inverting the orders of the finite subgroups of G.

Theorem 24.8 (Rational computation of topological K-theory for infinite groups).
Suppose that the group G satisfies the Baum–Connes Conjecture 20.4. Then there is an isomorphism

$$\bigoplus_{p+q=n} \bigoplus_{(C)\in(\mathcal{FCYC})} K_p(BZ_GC) \otimes_{\mathbb{Z}[W_GC]} \theta_C \cdot K_q(C_r^*(C)) \otimes_{\mathbb{Z}} \Lambda^G$$

$$\xrightarrow{\cong} K_n(C_r^*(G)) \otimes_{\mathbb{Z}} \Lambda^G.$$

If we tensor with $\mathbb{Q}$, we get an isomorphism

$$\bigoplus_{p+q=n} \bigoplus_{(C)\in(\mathcal{FCYC})} H_p(BZ_GC;\mathbb{Q}) \otimes_{\mathbb{Q}[W_GC]} \theta_C \cdot K_q(C_r^*(C)) \otimes_{\mathbb{Z}} \mathbb{Q}.$$

$$\xrightarrow{\cong} K_n(C_r^*(G)) \otimes_{\mathbb{Z}} \mathbb{Q}.$$

24.4.2   Rationalized Algebraic K-Theory for Infinite Groups

Recall that for the rational computation of the algebraic K-theory of the integral group ring we have already explained in Remark 21.23 that in the Farrell-Jones Conjecture we can reduce to the family of finite subgroups. A reduction to the family of finite subgroups also works if the coefficient ring is a regular $\mathbb{Q}$-algebra (see Lemma 21.24 (2)). The next result is a variation of [154, Theorem 0.4].

Theorem 24.9 (Rational computation of algebraic K-theory). *Suppose that the group G satisfies the Farrell–Jones Conjecture 20.3 for the algebraic K-theory of RG, where either $R = \mathbb{Z}$ or R is a regular ring with $\mathbb{Q} \subseteq R$. Then we get an isomorphism*

$$\bigoplus_{p+q=n} \bigoplus_{(C)\in(\mathcal{FCYC})} H_p(BZ_GC;\mathbb{Q}) \otimes_{\mathbb{Q}[W_GC]} \theta_C \cdot K_q(RC) \otimes_{\mathbb{Z}} \mathbb{Q}$$

$$\xrightarrow{\cong} K_n(RG) \otimes_{\mathbb{Z}} \mathbb{Q}.$$

Example 24.10 (The comparison map from algebraic to topological K-theory).
If we consider $R = \mathbb{C}$ as coefficient ring and apply $- \otimes_{\mathbb{Z}} \mathbb{C}$ instead of $- \otimes_{\mathbb{Z}} \mathbb{Q}$, the formulas simplify. Suppose that G satisfies the Baum–Connes Conjecture 20.4 and the Farrell–Jones Conjecture 20.3 for algebraic K-theory with $\mathbb{C}$ as coefficient ring. Let $\mathrm{con}(G)_f$ be the set of conjugacy classes (g) of elements $g \in G$ of finite order. We denote for $g \in G$ by $\langle g \rangle$ the cyclic subgroup generated by g. Then we get the following commutative square, whose horizontal maps are isomorphisms and whose vertical maps are induced by the obvious change of theory homomorphisms

(see [154, Theorem 0.5])

$$
\begin{array}{ccc}
\bigoplus_{p+q=n} \bigoplus_{(g)\in\mathrm{con}(G)_f} H_p(Z_G\langle g\rangle; \mathbb{C}) \otimes_{\mathbb{Z}} K_q(\mathbb{C}) & \xrightarrow{\;\cong\;} & K_n(\mathbb{C}G) \otimes_{\mathbb{Z}} \mathbb{C} \\
\downarrow & & \downarrow \\
\bigoplus_{p+q=n} \bigoplus_{(g)\in\mathrm{con}(G)_f} H_p(Z_G\langle g\rangle; \mathbb{C}) \otimes_{\mathbb{Z}} K_q^{\mathrm{top}}(\mathbb{C}) & \xrightarrow{\;\cong\;} & K_n(C_r^*(G)) \otimes_{\mathbb{Z}} \mathbb{C}
\end{array}
$$

Example 24.11 (A formula for $K_0(\mathbb{Z}G) \otimes_{\mathbb{Z}} \mathbb{Q}$). Suppose that the Farrell–Jones Conjecture is true rationally for $K_0(\mathbb{Z}G)$, i.e., the assembly map

$$
A_{\mathcal{V}\mathcal{C}\mathcal{Y}}\colon H_0^G(E_{\mathcal{V}\mathcal{C}\mathcal{Y}}(G); \mathbf{K}_{\mathbb{Z}}) \otimes_{\mathbb{Z}} \mathbb{Q} \to K_0(\mathbb{Z}G) \otimes_{\mathbb{Z}} \mathbb{Q}
$$

is an isomorphism. Then we obtain

$$
\widetilde{K}_0(\mathbb{Z}G) \otimes_{\mathbb{Z}} \mathbb{Q} \cong \bigoplus_{(C)\in(\mathcal{FCY}\mathcal{C})} H_1(BZ_G C; \mathbb{Q}) \otimes_{\mathbb{Q}[W_G C]} \theta_C \cdot K_{-1}(RC) \otimes_{\mathbb{Z}} \mathbb{Q}.
$$

Notice that $\widetilde{K}_0(\mathbb{Z}G) \otimes_{\mathbb{Z}} \mathbb{Q}$ contains only contributions from $K_{-1}(\mathbb{Z}C) \otimes_{\mathbb{Z}} \mathbb{Q}$ for finite cyclic subgroups $C \subseteq G$.

24.4.3 Rationalized Algebraic L-Theory for Infinite Groups

Here is the L-theory analogue of the results above. Compare [154, Theorem 0.4].

Theorem 24.12 (Rational computation of algebraic L-theory for infinite groups). *Suppose that the group G satisfies the Farrell–Jones Conjecture 20.3 for L-theory. Then we get for all $j \in \mathbb{Z}, j \leq 1$ an isomorphism*

$$
\bigoplus_{p+q=n} \bigoplus_{(C)\in(\mathcal{FCY}\mathcal{C})} H_p(BZ_G C; \mathbb{Q}) \otimes_{\mathbb{Q}[W_G C]} \theta_C \cdot L_q^{\langle j\rangle}(RC) \otimes_{\mathbb{Z}} \mathbb{Q}
$$

$$
\xrightarrow{\;\cong\;} L_n^{\langle j\rangle}(RG) \otimes_{\mathbb{Z}} \mathbb{Q}.
$$

Remark 24.13 (Separation of variables). Theorems 24.8, 24.9 and 24.12 support the following general principle *separation of variables* for the computation of K and L-groups of the group ring RG or reduced C^*-algebra $C_r^*(G)$ of a group G. Namely, there is a group homology part which is independent of the coefficient ring R and the K- or L-theory under consideration and a part depending only on the values of the theory under consideration on RC or $C_r^*(C)$ for all finite cyclic subgroups $C \subseteq G$.

24.5 Integral Computations

In contrast to the rational case no general pattern for integral calculations is known or expected. In algebraic K and L-theory the Nil-terms and UNil-terms are extremely hard to determine and the algebraic K-theory of the ring of integers is not yet fully understood.

Concrete calculations are usually based on a good understanding of the space $E_{\mathcal{FIN}}(G)$ usually coming from some geometric input. Notice however, that these spaces can be as complicated as possible in general. Namely, for any CW-complex X there exists a group G such that $G\backslash E_{\mathcal{FIN}}(G)$ and X are homotopy equivalent [146].

We mention at least one situation where a certain class of groups can be treated simultaneously.

Let $\mathcal{MFIN}$ be the subfamily of $\mathcal{FIN}$ consisting of elements in $\mathcal{FIN}$ which are maximal in $\mathcal{FIN}$. Consider the following assertions on the group G.

(M) $M_1, M_1 \in \mathcal{MFIN}, M_1 \cap M_2 \neq 1 \Rightarrow M_1 = M_2$;

(NM) $M \in \mathcal{MFIN} \Rightarrow N_G M = M$;

(VCL$_I$) If V is an infinite virtually cyclic subgroup of G, then V is of type I (see Lemma 21.19);

(FJK$_N$) The Isomorphism Conjecture of Farrell–Jones for algebraic K-theory 20.3 is true for $\mathbb{Z}G$ in the range $n \leq N$ for a fixed element $N \in \mathbb{Z} \amalg \{\infty\}$, i.e., the assembly map $A: \mathcal{H}_n^G(E_{\mathcal{VCYC}}(G); \mathbf{K}_R) \xrightarrow{\cong} K_n(RG)$ is bijective for $n \in \mathbb{Z}$ with $n \leq N$.

Let $\widetilde{K}_n(C_r^*(H))$ be the cokernel of the map $K_n(C_r^*(\{1\})) \to K_n(C_r^*(H))$ and $\overline{L}_n^{\langle j \rangle}(RG)$ be the cokernel of the map $L_n^{\langle j \rangle}(R) \to L_n^{\langle j \rangle}(RG)$. This coincides with $\widetilde{L}_n^{\langle j \rangle}(R)$, which is the cokernel of the map $L_n^{\langle j \rangle}(\mathbb{Z}) \to L_n^{\langle j \rangle}(R)$ if $R = \mathbb{Z}$ but not in general. Denote by $\mathrm{Wh}_n^R(G)$ the n-th Whitehead group of RG which is the $(n-1)$-th homotopy group of the homotopy fiber of the assembly map $BG_+ \wedge \mathbf{K}(R) \to \mathbf{K}(RG)$. It agrees with the previous defined Whitehead group $\mathrm{Wh}(G)$ if $R = \mathbb{Z}$ and $n = 1$. The next result is taken from [65, Theorem 4.1].

Theorem 24.14. *Let $\mathbb{Z} \subseteq \Lambda \subseteq \mathbb{Q}$ be a ring such that the order of any finite subgroup of G is invertible in Λ. Let $(\mathcal{MFIN})$ be the set of conjugacy classes (H) of subgroups of G such that H belongs to $\mathcal{MFIN}$. Then:*

(1) *If G satisfies (M), (NM) and the Baum–Connes Conjecture 20.4, then for $n \in \mathbb{Z}$ there is an exact sequence of topological K-groups*

$$0 \to \bigoplus_{(H) \in (\mathcal{MFIN})} \widetilde{K}_n(C_r^*(H)) \to K_n(C_r^*(G)) \to K_n(G\backslash E_{\mathcal{FIN}}(G)) \to 0,$$

which splits after applying $- \otimes_{\mathbb{Z}} \Lambda$.

(2) *If G satisfies (M), (NM), (VCL$_I$) and the L-theory part of the Farrell–Jones Conjecture 20.3, then for all $n \in \mathbb{Z}$ there is an exact sequence*

$$\ldots \to H_{n+1}(G \backslash E_{\mathcal{FIN}}(G); \mathbf{L}^{\langle -\infty \rangle}(R)) \to \bigoplus_{(H) \in (\mathcal{MFIN})} \overline{L}_n^{\langle -\infty \rangle}(RH)$$

$$\to L_n^{\langle -\infty \rangle}(RG) \to H_n(G \backslash E_{\mathcal{FIN}}(G); \mathbf{L}^{\langle -\infty \rangle}(R)) \to \ldots$$

It splits after applying $- \otimes_{\mathbb{Z}} \Lambda$, more precisely

$$L_n^{\langle -\infty \rangle}(RG) \otimes_{\mathbb{Z}} \Lambda \to H_n(G \backslash E_{\mathcal{FIN}}(G); \mathbf{L}^{\langle -\infty \rangle}(R)) \otimes_{\mathbb{Z}} \Lambda$$

is a split-surjective map of Λ-modules.

(3) *If G satisfies (M), (NM), and the Farrell–Jones Conjecture 20.3 for $L_n(RG)[1/2]$, then the conclusion of assertion 2 still holds if we invert 2 everywhere. Moreover, in the case $R = \mathbb{Z}$ the sequence reduces to a short exact sequence*

$$0 \to \bigoplus_{(H) \in (\mathcal{MFIN})} \widetilde{L}_n^{\langle j \rangle}(\mathbb{Z}H)[\tfrac{1}{2}] \to L_n^{\langle j \rangle}(\mathbb{Z}G)[\tfrac{1}{2}]$$

$$\to H_n(G \backslash E_{\mathcal{FIN}}(G); \mathbf{L}(\mathbb{Z})[\tfrac{1}{2}]) \to 0,$$

which splits after applying $- \otimes_{\mathbb{Z}[\frac{1}{2}]} \Lambda[\tfrac{1}{2}]$.

(4) *If G satisfies (M), (NM), and (FJK$_N$), then there is for $n \in \mathbb{Z}, n \leq N$ an isomorphism*

$$H_n(E_{\mathcal{VCYC}}(G), E_{\mathcal{FIN}}(G); \mathbf{K}_R) \oplus \bigoplus_{(H) \in (\mathcal{MFIN})} \mathrm{Wh}_n^R(H) \xrightarrow{\cong} \mathrm{Wh}_n^R(G),$$

where $\mathrm{Wh}_n^R(H) \to \mathrm{Wh}_n^R(G)$ is induced by the inclusion $H \to G$.

Remark 24.15. In [65] it is explained that the following classes of groups do satisfy the assumption appearing in Theorem 24.14 and what the conclusions are in the case $R = \mathbb{Z}$. Some of these cases have been treated earlier in [25], [162].

- Extensions $1 \to \mathbb{Z}^n \to G \to F \to 1$ for finite F such that the conjugation action of F on $\mathbb{Z}^n$ is free outside $0 \in \mathbb{Z}^n$;

- Fuchsian groups F;

- One-relator groups G.

Theorem 24.14 is generalized in [157] in order to treat for instance the semi-direct product of the discrete 3-dimensional Heisenberg group by $\mathbb{Z}/4$. For this group $G \backslash E_{\mathcal{FIN}}(G)$ is S^3.

A calculation for 2-dimensional crystallographic groups and more general cocompact NEC-groups is presented in [162] (see also [185]). For these groups the orbit spaces $G \backslash E_{\mathcal{FIN}}(G)$ are compact surfaces possibly with boundary.

Example 24.16. Let F be a cocompact Fuchsian group with presentation

$$F = \langle a_1, b_1, \ldots, a_g, b_g, c_1, \ldots, c_t \mid$$
$$c_1^{\gamma_1} = \ldots = c_t^{\gamma_t} = c_1^{-1} \cdots c_t^{-1}[a_1, b_1] \cdots [a_g, b_g] = 1 \rangle$$

for integers $g, t \geq 0$ and $\gamma_i > 1$. Then $G \backslash E_{\mathcal{FIN}}(G)$ is a closed orientable surface of genus g. The following is a consequence of Theorem 24.14 (see [162] for more details).

- There are isomorphisms

$$K_n(C_r^*(F)) \cong \begin{cases} \left(2 + \sum_{i=1}^{t}(\gamma_i - 1)\right) \cdot \mathbb{Z} & n = 0; \\ (2g) \cdot \mathbb{Z} & n = 1. \end{cases}$$

- The inclusions of the maximal subgroups $\mathbb{Z}/\gamma_i = \langle c_i \rangle$ induce an isomorphism

$$\bigoplus_{i=1}^{t} \mathrm{Wh}_n(\mathbb{Z}/\gamma_i) \xrightarrow{\cong} \mathrm{Wh}_n(F)$$

for $n \leq 1$.

- There are isomorphisms

$$L_n(\mathbb{Z}F)[1/2] \cong \begin{cases} \left(1 + \sum_{i=1}^{t} \left[\frac{\gamma_i}{2}\right]\right) \cdot \mathbb{Z}[1/2] & n \equiv 0 \ (4); \\ (2g) \cdot \mathbb{Z}[1/2] & n \equiv 1 \ (4); \\ \left(1 + \sum_{i=1}^{t} \left[\frac{\gamma_i - 1}{2}\right]\right) \cdot \mathbb{Z}[1/2] & n \equiv 2 \ (4); \\ 0 & n \equiv 3 \ (4), \end{cases}$$

where $[r]$ for $r \in \mathbb{R}$ denotes the largest integer less than or equal to r.

From now on suppose that each γ_i is odd. Then the number m above is odd and we get for $\varepsilon = p$ and s

$$L_n^\varepsilon(\mathbb{Z}F) \cong \begin{cases} \mathbb{Z}/2 \oplus \left(1 + \sum_{i=1}^{t} \frac{\gamma_i - 1}{2}\right) \cdot \mathbb{Z} & n \equiv 0 \ (4); \\ (2g) \cdot \mathbb{Z} & n \equiv 1 \ (4); \\ \mathbb{Z}/2 \oplus \left(1 + \sum_{i=1}^{t} \frac{\gamma_i - 1}{2}\right) \cdot \mathbb{Z} & q \equiv 2 \ (4); \\ (2g) \cdot \mathbb{Z}/2 & n \equiv 3 \ (4). \end{cases}$$

For $\varepsilon = h$ we do not know an explicit formula. The problem is that no general formula is known for the 2-torsion contained in $\widetilde{L}_{2q}^h(\mathbb{Z}[\mathbb{Z}/m])$, for m odd, since it is given by the term $\widehat{H}^2(\mathbb{Z}/2; \widetilde{K}_0(\mathbb{Z}[\mathbb{Z}/m]))$, see [12, Theorem 2].

Chapter 25

Exercises

0.1. Show that the manifold $N(M)$ appearing in Example 0.2 satisfies $\pi_1 \cong \mathbb{Z}\oplus\mathbb{Z}$, $\pi_2 = 0$ and $w_2 = 0$.

0.2. Prove (0.5).

0.3. Show that the total spaces of the two S^3-bundles over T^2 are not homotopy equivalent.

1.1. Let M be a closed orientable 8-dimensional smooth manifold. Suppose that $H^4(M;\mathbb{Q}) = 0$. Show that then all rational Pontrjagin classes vanish.

1.2. Let M and N be closed aspherical manifolds. Suppose that the Borel Conjecture 1.10 holds for $\pi_1(M)$. Show that the following assertions are equivalent for a map $f\colon M \to N$ and positive integer n:

(1) f is homotopic to a n-sheeted finite covering.

(2) The group homomorphism $\pi_1(f)$ is injective and its image has index n in $\pi_1(N)$.

1.3. Let M be a closed oriented $(4k + l)$-dimensional manifold and let B be an aspherical closed oriented l-dimensional manifold. Let $u\colon M \to B$ be a map. Choose a base point $y \in B$. Choose a map $f\colon M \to B$ which is transversal to y and homotopic to u. Then $N = f^{-1}(y)$ is a closed $4k$-dimensional manifold, which inherits an orientation from M and N.

Show that the higher signature $\text{sign}_{[B]}(M, u)$ associated to the cohomological fundamental class $[B] \in H_l(B)$ of B agrees with the signature $\text{sign}(N)$ of N.

2.1. Prove that $\Omega_n(X; E)$ is a group.

2.2.   Construct the exact sequence

$$\ldots \Omega_m(Y; E|_Y) \to \Omega_m(X; E) \to \Omega_m(X, Y; E) \to \Omega_{m-1}(Y; E|_Y) \to \ldots.$$

2.3.   Prove excision: for $A \subseteq Y$ such that $\bar{A}$ is contained in the interior of Y the inclusion induces an isomorphism

$$\Omega_m(X - A, Y - A; E|_{Y-A}) \cong \Omega_m(X, Y; E).$$

2.4.   Give the details of the proof of Proposition 2.3.

3.1.   Show that the homotopy type of a simply connected closed 4-manifold M is determined by the intersection form on $H_2(M)$.

3.2.   Show that the signature of the tensor product of two symmetric bilinear forms over $\mathbb{Q}$ is the product of the signatures.

4.1.   Prove that $\mathbb{CP}^2$ equipped with the opposite orientation has no complex structure.

4.2.   Prove the signature theorem in dimension 8.

4.3.   Prove that the signature of finite coverings $p \colon N \to M$ of degree k of closed oriented smooth manifolds is multiplicative:

$$\sigma(N) = k\sigma(M).$$

5.1.   Let F be a field and let V be the F-vector spaces $\bigoplus_{n \in \mathbb{Z}} F$. Let R be the ring of F-endomorphisms of V. Show that the R-modules $R \oplus R$ and R are R-isomorphic. Prove that $K_0(R) = 0$.

5.2.   Construct a ring R and a finitely generated projective R-module P such that for no natural numbers m and n there is an R-isomorphism $P \oplus R^m \cong R^n$.

5.3.   Let R_0 and R_1 be rings and $\mathrm{pr}_i \colon R_0 \times R_1 \to R_i$ be the projection for $i = 0, 1$. Show for $n = 0, 1$ that

$$(\mathrm{pr}_0)_* \times (\mathrm{pr}_2)_* \colon K_n(R_0 \times R_1) \xrightarrow{\cong} K_0(R_0) \times K_n(R_1)$$

is bijective.

5.4.   Show that $1 - t - t^{-1}$ is a unit in $\mathbb{Z}[\mathbb{Z}/5]$. Conclude that $\mathrm{Wh}(\mathbb{Z}/5)$ contains an element of infinite order.

6.1.   Let P_* be a finite projective R-chain complex. Define its finiteness obstruction $o(P_*) \in K_0(R)$ by $\sum_{n \geq 0}[P_n]$. Its reduced finiteness obstruction $\tilde{o}(P_*) \in \tilde{K}_0(R)$ is the image of $o(P_*)$ under the canonical projection. Prove:

(1) $o(P_*)$ depends only on the R-chain homotopy class of P_*.

(2) Let $0 \to P_* \to P'_* \to P''_* \to 0$ be an exact sequence of finite projective R-chain complexes. Then

$$o(P_*) - o(P'_*) + o(P''_*) = 0.$$

(3) The R-chain complex P_* is R-chain homotopy equivalent to a finite free R-chain complex if and only if $\tilde{o}(P_*) = 0$.

6.2.   Let $f\colon X \to Y$ be a homotopy equivalence of finite CW-complexes. Show that $f \times \mathrm{id}_{S^1}\colon X \times S^1 \to Y \times S^1$ is a simple homotopy equivalence.

6.3.   Let $f\colon X \to Y$ be a homotopy equivalence of finite CW-complexes. Let $i\colon X \to \mathrm{cyl}(f)$ and $j\colon Y \to \mathrm{cyl}(f)$ be the canonical inclusions and $p\colon \mathrm{cyl}(f) \to Y$ be the canonical projection. Show that $\tau(j) = 0$ and $p_*(\tau(i)) = \tau(f)$.

7.1.   Show that there exists a h-cobordism, which becomes trivial after crossing with S^n for odd n, but is not trivial after crossing with S^n for even n.

7.2.   Let $(W; M_0, f_0, M_1, f_1)$ and $(V; M_1, g_1, M_2, g_2)$ be two h-cobordisms. Let Z be obtained from W and V by glueing along the diffeomorphism $g_1 \circ f_1^{-1}\colon \partial_1 W \xrightarrow{\cong} \partial_0 V$. Show that (Z, M_0, f_0, M_2, g_2) is a h-cobordism and that we get in $\mathrm{Wh}(M_0)$

$$\tau(Z; M_0, f_0, M_2, g_2) = \tau(W; M_0, f_0, M_1, f_1) + u(\tau(V; M_1, g_1, M_2, g_2)),$$

where $u\colon \mathrm{Wh}(\pi_1(M_1)) \to \mathrm{Wh}(\pi_1(M_0))$ is the map induced by the various homotopy equivalences $f_1, \partial_1 W \to W, \partial_0 W \to W$ and f_0 in the obvious way.

7.3.   Put $N := \left(S^1 \times S^{n-1} \sharp S^1 \times S^{n-1}\right) \times \mathbb{CP}^2$ for some $n \geq 3$. Let M be a closed manifold which is h-cobordant to N. Show that M and N are diffeomorphic.

8.1.   Let M be a closed manifold which is obtained from the empty set by attaching two handles. Show that M is homeomorphic to a sphere.

8.2.   Let $\partial_0 W$ be a closed smooth manifold of dimension $(n-1)$. Suppose that the compact smooth n-dimensional manifold W is obtained from $\partial W_0 \times [0,1]$ by attaching one q-handle (φ^q), i.e., $W = \partial_0 W \times [0,1] + (\varphi^q)$. Let $\partial_1 W$ be $\partial W - \partial_0 W$.

Show that W can be obtained from $\partial_1 W \times [0,1]$ by attaching a $(n-q)$-handle, i.e., $W = \partial_1 W \times [0,1] + (\psi^{n-q})$.

9.1.   Compute w_2 of the Kummer surface. Show that the complement of a point in the Kummer surface is stably parallelizable.

9.2.   Compute the signature of the Kummer surface.

10.1. Add details to the computation of the fundamental group and the homology groups of the trace of surgery T.

10.2. Show (notations as in Chapter 10):

$$\langle x \cup p_1(V'(d)), [V'(d)]\rangle - 4 \cdot \langle x^3, [V'(d)]\rangle \;=\; d(5 - d^2) - 4d.$$

11.1. Prove that W is an h-cobordism if and only if for $i = 0, 1$:

$$i_* : \pi_1(M_i) \xrightarrow{\cong} \pi_1(W)$$

and if for all j:

$$i_* : H_j(\widetilde{M_i}) \xrightarrow{\cong} H_j(\widetilde{W}).$$

11.2. Prove that if $r > k$, a stably trivial r-dimensional vector bundle over S^k is trivial.

12.1. Show that
$$\mathrm{sign} : L^h_{4m}(1) \to \mathbb{Z}$$
is injective and non-trivial.

12.2. Show that each element in the image of
$$\mathrm{sign} : L^h_{4m}(1) \to \mathbb{Z}$$
is divisible by 2.

12.3. Prove that $L^h_{4m+2}(1)$ is non-trivial.

13.1. Prove that if all homology groups of the universal covering of a finite CW-complex X, except perhaps one, are trivial, this homology group is a stably free $\mathbb{Z}[\pi_1(W)]$-module.

13.2. Prove that if $R = \mathbb{Z}[\pi_1(W)]$, then $R/\{a - \bar{a}\} = \Sigma_{g^2=1}\mathbb{Z}g \oplus \Sigma_{\{g,g^{-1}\}_{g^2 \neq 1}}\mathbb{Z}$, a torsionfree abelian group.

13.3. Consider the kernel of the map

$$H_k(\widetilde{S}) \to H_k(\widetilde{T}, \widetilde{M_0} + \widetilde{M_1})$$

(notation as in Chapter 13, the definition of the surgery obstruction). Show that this is a free direct summand of half rank, on which the intersection form and quadratic refinement of the hyperbolic form vanish.

14.1. Show that a stably trivial $2k$-dimensional real vector bundle over S^{2k} is trivial if and only if the Euler class vanishes.

14.2. (Notation as in Chapter 14, the proof of the main theorem) Show that the map
$$H_k(W; \mathbb{Z}[\pi_1]) \to H_{k+1}(W, T; \mathbb{Z}[\pi_1]) \cong \mathbb{Z}[\pi_1(x)]$$
agrees with the map which maps $\alpha \in H_k(W; \mathbb{Z}[\pi_1])$ to $\lambda(\alpha, e_1)$.

14.3. Compute $\Omega_6^{Spin}(T)$, where T is the 2-torus.

15.1. Let Q be obtained from a closed n-manifold by removing two open disks. Show that there is a map $f \colon Q \to S^{n-1}$ whose restriction to both boundary components is the identity.

16.1. Show that the higher signatures of $M \times Q \cup_{\mathrm{id}+\mathrm{id}} M \times S^3 \times [0, 1]$ equipped with the obvious map to BG (as explained in Chapter 16) vanish if and only if the higher signatures of (M, f) vanish.

16.2. Let V be the total space of a mapping torus and f and g be maps from V to S^1, which agree on some fibre F. Show that $(V, f) - (V, g)$ is bordant to $(S^1 \times F, h)$ in $\Omega_*(S^1)$ for some map $h \colon S^1 \times F \to S^1$.

16.3. Let P be a simply connected manifold. Show that the signature of a fibre over a regular point of any smooth map from $S^1 \times S^1 \times P$ to S^1 vanishes.

16.4. (Notation as in the proof of Theorem 16.1 for $n = 1$) Show that the inclusion of A_+ to $\hat{\varphi}([0, \infty) \times P)$ is a homotopy equivalence and A_+ is an h-cobordism.

17.1. Let M be a connected oriented closed 3-manifold whose fundamental group has order 2. Compute the homology and cohomology of M with integral coefficients.

17.2. Let B_* be a chain complex of $R[\mathbb{Z}/2]$-modules. Check that there is a natural R-chain map $1 + T \colon W_* \otimes_{R[\mathbb{Z}/2]} B_* \to \hom_{R[\mathbb{Z}/2]}(W_*, B_*)$. Now let R be a ring with involution and C_* a chain complex of finitely generated projective R-modules. Show that the map $1 + T$ above for $B_* = \hom_R(C^{-*}, C_*)$ induces an isomorphism in homology if 2 is invertible in R, i.e., if R is a $\mathbb{Z}[\frac{1}{2}]$-algebra.

17.3. Let R be a ring with involution and $f \colon C_* \to D_*$ an R-chain map of chain complexes of finitely generated free R-modules. Show that the inclusion into W_* of the $\mathbb{Z}[\mathbb{Z}/2]$-chain complex concentrated in degree zero defined by the module $\mathbb{Z}[\mathbb{Z}/2]$ itself induces natural homomorphisms
$$\mathrm{ev} \colon Q^n(C_*) \to [C^{n-*}, C_*]_R \qquad \text{and} \qquad \mathrm{ev} \colon Q^{n+1}(f) \to [C(f)^{n+1-*}, D_*]_R.$$

17.4. Prove Theorem 17.33 (1), i.e., that the symmetric and quadratic algebraic L-groups are 4-periodic.

18.1. Give an example which shows that Lemma 18.3 does not hold without the assumption that each $\mathbf{E}_i$ is an Ω-spectrum.

18.2. Give examples showing that the pushout and the pullback do not satisfy homotopy invariance in contrast to the homotopy pushout and homotopy pullback (see Remarks 18.7 and 18.8).

18.3. Serre has shown that $\pi_n^s(\{\bullet\}_+) \otimes_{\mathbf{Z}} \mathbb{Q}$ is trivial for $n \geq 1$. Prove using this fact that there is a natural isomorphism

$$h_n(X, A)\colon \pi_n^s(X_+, A_+) \otimes_{\mathbf{Z}} \mathbb{Q} \xrightarrow{\cong} H_n(X, A; \mathbb{Q})$$

for all pairs of CW-complexes (X, A), where $H_n(X, A; \mathbb{Q})$ is singular homology with rational coefficients.

19.1. Suppose that there is a model for $\underline{E}G$, which is n-dimensional or which is of finite type respectively . Show that then the group homology $H_p(G; \mathbb{Q})$ and group cohomology $H^p(G; \mathbb{Q})$ of G vanishes for $p \geq n + 1$ or is finitely generated as $\mathbb{Q}$-module for all $p \geq 0$ respectively.

19.2. Let G and H be groups. Is $\underline{E}G \times \underline{E}H$ a model for $\underline{E}(G \times H)$? Is $E_{\mathcal{VCYC}}(G) \times E_{\mathcal{VCYC}}(H)$ a model for $E_{\mathcal{VCYC}}(G \times H)$?

19.3. Consider $G = H_0 * H_1$ for two finite groups H_0 and H_1. Construct a 1-dimensional model for $\underline{E}G$ and describe $G\backslash\underline{E}G$.

20.1. Let $D_\infty = \langle t, s \mid s^2 = 1, sts = t^{-1} \rangle$ be the infinite dihedral group which is the free amalgamated product $\mathbb{Z}/2 * \mathbb{Z}/2$ or, equivalently, the non-trivial semi-direct product $\mathbb{Z} \rtimes \mathbb{Z}/2$. Show that $\mathbb{R}$ with the obvious $\mathbb{Z}/2 * \mathbb{Z}/2$ coming from the translation action of $\mathbb{Z}$ and the $\mathbb{Z}/2$ action given by $-$ id is a model for $E_{\mathcal{FIN}}(D_\infty)$. Then compute $H_n^{D_\infty}(E_{\mathcal{FIN}}(D_\infty); \mathbf{K}^{\mathrm{top}})$ using the fact that $K_n(C_*^r(H))$ is isomorphic to the complex representation ring $R(H)$ of H for n even and $K_n(C_*^r(H))$ vanishes for n odd, if H is a finite group.

20.2. Let X and X' be contravariant pointed $\mathcal{C}$-spaces, let Y be a covariant pointed $\mathcal{C}$-space and let Z be a pointed space. Let $\mathrm{map}(Y, Z)$ be the contravariant pointed $\mathcal{C}$-space which sends an object c in $\mathcal{C}$ to $\mathrm{map}(Y(c), Z)$. Let $\hom_{\mathcal{C}}(X, X')$ be the topological space of a natural transformations $X \to X'$ with the subspace topology for the obvious embedding $i\colon \hom_{\mathcal{C}}(X, X') \to \prod_{c \in \mathrm{ob}(\mathcal{C})} \mathrm{map}(X(c), X'(c))$. Construct a homeomorphism, natural in X, Y and Z,

$$\mathrm{map}(X \wedge_{\mathcal{C}} Y, Z) \xrightarrow{\cong} \hom_{\mathcal{C}}(X, \mathrm{map}(Y, Z)).$$

20.3.   Let $\mathbf{F}$ be a spectrum and $H_*(-;\mathbf{F})$ be the associated homology theory (see Lemma 18.12). We constructed in Example 20.8 equivariant homology theories associated to $H_*(-;\mathbf{F})$ using the quotient space and the Borel construction, namely we considered for a G-CW-complex X the groups $H_*(G\backslash X;\mathbf{F})$ and $H_*(EG\times X;\mathbf{F})$. Construct for each of them a covariant functor $\mathbf{E}\colon$ GROUPOIDS $\to$ SPECTRA such that the equivariant homology theory is given by $H_*^?(-;\mathbf{E})$ in the sense of Lemma 20.14.

21.1.   Let $\mathcal{H}_*$ be a (non-equivariant) homology theory (satisfying the disjoint union axiom). Let G be a group and $f\colon EG \to \underline{E}G$ be a G-map. Show that the induced map

$$\mathcal{H}_n(G\backslash f) \otimes_{\mathbb{Z}} \mathrm{id}_{\mathbb{Q}}\colon \mathcal{H}_n(BG) \otimes_{\mathbb{Z}} \mathbb{Q} \to \mathcal{H}_n(G\backslash \underline{E}G) \otimes_{\mathbb{Z}} \mathbb{Q}$$

is bijective for all $n \in \mathbb{Z}$.

21.2.   Let l be an odd natural number. Show that the assembly map for the trivial family

$$A_{\mathcal{TR}}\colon H_1(B\mathbb{Z}/l;\mathbf{L}^{\langle -\infty\rangle}) \to L_1^{\langle -\infty\rangle}(\mathbb{Z}[\mathbb{Z}/l])$$

is not injective using the facts that $L_n^s(\mathbb{Z}[G]) = 0$ holds for any finite odd order group G and any odd integer n and the even dimensional L^s-groups of the ring $\mathbb{Z}$ are given by $L_0^s(\mathbb{Z}) \cong \mathbb{Z}$ and $L_2^s(\mathbb{Z}) = \mathbb{Z}/2$.

21.3.   Let G be the fundamental group of a compact orientable surface. Compute $\mathrm{Wh}(G)$, $\widetilde{K}_n(\mathbb{Z}[G])$ for $n \leq 0$ and $L_n^{\langle j\rangle}(\mathbb{Z}[G])$ for all $n \in \mathbb{Z}$ and decorations $j \in \{-\infty\}$ II $\{j \in \mathbb{Z} \mid j \leq 2\}$ using the fact that the Farrell–Jones Conjecture 20.3 is true for G and $L_{4n}^h(\mathbb{Z}) \cong \mathbb{Z}$, $L_{4n+2}^h(\mathbb{Z}) = \mathbb{Z}/2$ and $L_{2n+1}^h(\mathbb{Z}) = 0$ holds for $n \in \mathbb{Z}$.

22.1.   Let G be a group which is not torsionfree. Show that the rings $\mathbb{Q}G$, $\mathbb{C}G$ and $C_r^*(G)$ contain idempotents different from 0 and 1 and the image of $\mathrm{tr}_{C_r^*(G)}\colon K_0(C_r^*(G)) \to \mathbb{R}$ is not contained in $\mathbb{Z}$.

22.2.   Let G be the fundamental group of a compact orientable surface M. Compute $K_n(C_r^*(G \times \mathbb{Z}^k)$ for each $k \geq 0$ using the fact that the Baum–Connes Conjecture 20.4 holds for G

22.3.   Let G be a discrete subgroup of a connected Lie group L such that $G\backslash L$ is compact. Suppose that G satisfies the Baum–Connes Conjecture 20.4. Prove that $K_n(C_r^*(G))$ is finitely generated for $n \in \mathbb{Z}$.

23.1.   Let G be a finite group. Let $r(G)$ be the number of $\mathbb{R}$-conjugacy classes of G, where two elements of g_1 and g_2 are called $\mathbb{R}$-conjugate if g_1 and g_2 or g_1^{-1} and g_2 are conjugate. Show that

$$L_0^s(\mathbb{Z}G)[1/2] \cong \mathbb{Z}[1/2]^{r(g)}$$

using the fact that $r(g)$ is the number of isomorphism classes of irreducible real G-representations is $r(G)$.

23.2.   Let $T_*\colon \mathcal{H}_* \to \mathcal{K}_*$ be a transformation of (non-equivariant) homology theories satisfying the disjoint union axiom. Suppose that $T_n(\{\bullet\})$ is an isomorphism for all $n \in \mathbb{Z}$. Show that then $T(X)$ is bijective for all CW-complexes X.

24.1.   Let $\mathbb{Z}/3$ act on $\mathbb{Z}^3$ by permuting the coordinates. Let A be the quotient of $\mathbb{Z}^3$ by the $\mathbb{Z}/3$-fixed point set. Show that A is isomorphic to $\mathbb{Z}^2$ and the induced $\mathbb{Z}/3$-action is free outside $0 \in A$. Let G be the semi-direct product $A \rtimes \mathbb{Z}/3$. Compute $K_n(C_r^*(G))$ for $n \in \mathbb{Z}$.

24.2.   Let G be a group which satisfies the Farrell–Jones Conjecture 20.3 for $K_n(\mathbb{C}G)$ and the Baum–Connes Conjecture 20.4. Let $\widetilde{K}_n(\mathbb{C}G)$ and $\widetilde{K}_n(C_r^*(G))$ be the cokernels of the maps $K_n(\mathbb{C}) \to K_n(\mathbb{C}G)$ and $K_n(C_r^*(\{1\})) \to K_n(C_r^*(G))$ induced by the inclusion of the trivial group into G. Show that the following statements for a group G are equivalent:

(1) $\widetilde{K}_n(\mathbb{C}G) \otimes_{\mathbb{Z}} \mathbb{Q} = 0$ for all $n \in \mathbb{Z}$.

(2) $\widetilde{K}_n(C_r^*(G)) \otimes_{\mathbb{Z}} \mathbb{Q} = 0$ for all $n \in \mathbb{Z}$.

(3) G is torsionfree and $H_p(BG; \mathbb{Q}) = 0$ for $p \geq 1$.

24.3.   Show for a Fuchsian group

$$F = \langle a_1, b_1, \ldots, a_g, b_g, c_1, \ldots, c_t \mid$$
$$c_1^{\gamma_1} = \ldots = c_t^{\gamma_t} = c_1^{-1} \cdots c_t^{-1}[a_1, b_1] \cdots [a_g, b_g] = 1 \rangle$$

that $\mathrm{Wh}(F) = 0$ if and only if $t = 0$ or if $t \geq 1$ and $\gamma_i \in \{2, 3, 4, 6\}$ for $i = 1, 2, \ldots, t$.

Chapter 26

Hints to the Solutions of the Exercises

0.1. Apply the Seifert–van Kampen theorem and the Mayer–Vietoris sequence.

0.2. Show that $\langle x \cup p_1, [\ldots] \rangle$ is a bordism invariant.

0.3. Compute the second Stiefel–Whitney class and express this via the Wu formulas in terms of the Steenrod squares.

1.1. Since $H^4(N; \mathbb{Q}) = 0$, we get for the first rational Pontrjagin class $p_1(M; \mathbb{Q}) = 0$ and for the signature $\operatorname{sign}(M) = 0$. Now the Signature Theorem 1.3 implies $p_2(M; \mathbb{Q}) = 0$.

1.2. The implication $(1) \Rightarrow (2)$ follows from elementary covering theory. Suppose (2) holds. By elementary covering theory there is a covering $p \colon \overline{N} \to N$ and a map $\overline{f} \colon M \to \overline{N}$ such that the image of $\pi_1(p)$ and $\pi_1(f)$ agree and $p \circ \overline{f} = f$. The map $\pi_1(\overline{f})$ is bijective. Hence $\overline{f}$ is homotopic to a homeomorphism by the Borel Conjecture 1.10.

1.3. Let $t(\nu_{N \subseteq M}) \in H^l(D\nu_{N \subseteq M}, S\nu_{N \subseteq M})$ be the Thom class of the normal bundle $\nu_{N \subseteq M}$ with the orientation induced by the ones on M and N. Let φ be the composition of the maps induced by the obvious inclusions

$$\varphi \colon H^l(D\nu_{N \subseteq M}, S\nu_{N \subseteq M}) \xleftarrow{\cong} H^l(M, M - (D\nu_{N \subseteq M})^\circ) \to H^l(M).$$

Then one gets for $i \colon N \to M$ the inclusion

$$i_*([N]) = \varphi(t(\nu_{N \subseteq M})) \cap [M].$$

Now we consider the situation appearing in the exercise. Notice that f is covered by a bundle map $\nu_{N \subseteq M} \to \nu_{\{y\} \subseteq B}$. This implies $f^*(\varphi(t(\nu_{\{y\} \subseteq B}))) = \varphi(t(\nu_{N \subseteq M}))$. We get

$$
\begin{aligned}
\mathrm{sign}_{[B]}(M, u) &= \langle \mathcal{L}(M) \cup f^*([B]), [M] \rangle = \langle \mathcal{L}(M), f^*([B]) \cap [M] \rangle \\
&= \langle \mathcal{L}(M), f^*(\varphi(t(\nu_{\{y\} \subseteq B}))) \cap [M] \rangle = \langle \mathcal{L}(M), \varphi(t(\nu_{N \subseteq M})) \cap [M] \rangle \\
&= \langle \mathcal{L}(M), i_*([N]) \rangle = \langle i^*(\mathcal{L}(M)), [N] \rangle = \langle \mathcal{L}(N), [N] \rangle = \mathrm{sign}(N).
\end{aligned}
$$

2.1. Extend the proof for Ω_n.

2.2. The boundary operator associates to (W, f, α) the restriction to the boundary.

2.3. If (W, f, α) represents an element in $\Omega_m(X, Y; E)$ use partition of unity to decompose W as $W_1 \cup W_2$ with $\partial W_2 = \partial W + \partial W_1$, such that $f(W_1) \subseteq X - A$. Then consider $(W_1, f|_{W_1}, \alpha|_{W_1})$ and show that this gives an inverse to the homomorphism induced by inclusion.

2.4. Extend the proof for framed bordism of a point.

3.1. Choose a map f from M to a product K of infinite dimensional complex projective spaces inducing an isomorphism on H_2. Show that the intersection form on M is equivalent to the image of the fundamental class in $H_4(K)$. If M' is another manifold an g is an isometry from the intersection form of M to that of M', choose f' as for M such that it commutes with the maps to K up to homotopy. The proof is finished by showing that the gf can be lifted to a map from M to M' which is automatically a homotopy equivalence.

3.2. Standard linear algebra.

4.1. Apply the signature theorem in dimension 4. Use that the $p_1 = c_1^2 - 2c_2$, where c_i are the Chern classes, and that c_2 of a 4-manifold is the Euler characteristic.

4.2. Use that $\Omega_8 \otimes \mathbb{Q}$ has dimension 2 and find a basis. Check the signature theorem on the base elements.

4.3. Apply the signature theorem.

5.1. In the sequel we consider for any F-vector space W the abelian group $\hom_F(W, V)$ as a left R-module by composition.

Choose an R-isomorphism $u \colon \bigoplus_{i=1}^n V \to V$. It induces a bijection

$$
\psi \colon \bigoplus_{i=1}^n \hom_F(V, V) \xrightarrow{\cong} \hom_F\left(\bigoplus_{i=1}^n V, V\right) \xrightarrow{\hom_F(u, \mathrm{id}_V)} \hom_F(V, V).
$$

This is an isomorphism $\psi\colon R^n \to R$ of R-modules.

Let P be a finitely generated projective R-module. We can find an element $p \in R$ with $p^2 = p$ such that the R-map $r_p\colon R \to R$ given by right multiplication with p is R-isomorphic to P. Let Q be $\mathrm{im}(r_{1-p})$. There is a canonical R-isomorphism $P \oplus Q \overset{\cong}{\to} R$. The inclusions $\mathrm{im}(p) \to V$ and $P \to P \oplus Q \overset{\cong}{\to} R$ induce a R-map $\varphi_P\colon P \to \hom_F(\mathrm{im}(p), V)$ and analogously we get $\varphi_Q\colon Q \to \hom_F(\mathrm{im}(1 - p), V)$. One easily checks that $\varphi_P \oplus \varphi_Q$ is bijective and hence φ_P and φ_Q are bijective. At least one of the F-vector spaces $\mathrm{im}(p)$ or $\mathrm{im}(1 - p)$ possesses a countable infinite F-basis. Suppose that this is true for $\mathrm{im}(1 - p)$. Then there is a F-isomorphism $\mathrm{im}(1 - p) \overset{\cong}{\to} V$, which induces an R-isomorphism $Q \overset{\cong}{\to} R$. This implies in $K_0(R)$

$$[P] \;=\; [P] + [Q] - [Q] \;=\; [R] - [Q] \;=\; [R] - [R] = 0.$$

In the other case we conclude $P \cong R$ and this implies $[P] = [R]$. But obviously $2 \cdot [R] = [R \oplus R] = [R]$ and hence $[P] = 0$ in $K_0(R)$.

5.2. Take $R = \mathbb{Z}/2 \times \mathbb{Z}/2$ and $M = \mathbb{Z}/2 \times \{0\}$.

5.3. Let P_i be a finitely generated projective R_i-module for $i = 0, 1$. Then $P_0 \times P_1$ with the obvious $R_0 \times R_1$-module structure is a finitely generated projective $R_0 \times R_1$-module. Use this construction to define a map

$$K_n(R_0) \times K_0(R_1) \to K_n(R_0 \times R_1).$$

This map is the inverse of $(\mathrm{pr}_0)_* \times (\mathrm{pr}_1)_*$ because there is a natural R_i-isomorphism

$$(\mathrm{pr}_i)_*(P_0 \times P_1) \overset{\cong}{\to} P_i$$

for $i = 0, 1$ and for each finitely generated projective $R_0 \times R_1$-module M there is a $R_0 \times R_1$-isomorphism

$$M \overset{\cong}{\to} (\mathrm{pr}_0)_* M \times (\mathrm{pr}_1)_* M.$$

5.4. The inverse is $1 - t^2 - t^3$. The map $\mathbb{Z}[\mathbb{Z}/5] \to \mathbb{C}$ sending t to $\exp(2\pi i/5)$, the determinant over $\mathbb{C}$ and the map $\mathbb{C}^{\mathrm{inv}} \to (0, \infty), z \mapsto |z|$ together induce a map $\mathrm{Wh}(\mathbb{Z}/5) \to (0, \infty)$ of abelian groups, where we equip the target with the group structure given by multiplying positive real numbers. It sends the element given by the unit above to $|1 - \cos(2\pi/5)|$, which is different from 1.

6.1. (1) If $f\colon P_* \to Q_*$ is a R chain homotopy equivalence, then $\mathrm{cone}(f_*)$ is contractible. This implies $\mathrm{cone}(f)_{\mathrm{odd}} \cong \mathrm{cone}(f)_{\mathrm{ev}}$ and hence $o(P_*) = o(Q_*)$.

(2) is obvious.

(3) Suppose that $\widetilde{o(P_*)} = 0$. By adding elementary finite projective R-chain complexes $\ldots 0 \to 0 \to Q \overset{\mathrm{id}}{\to} Q \to 0 \to 0 \to \ldots$ to P_* one can change P_* within its

R-chain homotopy class such that all R-chain modules are finitely generated free except the top-dimensional one. But this top dimensional one must be stably free because of $o(\widetilde{P_*}) = 0$. Now add another appropriate elementary finite free R-chain complex to turn also this top-dimensional module into a finitely generated free one.

If P_* is up to homotopy finite free, $o(\widetilde{P_*}) = 0$ follows from assertion (1).

6.2.   Use the product formula for Whitehead torsion.

6.3.   We have $p \circ i = f$ and $p \circ j = \mathrm{id}_Y$. We use the homotopy equivalences i and p to identify $\pi = \pi_1(X) = \pi_1(\mathrm{cyl}(f)) = \pi_1(Y)$. We have the commutative diagram of based exact $\mathbb{Z}\pi$- chain complexes whose rows are based exact

$$
\begin{array}{ccccccccc}
0 & \longrightarrow & C_*(\widetilde{Y}) & \longrightarrow & C_*(\widetilde{Y}) & \longrightarrow & 0_* & \longrightarrow & 0 \\
& & \mathrm{id}\downarrow & & C_*(\widetilde{j})\downarrow & & 0_*\downarrow & & \\
0 & \longrightarrow & C_*(\widetilde{Y}) & \longrightarrow & C_*(\widetilde{\mathrm{cyl}(f)}) & \longrightarrow & \mathrm{cone}_*(C_*(\widetilde{X})) & \longrightarrow & 0
\end{array}
$$

It is easy to check that the left and the right vertical arrow have trivial Whitehead torsion. Hence the same is true for the middle vertical arrow. This shows $\tau(j) = 0$. Finally apply the composition formula for Whitehead torsion.

7.1.   Use the s-Cobordism theorem, the fact that $\mathrm{Wh}(\mathbb{Z}/5)$ contains an element of infinite order and the product formula for Whitehead torsion.

7.2.   Use the sum and the composition formula for Whitehead torsion.

7.3.   Show that $\pi_1(N) = \mathbb{Z} * \mathbb{Z}$. Conclude using the formulas for free amalgamated products and the Bass–Heller–Swan decomposition that $\mathrm{Wh}(\pi_1(N)) = 0$. Finally apply the s-Cobordism Theorem.

8.1.   One can attach to the empty set only a 0-handle $D^0 \times D^n$ since for a q-handle $D^q \times D^{n-q}$ there exists an embedding $S^{q-1} \times D^{n-q} \to \emptyset$ only for $q = 0$. Since ∂M is empty by assumption, the attaching embedding $S^{q-1} \times D^{n-q} \to S^{n-1} = \partial D^n$ for the second handle must be surjective. Hence we must have $q = n$ and M looks like $D^n \cup_f D^n$ for some diffeomorphism $f \colon S^{n-1} \to S^{n-1}$. This implies using the Alexander trick that M is homeomorphic to S^n.

8.2.   Let M be the manifold with boundary $S^{q-1} \times S^{n-1-q}$ obtained from $\partial_0 W$ by removing the interior of $\varphi^q(S^{q-1} \times D^{n-q})$. We get

$$
\begin{aligned}
W & \cong \partial_0 W \times [0,1] \cup_{S^{q-1} \times D^{n-q}} D^q \times D^{n-q} \\
& = M \times [0,1] \cup_{S^{q-1} \times S^{n-2-q} \times [0,1]} \\
& \qquad \left(S^{q-1} \times D^{n-1-q} \times [0,1] \cup_{S^{q-1} \times D^{n-q} \times \{1\}} D^q \times D^{n-q} \right).
\end{aligned}
$$

Inside $S^{q-1} \times D^{n-1-q} \times [0,1] \cup_{S^{q-1} \times D^{n-q} \times \{1\}} D^q \times D^{n-q}$ we have the following submanifolds

$$X := S^{q-1} \times 1/2 \cdot D^{n-1-q} \times [0,1] \cup_{S^{q-1} \times 1/2 \cdot D^{n-q} \times \{1\}} D^q \times 1/2 \cdot D^{n-q};$$
$$Y := S^{q-1} \times 1/2 \cdot S^{n-1-q} \times [0,1] \cup_{S^{q-1} \times 1/2 \cdot S^{n-q} \times \{1\}} D^q \times 1/2 \cdot S^{n-q}.$$

The pair (X, Y) is diffeomorphic to $(D^q \times D^{n-q}, D^q \times S^{n-1-q})$, i.e., it is a handle of index $(n-q)$. Let N be obtained from W by removing the interior of X. Then W is obtained from N by adding a $(n-q)$-handle, the so-called *dual handle*. One easily checks that N is diffeomorphic to $\partial_1 W \times [0,1]$ relative $\partial_1 W \times \{1\}$.

9.1.   i) Show that the Whitney sum of the stable tangent bundle of a hypersurface V of degree d in $\mathbb{CP}^{n+1}$ with the restriction of H^d to V is the restriction of $\bigoplus_{n+2} H$ to V, where H^d is the d-fold tensor product of the Hopf bundle H.
ii) Show that on a 3-dimensional CW-complex all oriented vector bundles with trivial second Stiefel–Whitney class are stably parallelizable.
iii) Use that a compact n-dimensional manifold with non-empty boundary is homotopy equivalent to a $n-1$-dimensional CW-complex.

9.2.   Apply the hint i) to exercise 9.1 and use the signature Theorem.

10.1.   Apply the Seifert–van Kampen theorem and the Mayer–Vietoris sequence.

10.2.   Apply the hint i) to exercise 9.1.

11.1.   Apply the Hurewicz Theorem

11.2.   This can be reduced to Sard's Theorem.

12.1.   We have seen already that if the signature of a unimodular bilinear form vanishes, then there is a half rank summand on which the form vanishes. Use this to produce a half rank direct summand with the same properties. This leads to injectivity. To show that the signature is non-trivial one has to find an even unimodular form with non-trivial signature. This is surprisingly not so easy, look at a book about Lie groups (Dynkin diagrams) to see a candidate.

12.2.   Consider the reduction of an unimodular even form   mod 2.

12.3.   List all elements with rank ≤ 2 (don't forget the quadratic refinement).

13.1.   Relate the statement to the cellular chain complex of the universal covering.

13.2.   Elementary linear algebra.

13.3.   Apply the Lefschetz duality theorem.

14.1. Show that n-dimensional vector bundles over S^n are classified by $\pi_{n-1}(O(n))$. Apply the exact homotopy sequence of the fibre bundle $O(n+1) \to S^n$ with fibre $O(n)$ and relate the boundary map $d : \pi_n(S^n) \to \pi_{n-1}(O(n))$ to the Euler class of the vector bundle given by $d(1)$.

14.2. Apply excision.

14.3. Apply the Mayer–Vietoris sequence and use that the 4-dimensional spin bordism group is isomorphic to $\mathbb{Z}$ (the isomorphism is given by the signature divided by 16, and that the spin bordism groups are 0 in dimension 5 and 6 ([231].

15.1. Show that the identity on one of the two boundary components can be extended to Q. Since the maps of an $n-1$-dimensional manifold to S^n are classified by the degree, one has to show that the restriction to the other boundary component has degree one.

16.1. Prove that the manifold is bordant to $M \times K$, where K is the Kummer surface, in such a way that the maps to BG can be extended.

16.2. After a homotopy achieve that the maps agree on a tubular neighborhood of F. Then construct a bordism between $V+(-V)$ and $S^1 \times F$ such that the maps f on V and g on $-V$ can be extended to this bordism.

16.3. Since P is simply connected a map from $S^1 \times S^1 \times P$ to S^1 factors through $S^1 \times S^1$. Then the fibre is a product of a 1-dimensional manifold and P.

16.4. Construct a homotopy inverse as the identity on A_+ and the projection to $k \times P$ on the rest.

17.1. Since M is connected, $H_0(M)$ and $H^0(M)$ are $\mathbb{Z}$. By Poincaré duality $H_3(M)$ and $H^3(M)$ are $\mathbb{Z}$. Since $\pi_1(M)$ is abelian, it is isomorphic to $H_1(M)$. Hence $H_1(M)$ is $\mathbb{Z}/2$. By the universal coefficient theorem $H^1(M) = 0$. By Poincaré duality $H_2(M) = 0$ and $H^2(M) = \mathbb{Z}/2$. Since M is 3-dimensional $H_p(M)$ and $H^p(M)$ vanish for $p \geq 4$.

17.2. Equip R with the trivial $\mathbb{Z}/2$-action. Denote by R also the $R[\mathbb{Z}/2]$-chain complex which is concentrated in dimension zero and given there by R. Let $e_* : W_* \to R$ be the obvious chain map inducing an isomorphism on H_0. Let $u_* : R \otimes_{R[\mathbb{Z}/2]} B_* \to (B_*)^{\mathbb{Z}/2}$ be the R-chain map sending $r \otimes b$ to $r \cdot \frac{1+T}{2} \cdot b$ for $T \in \mathbb{Z}/2$ the generator.

The map $1+T$ is given by the composition

$$W_* \otimes_{R[\mathbb{Z}/2]} B_* \xrightarrow{\ e_* \otimes_{R[\mathbb{Z}/2]} \mathrm{id}_{B_*}\ } R \otimes_{R[\mathbb{Z}/2]} B_* \xrightarrow{\ u_*\ } (B_*)^{\mathbb{Z}/2} = \hom_{R[\mathbb{Z}/2]}(R, B_*)$$

$$\xrightarrow{\ \hom_R(e_*, \mathrm{id}_{B_*})\ } \hom_{R[\mathbb{Z}/2]}(W_*, B_*).$$

If R contains $1/2$, the $R[\mathbb{Z}/2]$-module R with the trivial $\mathbb{Z}/2$-action is projective. Hence $e_* \colon W_* \to R$ is a $R[\mathbb{Z}/2]$-chain equivalence. If R contains $1/2$, the R-chain map $u_* \colon R \otimes_{R[\mathbb{Z}/2]} B_* \to (B_*)^{\mathbb{Z}/2}$ is an isomorphism.

17.3. One has a natural identification of $[C^{n-*}, C_*]$ with $H_n\left(\hom_R(C^{n-*}, C_*)\right)$.

17.4. The periodicity isomorphism is given by the double suspension. Since we make no restrictions on the dimensions and only require boundedness for the chain complexes underlying a symmetric or quadratic Poincaré chain complex, this is already an isomorphism on the level of categories of chain complexes. See [200, Proposition 1.10].

18.1. Let $\mathbf{E}$ be an Ω-spectrum such that $A = \pi_0(\mathbf{E}) \cong \pi_k(E(k))$ is a nontrivial abelian group. Denote by $\mathbf{E}_{(i,\infty)}$ the spectrum obtained from $\mathbf{E}$ by replacing the spaces $E(0)$, $E(1)$, ..., $E(i)$ by $\{\bullet\}$. We have

$$\pi_k(E_{(i,\infty)}(k)) = \begin{cases} 0 & \text{if } k \leq i \\ A & \text{if } k > i \end{cases}$$

and the maps $\pi_k(E_{(i,\infty)}(k)) \to \pi_{k+1}(E_{(i,\infty)}(k+1))$ are isomorphisms for $k > i$. We see that for all i we have $\pi_0(\mathbf{E}_{(i,\infty)}) = A$ and

$$\pi_0(\prod_{i\in\mathbb{N}} E_{(i,\infty)}) = \operatorname{colim}_k \pi_k(\prod_{i\in\mathbb{N}} E_{(i,\infty)}(k)) = \operatorname{colim}_k \bigoplus_{i=1}^{k} A = \bigoplus_{i=1}^{\infty} A.$$

The natural map

$$\pi_0(\prod_{i\in\mathbb{N}} \mathbf{E}_{(i,\infty)}) \to \prod_{i\in\mathbb{N}} \pi_0(\mathbf{E}_{(i,\infty)})$$

can be identified with the natural inclusion $\bigoplus_{i\in\mathbb{N}} A \to \prod_{i\in\mathbb{N}} A$ and is not an isomorphism.

18.2. For the pushout consider the following commutative diagram

$$
\begin{array}{ccccc}
S^n_- & \xleftarrow{\ i_-\ } & S^{n-1} & \xrightarrow{\ i_+\ } & S^n_+ \\
\downarrow & & {\scriptstyle \mathrm{id}_{S^{n-1}}}\downarrow & & \downarrow \\
\{\bullet\} & \xleftarrow{\ k_1\ } & S^{n-1} & \xrightarrow{\ k_2\ } & \{\bullet\}
\end{array}
$$

where S^n_+ and S^n_- are the upper and lower hemisphere in S^n and S^{n-1} is their intersection and the maps i_- and i_+ are inclusions. Notice that the pushout of the upper row is S^n and of the lower row is $\{\bullet\}$.

For the pullback consider for a pointed space $X = (X, x)$ the commutative diagram

$$\text{map}([0, 1/2], X) \xrightarrow{\;i_1\;} X \xleftarrow{\;i_2\;} \text{map}([1/2, 1], X)$$

$$\downarrow \qquad\qquad \text{id}\downarrow \qquad\qquad \downarrow$$

$$\{\bullet\} \xrightarrow[\;k_1\;]{} X \xleftarrow[\;k_2\;]{} \{\bullet\}$$

where the mapping spaces are pointed mapping spaces with $0 \in [0, 1/2]$ and $1 \in [1/2, 1]$ as base points and the maps i_1 and i_2 are given by evaluation at $1/2$. Notice that the pullback of the upper row is ΩX and of the lower row is $\{\bullet\}$.

18.3. Tensoring over $\mathbb{Z}$ with $\mathbb{Q}$ is an exact functor and respects direct sums over arbitrary index sets. Hence we get a homology theory on the category of CW-pairs with values in $\mathbb{Q}$-modules satisfying the disjoint union axiom and the dimension axiom by $\pi_*^s \otimes_{\mathbb{Z}} \mathbb{Q}$. The same is true for singular homology with coefficients in $\mathbb{Q}$. Now apply the general result that any such homology theory is the same as cellular homology with coefficients in $\mathbb{Q}$.

19.1. The cellular $\mathbb{Q}$-chain complex of $C_*(\underline{E}G)$ with the obvious G-action is a projective $\mathbb{Q}G$-chain complex, since its n-th chain module looks like $\bigoplus_{i \in I_n} \mathbb{Q}[G/H_i]$ if the equivariant n-cells are given by $\{G/H_i \times D^n \mid i \in I\}$.

19.2. The answer is yes for the family $\mathcal{FIN}$ by the following argument. First of all one must show that the product of a G-CW-complex with a H-CW-complex is a $G \times H$-CW-complex. This is true in general since we are working in the category of compactly generated spaces. Obviously all isotropy groups of $\underline{E}G \times \underline{E}H$ are finite. Let $K \subseteq G \times H$ be a subgroup. Let K_G and K_H respectively be the image of K under the projection $G \times H$ to G and H respectively. Then

$$(\underline{E}G \times \underline{E}H)^K = \underline{E}G^{K_G} \times \underline{E}G^{K_H}$$

and K is finite if and only if both K_G and K_H are finite. Now the claim follows.

The answer is no for the family $\mathcal{VCYC}$ as the example $G = H = \mathbb{Z}$ shows. Obviously $E_{\mathcal{VCYC}}(\mathbb{Z})$ can be chosen to be a point, but $E_{\mathcal{VCYC}}(\mathbb{Z} \times \mathbb{Z})$ cannot be $\mathbb{Z} \times \mathbb{Z}$-homotopy equivalent to a point since $\mathbb{Z} \times \mathbb{Z}$ is not virtually cyclic.

19.3. This is a special case of the result of Section 19.3.7.

20.1. A concrete D_∞-CW-structure for $E_{\mathcal{FIN}}(D_\infty)$ is given by

$$D_\infty \times S^0 \longrightarrow D_\infty/\{1\} * \mathbb{Z}/2 \amalg D_\infty/\mathbb{Z}/2 * \{1\}$$

$$\downarrow \qquad\qquad\qquad\qquad \downarrow$$

$$D_\infty \times D^1 \longrightarrow \qquad\qquad \mathbb{R}$$

The associated Mayer–Vietoris sequence together with G-homotopy invariance and the induction structure yields a long exact sequence

$$\ldots \to K_n^{\text{top}}(C_r^*(\{1\})) \oplus K_n^{\text{top}}(C_r^*(\{1\})$$
$$\to K_n^{\text{top}}(C_r^*(\{1\})) \oplus K_n(C_r^*(\mathbb{Z}/2)) \oplus K_n(C_r^*(\mathbb{Z}/2)) \to H_n^{D_\infty}(E_{\mathcal{FIN}}(D_\infty); \mathbf{K}^{\text{top}})$$
$$\to K_n^{\text{top}}(C_r^*(\{1\})) \oplus K_n^{\text{top}}(C_r^*(\{1\})) \to \ldots$$

This yields $H_n^{D_\infty}(E_{\mathcal{FIN}}(D_\infty); \mathbf{K}^{\text{top}}) = 0$ for n odd and the short exact sequence

$$0 \to R(\{1\}) \to R(\mathbb{Z}/2) \oplus R(\mathbb{Z}/2) \to H_n^{D_\infty}(E_{\mathcal{FIN}}(D_\infty); \mathbf{K}^{\text{top}}) \to 0$$

and hence $H_n^{D_\infty}(E_{\mathcal{FIN}}(D_\infty); \mathbf{K}^{\text{top}}) \cong \mathbb{Z}^3$ for n even.

20.2. Let $p: \bigvee_{c \in \text{ob}(\mathcal{C})} X(c) \wedge Y(c) \to X \wedge_{\mathcal{C}} Y$ be the canonical identification. There is a canonical homeomorphism

$$h: \text{map}\left(\bigvee_{c \in \text{ob}(\mathcal{C})} X(c) \wedge Y(c), Z \right) \xrightarrow{\cong} \prod_{c \in \text{ob}(\mathcal{C})} \text{map}(X(c) \wedge Y(c), Z).$$

Define α as the map which makes the following diagram commutative

$$
\begin{array}{ccc}
\text{map}(X \wedge_{\mathcal{C}} Y, Z) & \xrightarrow{\ \alpha\ } & \hom_{\mathcal{C}}(X, \text{map}(Y, Z)) \\
{\scriptstyle j}\downarrow & & \downarrow{\scriptstyle i} \\
\text{map}(X(c) \wedge Y(c), Z) & \xrightarrow[\prod_{c \in \text{ob}(\mathcal{C})} \beta_c]{} & \prod_{c \in \text{ob}(\mathcal{C})} \text{map}(X(c), \text{map}(Y(c), Z))
\end{array}
$$

where j is the embedding of topological spaces coming from the composition $h: \text{map}(p, \text{id}_Z)$, i is the embedding of topological spaces used to define a topology on $\hom_{\mathcal{C}}(X, \text{map}(Y, Z))$ and each map β_c is the canonical homeomorphism (18.2).

20.3. The solution is given by the covariant functors

$$\mathbf{E}_1: \text{GROUPOIDS} \to \text{SPECTRA}, \qquad \mathcal{G} \mapsto \mathbf{F},$$
$$\mathbf{E}_1: \text{GROUPOIDS} \to \text{SPECTRA}, \qquad \mathcal{G} \mapsto B\mathcal{G}_+ \wedge \mathbf{F},$$

where $B\mathcal{G}$ is the classifying space of the groupoid $\mathcal{G}$, i.e., the geometric realization of the simplicial set given by the nerve of $\mathcal{G}$. (If G is a group and we consider G as a groupoid with one object, this is the same as $BG = G \backslash EG$.) One has to check for a G-CW-complex X, that we get for the associated $\text{Or}(G)$-space $\text{map}_G(-, X)$ a homotopy equivalence

$$\text{map}_G(-, X)_+ \wedge_{\text{Or}(G)} \mathbf{E}_1(\mathcal{G}^G(-)) \to (G \backslash X)_+ \wedge \mathbf{F},$$
$$\text{map}_G(-, X)_+ \wedge_{\text{Or}(G)} \mathbf{E}_2(\mathcal{G}^G(-)) \to (EG \times_G X)_+ \wedge \mathbf{F}.$$

For this purpose one constructs for a G-CW-complex Z a homeomorphism of (unpointed) spaces

$$\mathrm{map}_G(-,X) \wedge_{\mathrm{Or}(G)} - \times_G Z \;\to\; X \times_G Z,$$

where $- \times_G Z$ is the covariant $\mathrm{Or}(G)$-space sending G/H to $G/H \times_G Z = H\backslash Z$, and applies this to the case $Z = \{\bullet\}$ and $Z = EG$.

21.1. For any CW-complex X the Atiyah–Hirzebruch spectral sequence converges to $\mathcal{H}_{p+q}(X)$ in the strong sense, since we are dealing with a homology theory satisfying the disjoint union axiom. It has as E^2-term the cellular homology $H_p(X; \mathcal{H}_q(\{\bullet\}))$ with coefficients in $\mathcal{H}_q(\{\bullet\})$. Recall that $\mathbb{Q}$ is flat as $\mathbb{Z}$-module. By a spectral sequence comparison argument it suffices to prove the claim in the special case, where $\mathcal{H}_*$ is cellular homology with rational coefficients $H_*(-;\mathbb{Q})$. The cellular $\mathbb{Q}$-chain complexes $C_*(EG;\mathbb{Q})$ and $C_*(\underline{E}G;\mathbb{Q})$ inherit the structure of $\mathbb{Q}G$-chain complexes. They both are projective $\mathbb{Q}G$-resolution of the trivial $\mathbb{Q}G$-module $\mathbb{Q}$ since EG and $\underline{E}G$ are contractible. By the fundamental lemma of homological algebra the chain map $C_*(f; M\mathbb{Q}): C_*(EG;\mathbb{Q}) \to C_*(\underline{E}G;\mathbb{Q})$ is a $\mathbb{Q}G$-chain equivalence since it induces an isomorphism on the zero-th homology. Hence also the $\mathbb{Q}$-chain map

$$\mathrm{id}_\mathbb{Q} \otimes_{\mathbb{Q}G} C_*(f; M\mathbb{Q}): \mathbb{Q} \otimes_{\mathbb{Q}G} C_*(EG;\mathbb{Q}) \to \mathbb{Q} \otimes_{\mathbb{Q}G} C_*(\underline{E}G)$$

is a $\mathbb{Q}$-chain equivalence and hence induces an isomorphism on homology. It can be identified with the chain map $C_*(G\backslash f;\mathbb{Q}): C_*(BG;\mathbb{Q}) \to C_*(G\backslash \underline{E}G;\mathbb{Q})$.

21.2. It suffices to prove that $H_1(B\mathbb{Z}/l; \mathbf{L}^{\langle -\infty\rangle})[1/2]$ is different from zero. Inverting 2 has the advantage that we can ignore the decorations. One easily computes $H_n(B\mathbb{Z}/l;\mathbb{Z}) = \mathbb{Z}/l$ for odd n and $H_n(B\mathbb{Z}/l;\mathbb{Z}) = 0$ for even $n \geq 2$. Now the claim follows from the Atiyah Hirzebruch spectral sequence since its E^2-term has a checkerboard pattern which forces all differentials to be zero.

21.3. The group G is torsionfree. Hence we can use the version of the Farrell–Jones Conjecture for torsionfree groups 21.11 and 21.16. This implies $\mathrm{Wh}(G) = 0$ and $\widetilde{K}_n(\mathbb{Z}G) = 0$ for $n \leq 0$. We conclude from the Rothenberg sequences that $L_n^{\langle j\rangle}(\mathbb{Z}G)$ is independent of j so that we can pick $j = -\infty$. Notice that G is a finitely generated free group if the surface has boundary and the surface itself is a model for BG if the surface is closed of genus ≥ 1. Now an easy calculation with the Atiyah–Hirzebruch spectral sequence finishes the computation. The result is

$$L_n^{\langle j\rangle}(\mathbb{Z}G) \cong \begin{cases} L_0(\mathbb{Z}) \oplus H_2(BG; L_2(\mathbb{Z})) & n \equiv 0 \mod 4; \\ H_1(F_g; L_0(\mathbb{Z})) & n \equiv 1 \mod 4; \\ L_2(\mathbb{Z}) \oplus H_2(BG; L_0(\mathbb{Z})) & n \equiv 2 \mod 4; \\ H_1(F_g; L_2(\mathbb{Z})) & n \equiv 3 \mod 4. \end{cases}$$

22.1. Let g be an element in G with $g^n = 1$ and $g \neq 1$ for some natural number $n \geq 2$. Put $p = n^{-1} \cdot \sum_{k=1}^n g^k$. This is an element in $\mathbb{Q}G$ with $p^2 = p$ which is different from 0 and 1. It satisfies $\mathrm{tr}_{C_r^*(G)}(p) = \frac{1}{n}$.

22.2. We get from the Pimsner–Voiculescu splitting for $k \geq 1$ an isomorphism

$$K_n^r(C_r^*(G \times \mathbb{Z}^k)) \cong K_0(C_r^*(G))^{k-1} \oplus K_1(C_r^*(G))^{k-1}$$

for all $n \in \mathbb{Z}$. Hence it suffices to treat the case $k = 0$. If M is S^2, then $G = \{1\}$ and $K_n(C_r^*(\{1\})) = K_n(\{\bullet\})$ is $\mathbb{Z}$ for $n = 0$ and vanishes for $n = 1$. Suppose that M is different from S^2. Then $\pi_1(M)$ is torsionfree and M is a model for BG. The Baum–Connes Conjecture for torsion free groups 22.2 implies $K_n(C_r^*(G)) = K_n(BG) = K_n(M)$ for $n \in \mathbb{Z}$. The Atiyah–Hirzebruch spectral sequence implies

$$
\begin{aligned}
K_0(BG) &= H_0(BG) \oplus H_2(BG), \\
K_1(BG) &= H_1(BG).
\end{aligned}
$$

So we get

$$
\begin{aligned}
K_0(C_r^*(G)) &= H_0(BG) \oplus H_2(BG), \\
K_1(C_r^*(G)) &= H_1(BG).
\end{aligned}
$$

and for $k \geq 1$

$$K_n(C_r^*(G \times \mathbb{Z}^k)) = (H_0(BG) \oplus H_1(BG) \oplus H_2(BG))^{k-1}.$$

22.3. Let $K \subseteq L$ be a maximal compact subgroup. Then L/K with the obvious G-action is a model for $E_{\mathcal{FIN}}(G)$ by Theorem 19.11. Since $G \backslash L/K$ is compact, there is a finite G-CW-complex model for $E_{\mathcal{FIN}}(G)$. Now one proves for every finite G-CW-complex X by induction over the number of equivariant cells, that $H_n^G(X; \mathbf{K}^{\mathrm{top}})$ is finitely generated. Use the facts that $K_n(C_r^*(H))$ is the complex representation ring of H for $n = 0$ and vanishes for $n = 1$ and hence is finitely generated for all $n \in \mathbb{Z}$.

23.1. There are isomorphisms

$$L_0^s(\mathbb{Z}G)[1/2] \cong L_0^p(\mathbb{Z}G)[1/2] \cong L_0^p(\mathbb{R}G)[1/2] \cong K_0(\mathbb{R}G)[1/2].$$

The group $K_0(\mathbb{R}G)$ is isomorphic to the real representation ring of G which is isomorphic to the free abelian group generated by the isomorphism classes of irreducible real G-representations.

23.2. Because of the non-equivariant version of Lemma 20.5 it suffices to prove the claim for a finite-dimensional CW-complex. Now use induction over the dimension.

The induction begin $n = 0$ follows from the disjoint union axiom because a zero-dimensional CW-complex is a disjoint union of points. In the induction step apply the Mayer–Vietoris sequence to a pushout

$$
\begin{array}{ccc}
\coprod_{i \in I_n} S^{n-1} & \longrightarrow & X_{n-1} \\
\downarrow & & \downarrow \\
\coprod_{i \in I_n} D^n & \longrightarrow & X_n
\end{array}
$$

The claim is true for all corners except for X_n by induction hypothesis and homotopy invariance. Now apply the 5-Lemma.

24.1.   Obviously the $\mathbb{Z}/3$-fixed point set is $\{(n, n, n) \mid n \in \mathbb{Z}\} \subseteq \mathbb{Z}^3$. The composition of the inclusion $\mathbb{Z}^2 = \mathbb{Z}^2 \times \{0\} \to \mathbb{Z}^3$ with the projection $\mathbb{Z}^3 \to A$ induces an isomorphism. If $\{e_1, e_2\}$ is the standard basis of $\mathbb{Z}^2$, the generator of $t \in \mathbb{Z}^3$ acts by $e_1 \mapsto e_2$ and $e_2 \mapsto e_1^{-1} e_2^{-1}$ and t^2 acts by $e_1 \mapsto e_1^{-1} e_2^{-1}$ and $e_2 \mapsto e_1$. This action is obviously free outside the origin. We want to apply Theorem 24.14 which is possible by Remark 24.15.

Each finite subgroup of G must have order 1 or 3 since we can write G as an extension $1 \to \mathbb{Z}^2 \to G \to \mathbb{Z}/3 \to 1$. An element of order 3 must be of the shape $t e_1^a e_2^b$ or $t^2 e_1^a e_2^b$. One easily checks that each element of this shape has indeed order 3,

$$
\begin{aligned}
(t e_1^a e_2^b)^3 &= t e_1^a e_2^b t^{-1} t^2 e_1^a e_2^b t^{-2} e_1^a e_2^b = e_2^a e_1^{-b} e_2^{-b} e_1^{-a} e_2^{-a} e_1^b e_1^a e_2^b = 1; \\
(t^2 e_1^a e_2^b)^3 &= t^2 e_1^a e_2^b t^{-2} t e_1^a e_2^b t^{-1} e_1^a e_2^b = e_1^{-a} e_2^{-a} e_1^b e_2^a e_1^{-b} e_2^{-b} e_1^a e_2^b = 1;
\end{aligned}
$$

Next we determine how many conjugacy classes of elements of order 3 of the shape $t e_1 e_2^b$ exists in G. One easily checks

$$
\begin{aligned}
t(t e_1^a e_2^b) t^{-1} &= t e_2^a e_1^{-b} e_2^{-b} = t e_1^{-b} e_2^{a-b}; \\
t^2 (t e_1^a e_2^b) t^{-2} &= t e_1^{-a} e_2^{-a} e_1^b = t e_1^{-a+b} e_2^{-a}; \\
e_1^m e_2^n (t e_1^a e_2^b)(e_1^m e_2^n)^{-1} &= t t^2 e_1^m e_2^n t^{-2} e_1^a e_2^b e_1^{-m} e_2^{-n} = t e_1^{-m} e_2^{-m} e_1^n e_1^a e_2^b e_1^{-m} e_2^{-n} \\
&= t e_1^{a-2m+n} e_2^{b-m-n}.
\end{aligned}
$$

This implies that the elements of order 3 of the shape $t e_1^a e_2^b$ fall into three conjugacy classes, namely the conjugacy classes of t, $t e_1$ and $t e_1^2$. We conclude that we have the following conjugacy classes of subgroups of order three: $\langle t \rangle$, $\langle t e_1 \rangle$ and $\langle t e_1^2 \rangle$.

Next we have to figure out the homotopy type of $G \backslash E_{\mathcal{FIN}}(G)$. There is an obvious action of G on $\mathbb{R}^2$ which combines the $\mathbb{Z}^2$-action on $\mathbb{R}^2$ by translation and the $\mathbb{R}$-linear $\mathbb{Z}/3$-action coming from the $\mathbb{Z}$-linear $\mathbb{Z}/3$-action on $\mathbb{Z}^2$ described above by extending from $\mathbb{Z}^2$ to $\mathbb{R}^2$. One easily checks that all isotropy groups are finite and the fixed point set for any finite group is a affine subspace of $\mathbb{R}^2$ and hence contractible. Hence $\mathbb{R}^2$ with this G-action is a model for $E_{\mathcal{FIN}}(G)$. Identify $\mathbb{Z}^2 \backslash \mathbb{R}^2$ with the 2-torus $T^2 = S^1 \times S^1$. Define a $\mathbb{Z}/3$-action on T^2 by letting the generator $t \in \mathbb{Z}/3$ act by $(z_1, z_2) \mapsto (z_2, z_1^{-1} z_2^{-1})$, where we consider S^1 as a subset of $\mathbb{C}$. One easily checks $G \backslash E_{\mathcal{FIN}}(G) = (\mathbb{Z}/3) \backslash T^2$. One easily checks that

$$H_n((\mathbb{Z}/3) \backslash T^2; \mathbb{Q}) \cong \mathbb{Q} \otimes_{\mathbb{Q}[\mathbb{Z}/3]} H_n(T^2; \mathbb{Q}) = H_2(S^2; \mathbb{Q}).$$

Although the $\mathbb{Z}/3$-action is not free, the quotient space $(\mathbb{Z}/3) \backslash T^2$ is a manifold, since T^2 is two-dimensional. Hence

$$G \backslash E_{\mathcal{FIN}}(G) = (\mathbb{Z}/3) \backslash T^2 = S^2.$$

We have

$$\widetilde{K}_n(C_r^*(\mathbb{Z}/3)) = \begin{cases} \mathbb{Z}^2 & n \text{ even;} \\ \{0\} & n \text{ odd;} \end{cases}$$

$$K_n(S^2) = \begin{cases} \mathbb{Z}^2 & n \text{ even;} \\ \{0\} & n \text{ odd.} \end{cases}$$

Theorem 24.14 implies

$$\widetilde{K}_n(C_r^*(G)) = \begin{cases} \mathbb{Z}^8 & n \text{ even;} \\ \{0\} & n \text{ odd.} \end{cases}$$

24.2. Let A be an abelian group. Then $A \otimes_{\mathbb{Z}} \mathbb{Q}$ vanishes if and only if $A \otimes_{\mathbb{Z}} \mathbb{C}$ vanishes. Hence it suffices to prove the claim in the situation, where we replace $- \otimes_{\mathbb{Z}} \mathbb{Q}$ everywhere by $- \otimes_{\mathbb{Z}} \mathbb{C}$. We conclude from Example 24.10 that $\widetilde{K}_n(\mathbb{C}G) \otimes_{\mathbb{Z}} \mathbb{C}$ vanishes for all $n \in \mathbb{Z}$ if and only if $H_p(Z_G\langle g \rangle; \mathbb{C}) \otimes_{\mathbb{Z}} K_q(\mathbb{C})$ vanishes for all $g \in G$ of finite order and $p \geq 0$ and $q \in \mathbb{Z}$ unless $g = 1$ and $p = 0$. Since $K_0(\mathbb{C}) \cong \mathbb{Z}$ and $H_0(Z_G\langle g \rangle; \mathbb{C}) = \mathbb{Z}$, this is equivalent to the condition that each $g \in G$ of finite order satisfies $g = 1$ and $H_p(Z_G\langle 1 \rangle; \mathbb{C}) = H_p(G; \mathbb{C}) = 0$ for $p \geq 1$. The proof for $K_*(C_r^*(G))$ is analogous.

24.3. Because of Example 24.16 it suffices to prove $\mathrm{Wh}(\mathbb{Z}/n) = 0$ if and only if $n \in \{2, 3, 4, 6\}$. This follows from Theorem 24.4 (4), provided we can show $r_{\mathbb{R}}(\mathbb{Z}/n) - r_{\mathbb{Q}}(\mathbb{Z}/n) \Leftrightarrow n \in \{1, 2, 3, 4, 6\}$. This follows easily from the fact that $r_{\mathbb{R}}(\mathbb{Z}/n)$ is the number of $\mathbb{R}$-conjugacy classes of elements in $\mathbb{Z}/n$ and $r_{\mathbb{Q}}(\mathbb{Z}/n)$ is the number of conjugacy classes of cyclic subgroups of $\mathbb{Z}/n$.

Bibliography

[1] H. Abels. A universal proper G-space. *Math. Z.*, 159(2):143–158, 1978.

[2] U. Abresch. Über das Glätten Riemannscher Metriken. Habilitationsschrift, Bonn, 1988.

[3] J. F. Adams. *Stable homotopy and generalised homology*. University of Chicago Press, Chicago, Ill., 1974. Chicago Lectures in Mathematics.

[4] A. Adem and Y. Ruan. Twisted orbifold K-theory. Preprint, to appear in Comm. Math. Phys., 2001.

[5] C. Anantharaman and J. Renault. Amenable groupoids. In *Groupoids in analysis, geometry, and physics (Boulder, CO, 1999)*, volume 282 of *Contemp. Math.*, pages 35–46. Amer. Math. Soc., Providence, RI, 2001.

[6] C. Anantharaman-Delaroche and J. Renault. *Amenable groupoids*, volume 36 of *Monographies de L'Enseignement Mathématique*. Geneva, 2000. With a foreword by Georges Skandalis and Appendix B by E. Germain.

[7] C. S. Aravinda, F. T. Farrell, and S. K. Roushon. Surgery groups of knot and link complements. *Bull. London Math. Soc.*, 29(4):400–406, 1997.

[8] M. F. Atiyah. *K-theory*. W. A. Benjamin, Inc., New York–Amsterdam, 1967.

[9] M. F. Atiyah. Elliptic operators, discrete groups and von Neumann algebras. *Astérisque*, 32–33:43–72, 1976.

[10] M. F. Atiyah and I. M. Singer. The index of elliptic operators. III. *Ann. of Math. (2)*, 87:546–604, 1968.

[11] A. Bak. Odd dimension surgery groups of odd torsion groups vanish. *Topology*, 14(4):367–374, 1975.

[12] A. Bak. The computation of surgery groups of finite groups with abelian 2-hyperelementary subgroups. In *Algebraic K-theory (Proc. Conf., Northwestern Univ., Evanston, Ill., 1976)*, pages 384–409. Lecture Notes in Math., Vol. 551. Springer-Verlag, Berlin, 1976.

[13] M. Banagl and A. Ranicki. Generalized Arf invariants in algebraic L-theory. arXiv: math.AT/0304362, 2003.

[14] A. Bartels. On the left hand side of the assembly map in algebraic K-theory. *ATG*, 3:1037–1050, 2003.

[15] A. Bartels. Squeezing and higher algebraic K-theory. *K-Theory*, 28(1):19–37, 2003.

[16] A. Bartels and W. Lück. Induction theorems and isomorphism conjectures for K- and L-theory. Preprintreihe SFB 478 — Geometrische Strukturen in der Mathematik, Heft 331, Münster, arXiv:math.KT/0404486, 2004.

[17] A. Bartels and W. Lück. Isomorphism conjectures for homotopy K-theory and group actions on trees. Preprintreihe SFB 478 — Geometrische Strukturen in der Mathematik, Heft 343, Münster, arXiv:math.KT/0407489, 2004.

[18] A. Bartels and H. Reich. On the Farrell–Jones conjecture for higher algebraic K-theory. Preprintreihe SFB 478 — Geometrische Strukturen in der Mathematik, Heft 253, Münster. Preprint, 2003.

[19] H. Bass. *Algebraic K-theory*. W. A. Benjamin, Inc., New York–Amsterdam, 1968.

[20] H. Bass, A. Heller, and R. G. Swan. The Whitehead group of a polynomial extension. *Inst. Hautes Études Sci. Publ. Math.*, (22):61–79, 1964.

[21] P. Baum and A. Connes. Geometric K-theory for Lie groups and foliations. *Enseign. Math. (2)*, 46(1–2):3–42, 2000.

[22] P. Baum, A. Connes, and N. Higson. Classifying space for proper actions and K-theory of group C^*-algebras. In C^*-algebras: 1943–1993 (San Antonio, TX, 1993), pages 240–291. Amer. Math. Soc., Providence, RI, 1994.

[23] P. Baum and R. G. Douglas. K-homology and index theory. In *Operator algebras and applications, Part I (Kingston, Ont., 1980)*, pages 117–173. Amer. Math. Soc., Providence, R.I., 1982.

[24] P. Baum, M. Karoubi, and J. Roe. On the Baum–Connes conjecture in the real case. Preprint, 2002.

[25] E. Berkove, D. Juan-Pineda, and K. Pearson. The lower algebraic K-theory of Fuchsian groups. *Comment. Math. Helv.*, 76(2):339–352, 2001.

[26] B. Blackadar. *K-theory for operator algebras*. Springer-Verlag, New York, 1986.

[27] M. Bökstedt, W. C. Hsiang, and I. Madsen. The cyclotomic trace and algebraic K-theory of spaces. *Invent. Math.*, 111(3):465–539, 1993.

[28] A. Borel. Stable real cohomology of arithmetic groups. *Ann. Sci. École Norm. Sup. (4)*, 7:235–272 (1975), 1974.

[29] A. Borel and J.-P. Serre. Corners and arithmetic groups. *Comment. Math. Helv.*, 48:436–491, 1973. Avec un appendice: Arrondissement des variétés à coins, par A. Douady et L. Hérault.

[30] M. R. Bridson and A. Haefliger. *Metric spaces of non-positive curvature.* Springer-Verlag, Berlin, 1999. Die Grundlehren der mathematischen Wissenschaften, Band 319.

[31] M. R. Bridson and K. Vogtmann. The symmetries of outer space. *Duke Math. J.*, 106(2):391–409, 2001.

[32] T. Bröcker and T. tom Dieck. *Kobordismentheorie.* Springer-Verlag, Berlin, 1970.

[33] W. Browder. *Surgery on simply-connected manifolds.* Springer-Verlag, New York, 1972. Ergebnisse der Mathematik und ihrer Grenzgebiete, Band 65.

[34] W. Browder. Homotopy type of differentiable manifolds. In *Novikov conjectures, index theorems and rigidity, Vol. 1 (Oberwolfach, 1993)*, pages 97–100. Cambridge Univ. Press, Cambridge, 1995.

[35] W. Browder and J. Levine. Fibering manifolds over a circle. *Comment. Math. Helv.*, 40:153–160, 1966.

[36] K. S. Brown. *Cohomology of groups*, volume 87 of *Graduate Texts in Mathematics*. Springer-Verlag, New York, 1982.

[37] G. Brumfiel. On the homotopy groups of BPL and PL/O. *Annals of Math.*, 22:73–79, 1968.

[38] S. Cappell, A. Ranicki, and J. Rosenberg, editors. *Surveys on surgery theory. Vol. 1.* Princeton University Press, Princeton, NJ, 2000. Papers dedicated to C. T. C. Wall.

[39] S. Cappell, A. Ranicki, and J. Rosenberg, editors. *Surveys on surgery theory. Vol. 2.* Princeton University Press, Princeton, NJ, 2001. Papers dedicated to C. T. C. Wall.

[40] S. E. Cappell. Mayer–Vietoris sequences in hermitian K-theory. In *Algebraic K-theory, III: Hermitian K-theory and geometric applications (Proc. Conf., Battelle Memorial Inst., Seattle, Wash., 1972)*, pages 478–512. Lecture Notes in Math., Vol. 343. Springer, Berlin, 1973.

[41] S. E. Cappell. Splitting obstructions for Hermitian forms and manifolds with $Z_2 \subset \pi_1$. *Bull. Amer. Math. Soc.*, 79:909–913, 1973.

[42] S. E. Cappell. Manifolds with fundamental group a generalized free product. I. *Bull. Amer. Math. Soc.*, 80:1193–1198, 1974.

[43] S. E. Cappell. Unitary nilpotent groups and Hermitian K-theory. I. *Bull. Amer. Math. Soc.*, 80:1117–1122, 1974.

[44] S. E. Cappell. On homotopy invariance of higher signatures. *Invent. Math.*, 33(2):171–179, 1976.

[45] S. E. Cappell and J. L. Shaneson. On 4-dimensional *s*-cobordisms. *J. Differential Geom.*, 22(1):97–115, 1985.

[46] M. Cárdenas and E. K. Pedersen. On the Karoubi filtration of a category. *K-Theory*, 12(2):165–191, 1997.

[47] G. Carlsson and E. K. Pedersen. Controlled algebra and the Novikov conjectures for *K*- and *L*-theory. *Topology*, 34(3):731–758, 1995.

[48] D. W. Carter. Localization in lower algebraic *K*-theory. *Comm. Algebra*, 8(7):603–622, 1980.

[49] D. W. Carter. Lower *K*-theory of finite groups. *Comm. Algebra*, 8(20):1927–1937, 1980.

[50] T. A. Chapman. Compact Hilbert cube manifolds and the invariance of Whitehead torsion. *Bull. Amer. Math. Soc.*, 79:52–56, 1973.

[51] T. A. Chapman. Topological invariance of Whitehead torsion. *Amer. J. Math.*, 96:488–497, 1974.

[52] P.-A. Cherix, M. Cowling, P. Jolissaint, P. Julg, and A. Valette. *Groups with the Haagerup property*, volume 197 of *Progress in Mathematics*. Birkhäuser Verlag, Basel, 2001.

[53] M. M. Cohen. *A course in simple-homotopy theory*. Springer-Verlag, New York, 1973. Graduate Texts in Mathematics, Vol. 10.

[54] P. E. Conner and E. E. Floyd. *Differentiable periodic maps*. Academic Press Inc., Publishers, New York, 1964.

[55] A. Connes. *Noncommutative geometry*. Academic Press Inc., San Diego, CA, 1994.

[56] F. X. Connolly and M. O. M. da Silva. The groups $N^r K_0(\mathbf{Z}\pi)$ are finitely generated $\mathbf{Z}[\mathbf{N}^r]$-modules if π is a finite group. *K-Theory*, 9(1):1–11, 1995.

[57] F. X. Connolly and T. Koźniewski. Nil groups in *K*-theory and surgery theory. *Forum Math.*, 7(1):45–76, 1995.

[58] F. X. Connolly and S. Prassidis. On the exponent of the cokernel of the forget-control map on K_0-groups. *Fund. Math.*, 172(3):201–216, 2002.

[59] F. X. Connolly and S. Prassidis. On the exponent of the NK_0-groups of virtually infinite cyclic groups. *Canad. Math. Bull.*, 45(2):180–195, 2002.

[60] F. X. Connolly and A. Ranicki. On the calculation of UNIL$_*$. arXiv:math.AT/0304016v1, 2003.

[61] M. Culler and K. Vogtmann. Moduli of graphs and automorphisms of free groups. *Invent. Math.*, 84(1):91–119, 1986.

[62] K. R. Davidson. *C*-algebras by example*, volume 6 of *Fields Institute Monographs*. American Mathematical Society, Providence, RI, 1996.

[63] J. F. Davis. Manifold aspects of the Novikov conjecture. In *Surveys on surgery theory, Vol. 1*, pages 195–224. Princeton Univ. Press, Princeton, NJ, 2000.

[64] J. F. Davis and W. Lück. Spaces over a category and assembly maps in isomorphism conjectures in *K*- and *L*-theory. *K-Theory*, 15(3):201–252, 1998.

[65] J. F. Davis and W. Lück. The *p*-chain spectral sequence. *K-Theory*, 30:71–104, 2003.

[66] W. Dicks and M. J. Dunwoody. *Groups acting on graphs*. Cambridge University Press, Cambridge, 1989.

[67] A. Dold. Relations between ordinary and extraordinary homology. Colloq. alg. topology, Aarhus 1962, 2–9, 1962.

[68] A. Dold. *Lectures on algebraic topology*. Springer-Verlag, Berlin, second edition, 1980.

[69] S. K. Donaldson. An application of gauge theory to four-dimensional topology. *J. Differential Geom.*, 18(2):279–315, 1983.

[70] S. K. Donaldson. Irrationality and the *h*-cobordism conjecture. *J. Differential Geom.*, 26(1):141–168, 1987.

[71] A. N. Dranishnikov, G. Gong, V. Lafforgue, and G. Yu. Uniform embeddings into Hilbert space and a question of Gromov. *Canad. Math. Bull.*, 45(1):60–70, 2002.

[72] W. Dywer, T. Schick, and S. Stolz. Remarks on a conjecture of Gromov and Lawson. arXiv:math.GT/0208011, 2002.

[73] D. S. Farley. A proper isometric action of Thompson's group V on Hilbert space. Preprint, 2001.

[74] F. T. Farrell. The nonfiniteness of Nil. *Proc. Amer. Math. Soc.*, 65(2):215–216, 1977.

[75] F. T. Farrell. The exponent of UNil. *Topology*, 18(4):305–312, 1979.

[76] F. T. Farrell and W. C. Hsiang. The topological-Euclidean space form problem. *Invent. Math.*, 45(2):181–192, 1978.

[77] F. T. Farrell and L. E. Jones. *K*-theory and dynamics. I. *Ann. of Math. (2)*, 124(3):531–569, 1986.

[78] F. T. Farrell and L. E. Jones. Negatively curved manifolds with exotic smooth structures. *J. Amer. Math. Soc.*, 2(4):899–908, 1989.

[79] F. T. Farrell and L. E. Jones. Stable pseudoisotopy spaces of compact non-positively curved manifolds. *J. Differential Geom.*, 34(3):769–834, 1991.

[80] F. T. Farrell and L. E. Jones. Isomorphism conjectures in algebraic K-theory. *J. Amer. Math. Soc.*, 6(2):249–297, 1993.

[81] F. T. Farrell and L. E. Jones. The lower algebraic K-theory of virtually infinite cyclic groups. *K-Theory*, 9(1):13–30, 1995.

[82] F. T. Farrell and L. E. Jones. Rigidity for aspherical manifolds with $\pi_1 \subset GL_m(\mathbb{R})$. *Asian J. Math.*, 2(2):215–262, 1998.

[83] F. T. Farrell and P. A. Linnell. K-theory of solvable groups. *Proc. London Math. Soc. (3)*, 87(2):309–336, 2003.

[84] F. T. Farrell and S. K. Roushon. The Whitehead groups of braid groups vanish. *Internat. Math. Res. Notices*, (10):515–526, 2000.

[85] T. Farrell. The Borel conjecture. In T. Farrell, L. Göttsche, and W. Lück, editors, *High dimensional manifold theory*, number 9 in ICTP Lecture Notes, pages 225–298. Abdus Salam International Centre for Theoretical Physics, Trieste, 2002. Proceedings of the summer school "High dimensional manifold theory" in Trieste May/June 2001, Number 1. http://www.ictp.trieste.it/~pub_off/lectures/vol9.html.

[86] T. Farrell, L. Göttsche, and W. Lück, editors. *High dimensional manifold theory*. Number 9 in ICTP Lecture Notes. Abdus Salam International Centre for Theoretical Physics, Trieste, 2002. Proceedings of the summer school "High dimensional manifold theory" in Trieste May/June 2001, Number 1. http://www.ictp.trieste.it/~pub_off/lectures/vol9.html.

[87] T. Farrell, L. Göttsche, and W. Lück, editors. *High dimensional manifold theory*. Number 9 in ICTP Lecture Notes. Abdus Salam International Centre for Theoretical Physics, Trieste, 2002. Proceedings of the summer school "High dimensional manifold theory" in Trieste May/June 2001, Number 2. http://www.ictp.trieste.it/~pub_off/lectures/vol9.html.

[88] T. Farrell, L. Jones, and W. Lück. A caveat on the isomorphism conjecture in L-theory. *Forum Math.*, 14(3):413–418, 2002.

[89] S. Ferry. A simple-homotopy approach to the finiteness obstruction. In *Shape theory and geometric topology (Dubrovnik, 1981)*, pages 73–81. Springer-Verlag, Berlin, 1981.

[90] S. Ferry and A. Ranicki. A survey of Wall's finiteness obstruction. In *Surveys on surgery theory, Vol. 2*, volume 149 of *Ann. of Math. Stud.*, pages 63–79. Princeton Univ. Press, Princeton, NJ, 2001.

[91] S. C. Ferry, A. A. Ranicki, and J. Rosenberg. A history and survey of the
      Novikov conjecture. In *Novikov conjectures, index theorems and rigidity,
      Vol. 1 (Oberwolfach, 1993)*, pages 7–66. Cambridge Univ. Press, Cambridge,
      1995.

[92] S. C. Ferry and S. Weinberger. Curvature, tangentiality, and controlled
      topology. *Invent. Math.*, 105(2):401–414, 1991.

[93] Z. Fiedorowicz. The Quillen–Grothendieck construction and extension
      of pairings. In *Geometric applications of homotopy theory (Proc. Conf.,
      Evanston, Ill., 1977)*, I, pages 163–169. Springer, Berlin, 1978.

[94] M. H. Freedman. The topology of four-dimensional manifolds. *J. Differential
      Geom.*, 17(3):357–453, 1982.

[95] M. H. Freedman. The disk theorem for four-dimensional manifolds. In *Pro-
      ceedings of the International Congress of Mathematicians, Vol. 1, 2 (War-
      saw, 1983)*, pages 647–663, Warsaw, 1984. PWN.

[96] D. Gabai. On the geometric and topological rigidity of hyperbolic 3-
      manifolds. *J. Amer. Math. Soc.*, 10(1):37–74, 1997.

[97] S. M. Gersten. On the spectrum of algebraic K-theory. *Bull. Amer. Math.
      Soc.*, 78:216–219, 1972.

[98] E. Ghys. Groupes aléatoires. preprint, Lyon, to appear in Séminaire Bour-
      baki, 55ème année, 2002–2003, no. 916, 2003.

[99] T. G. Goodwillie. Calculus. II. Analytic functors. *K-Theory*, 5(4):295–332,
      1991/92.

[100] D. Grayson. Higher algebraic K-theory. II (after Daniel Quillen). In *Al-
      gebraic K-theory (Proc. Conf., Northwestern Univ., Evanston, Ill., 1976)*,
      pages 217–240. Lecture Notes in Math., Vol. 551. Springer-Verlag, Berlin,
      1976.

[101] M. Gromov. Asymptotic invariants of infinite groups. In *Geometric group
      theory, Vol. 2 (Sussex, 1991)*, pages 1–295. Cambridge Univ. Press, Cam-
      bridge, 1993.

[102] M. Gromov. Geometric reflections on the Novikov conjecture. In *Novikov
      conjectures, index theorems and rigidity, Vol. 1 (Oberwolfach, 1993)*, pages
      164–173. Cambridge Univ. Press, Cambridge, 1995.

[103] M. Gromov. Spaces and questions. *Geom. Funct. Anal.*, (Special Volume,
      Part I):118–161, 2000. GAFA 2000 (Tel Aviv, 1999).

[104] M. Gromov. Random walk in random groups. *Geom. Funct. Anal.*, 13(1):73–
      146, 2003.

[105] E. Guentner, N. Higson, and S. Weinberger. The Novikov conjecture for
      linear groups. Preprint, 2003.

[106] A. Haefliger. Differential embeddings of s^n in s^{n+q} for $q > 2$. *Ann. of Math.*, 83(2):402–436, 1966.

[107] I. Hambleton and E. K. Pedersen. Identifying assembly maps in K- and L-theory. *Math. Annalen*, 328(1):27–58, 2004.

[108] I. Hambleton and L. R. Taylor. A guide to the calculation of the surgery obstruction groups for finite groups. In *Surveys on surgery theory, Vol. 1*, pages 225–274. Princeton Univ. Press, Princeton, NJ, 2000.

[109] L. Hesselholt and I. Madsen. On the K-theory of nilpotent endomorphisms. In *Homotopy methods in algebraic topology (Boulder, CO, 1999)*, volume 271 of *Contemp. Math.*, pages 127–140. Amer. Math. Soc., Providence, RI, 2001.

[110] N. Higson. Bivariant K-theory and the Novikov conjecture. *Geom. Funct. Anal.*, 10(3):563–581, 2000.

[111] N. Higson and G. Kasparov. E-theory and KK-theory for groups which act properly and isometrically on Hilbert space. *Invent. Math.*, 144(1):23–74, 2001.

[112] N. Higson and J. Roe. Amenable group actions and the Novikov conjecture. *J. Reine Angew. Math.*, 519:143–153, 2000.

[113] N. Higson and J. Roe. *Analytic K-homology*. Oxford University Press, Oxford, 2000. Oxford Science Publications.

[114] M. W. Hirsch. *Differential topology*. Springer-Verlag, New York, 1976. Graduate Texts in Mathematics, No. 33.

[115] F. Hirzebruch. *Neue topologische Methoden in der algebraischen Geometrie*, volume 131 of *Grundlehren der math. Wissenschaften*. Springer, 1966.

[116] F. Hirzebruch. The signature theorem: reminiscences and recreation. In *Prospects in mathematics (Proc. Sympos., Princeton Univ., Princeton, N.J., 1970)*, pages 3–31. Ann. of Math. Studies, No. 70. Princeton Univ. Press, Princeton, N.J., 1971.

[117] S. Illman. Existence and uniqueness of equivariant triangulations of smooth proper G-manifolds with some applications to equivariant Whitehead torsion. *J. Reine Angew. Math.*, 524:129–183, 2000.

[118] M. Joachim. K-homology of C^*-categories and symmetric spectra representing K-homology. *Math. Annalen*, 328:641–670, 2003.

[119] L. Jones. Foliated control theory and its applications. In T. Farrell, L. Göttsche, and W. Lück, editors, *High dimensional manifold theory*, number 9 in ICTP Lecture Notes, pages 405–460. Abdus Salam International Centre for Theoretical Physics, Trieste, 2002. Proceedings of the summer school "High dimensional manifold theory" in Trieste May/June 2001, Number 2. http://www.ictp.trieste.it/~pub_off/lectures/vol9.html.

[120] L. Jones. A paper for F.T. Farrell on his 60-th birthday. In T. Farrell, L. Göttsche, and W. Lück, editors, *High dimensional manifold theory*, pages 200–260. 2003. Proceedings of the summer school "High dimensional manifold theory" in Trieste May/June 2001.

[121] M. Karoubi. Bott periodicity, generalizations and variants. To appear in the handbook of K-theory, 2004.

[122] G. Kasparov. Novikov's conjecture on higher signatures: the operator K-theory approach. In *Representation theory of groups and algebras*, pages 79–99. Amer. Math. Soc., Providence, RI, 1993.

[123] G. Kasparov and G. Skandalis. Groupes "boliques" et conjecture de Novikov. *C. R. Acad. Sci. Paris Sér. I Math.*, 319(8):815–820, 1994.

[124] G. G. Kasparov. Equivariant KK-theory and the Novikov conjecture. *Invent. Math.*, 91(1):147–201, 1988.

[125] G. G. Kasparov. K-theory, group C^*-algebras, and higher signatures (conspectus). In *Novikov conjectures, index theorems and rigidity, Vol. 1 (Oberwolfach, 1993)*, pages 101–146. Cambridge Univ. Press, Cambridge, 1995.

[126] G. G. Kasparov and G. Skandalis. Groups acting on buildings, operator K-theory, and Novikov's conjecture. *K-Theory*, 4(4):303–337, 1991.

[127] S. P. Kerckhoff. The Nielsen realization problem. *Ann. of Math. (2)*, 117(2):235–265, 1983.

[128] M. A. Kervaire. Le théorème de Barden–Mazur–Stallings. *Comment. Math. Helv.*, 40:31–42, 1965.

[129] M. A. Kervaire and J. Milnor. Groups of homotopy spheres. I. *Ann. of Math. (2)*, 77:504–537, 1963.

[130] R. C. Kirby and L. C. Siebenmann. *Foundational essays on topological manifolds, smoothings, and triangulations*. Princeton University Press, Princeton, N.J., 1977. With notes by J. Milnor and M. F. Atiyah, Annals of Mathematics Studies, No. 88.

[131] S. Klaus and M. Kreck. A quick proof of the rational Hurewicz theorem and a computation of the rational homotopy groups of spheres. *Math. Proc. Cambridge Philos. Soc.*, 136(3):617–623, 2004.

[132] M. Kolster. K-theory and arithmetics. Lectures – ICTP Trieste 2002.

[133] M. Kreck. Surgery and duality. *Ann. of Math. (2)*, 149(3):707–754, 1999.

[134] M. Kreck. *Differential Algebraic Topology*. Preprint, 2004.

[135] M. Kreck. On the topological invariance of Pontrjagin classes. In preparation, 2004.

[136] M. Kreck and W. Lück. Topological rigidity for non-aspherical manifolds. In preparation, 2004.

[137] S. Krstić and K. Vogtmann. Equivariant outer space and automorphisms of free-by-finite groups. *Comment. Math. Helv.*, 68(2):216–262, 1993.

[138] A. O. Kuku. K_n, SK_n of integral group-rings and orders. In *Applications of algebraic K-theory to algebraic geometry and number theory, Part I, II (Boulder, Colo., 1983)*, pages 333–338. Amer. Math. Soc., Providence, RI, 1986.

[139] A. O. Kuku and G. Tang. Higher K-theory of group-rings of virtually infinite cyclic groups. *Math. Ann.*, 325(4):711–726, 2003.

[140] V. Lafforgue. Une démonstration de la conjecture de Baum–Connes pour les groupes réductifs sur un corps p-adique et pour certains groupes discrets possédant la propriété (T). *C. R. Acad. Sci. Paris Sér. I Math.*, 327(5):439–444, 1998.

[141] V. Lafforgue. Banach KK-theory and the Baum–Connes conjecture. In *European Congress of Mathematics, Vol. II (Barcelona, 2000)*, volume 202 of *Progr. Math.*, pages 31–46. Birkhäuser, Basel, 2001.

[142] V. Lafforgue. K-théorie bivariante pour les algèbres de Banach et conjecture de Baum–Connes. *Invent. Math.*, 149(1):1–95, 2002.

[143] E. C. Lance. *Hilbert C^*-modules*. Cambridge University Press, Cambridge, 1995. A toolkit for operator algebraists.

[144] T. Lance. Differentiable structures on manifolds. In *Surveys on surgery theory, Vol. 1*, pages 73–104. Princeton Univ. Press, Princeton, NJ, 2000.

[145] P. S. Landweber. Homological properties of comodules over $Mu_*(Mu)$ and $BP_*(BP)$. *Amer. J. Math.*, 98(3):591–610, 1976.

[146] I. J. Leary and B. E. A. Nucinkis. Every CW-complex is a classifying space for proper bundles. *Topology*, 40(3):539–550, 2001.

[147] R. Lee and R. H. Szczarba. The group $K_3(Z)$ is cyclic of order forty-eight. *Ann. of Math. (2)*, 104(1):31–60, 1976.

[148] E. Leichtnam, W. Lück, and M. Kreck. On the cut-and-paste property of higher signatures of a closed oriented manifold. *Topology*, 41(4):725–744, 2002.

[149] J. P. Levine. Lectures on groups of homotopy spheres. In *Algebraic and geometric topology (New Brunswick, N.J., 1983)*, pages 62–95. Springer, Berlin, 1985.

[150] E. Lluis-Puebla, J.-L. Loday, H. Gillet, C. Soulé, and V. Snaith. *Higher algebraic K-theory: an overview*, volume 1491 of *Lecture Notes in Mathematics*. Springer-Verlag, Berlin, 1992.

[151] W. Lück. The geometric finiteness obstruction. *Proc. London Math. Soc. (3)*, 54(2):367–384, 1987.

[152] W. Lück. *Transformation groups and algebraic K-theory.* Springer-Verlag, Berlin, 1989.

[153] W. Lück. A basic introduction to surgery theory. In T. Farrell, L. Göttsche, and W. Lück, editors, *High dimensional manifold theory*, number 9 in ICTP Lecture Notes, pages 1–224. Abdus Salam International Centre for Theoretical Physics, Trieste, 2002. Proceedings of the summer school "High dimensional manifold theory" in Trieste May/June 2001, Number 1. http://www.ictp.trieste.it/~pub_off/lectures/vol9.html.

[154] W. Lück. Chern characters for proper equivariant homology theories and applications to K- and L-theory. *J. Reine Angew. Math.*, 543:193–234, 2002.

[155] W. Lück. *L^2-invariants: theory and applications to geometry and K-theory*, volume 44 of *Ergebnisse der Mathematik und ihrer Grenzgebiete. 3. Folge. A Series of Modern Surveys in Mathematics [Results in Mathematics and Related Areas. 3rd Series. A Series of Modern Surveys in Mathematics].* Springer-Verlag, Berlin, 2002.

[156] W. Lück. The relation between the Baum–Connes conjecture and the trace conjecture. *Invent. Math.*, 149(1):123–152, 2002.

[157] W. Lück. K-and L-theory of the semi-direct product discrete three-dimensional Heisenberg group by $\mathbb{Z}/4$. In preparation, 2004.

[158] W. Lück. Survey on classifying spaces for families of subgroups. Preprintreihe SFB 478 — Geometrische Strukturen in der Mathematik, Heft 308, Münster, arXiv:math.GT/0312378 v1, 2004.

[159] W. Lück and H. Reich. The Baum–Connes and the Farrell–Jones conjectures in K- and L-theory. Preprintreihe SFB 478 — Geometrische Strukturen in der Mathematik, Heft 324, Münster, arXiv:math.GT/0402405, to appear in the handbook of K-theory, 2004.

[160] W. Lück, H. Reich, J. Rognes, and M. Varisco. Algebraic K-theory of integral group rings and topological cyclic homology. In preparation, 2004.

[161] W. Lück, H. Reich, and M. Varisco. Commuting homotopy limits and smash products. *K-Theory*, 30:137–165, 2003.

[162] W. Lück and R. Stamm. Computations of K- and L-theory of cocompact planar groups. *K-Theory*, 21(3):249–292, 2000.

[163] G. Lusztig. Novikov's higher signature and families of elliptic operators. *J. Differential Geometry*, 7:229–256, 1972.

[164] R. C. Lyndon and P. E. Schupp. *Combinatorial group theory.* Springer-Verlag, Berlin, 1977. Ergebnisse der Mathematik und ihrer Grenzgebiete, Band 89.

[165] J. Milnor. On manifolds homeomorphic to the 7-sphere. *Ann. of Math. (2)*, 64:399–405, 1956.

[166] J. Milnor. *Morse theory*. Princeton University Press, Princeton, N.J., 1963.

[167] J. Milnor. *Lectures on the h-cobordism theorem*. Princeton University Press, Princeton, N.J., 1965.

[168] J. Milnor. *Topology from the differentiable viewpoint*. The University Press of Virginia, Charlottesville, Va., 1965.

[169] J. Milnor. *Introduction to algebraic K-theory*. Princeton University Press, Princeton, N.J., 1971. Annals of Mathematics Studies, No. 72.

[170] J. Milnor and D. Husemoller. *Symmetric bilinear forms*. Springer-Verlag, New York, 1973. Ergebnisse der Mathematik und ihrer Grenzgebiete, Band 73.

[171] J. Milnor and J. D. Stasheff. *Characteristic classes*. Princeton University Press, Princeton, N. J., 1974. Annals of Mathematics Studies, No. 76.

[172] I. Mineyev and G. Yu. The Baum–Connes conjecture for hyperbolic groups. *Invent. Math.*, 149(1):97–122, 2002.

[173] A. S. Miščenko. Homotopy invariants of multiply connected manifolds. III. Higher signatures. *Izv. Akad. Nauk SSSR Ser. Mat.*, 35:1316–1355, 1971.

[174] A. S. Miščenko and A. T. Fomenko. The index of elliptic operators over C^*-algebras. *Izv. Akad. Nauk SSSR Ser. Mat.*, 43(4):831–859, 967, 1979. English translation in *Math. USSR-Izv.* 15 (1980), no. 1, 87–112.

[175] G. Mislin. Wall's finiteness obstruction. In *Handbook of algebraic topology*, pages 1259–1291. North-Holland, Amsterdam, 1995.

[176] G. Mislin and A. Valette. *Proper group actions and the Baum–Connes Conjecture*. Advanced Courses in Mathematics CRM Barcelona. Birkhäuser, 2003.

[177] S. A. Mitchell. On the Lichtenbaum–Quillen conjectures from a stable homotopy-theoretic viewpoint. In *Algebraic topology and its applications*, volume 27 of *Math. Sci. Res. Inst. Publ.*, pages 163–240. Springer, New York, 1994.

[178] G. J. Murphy. C^*-algebras and operator theory. Academic Press Inc., Boston, MA, 1990.

[179] S. P. Novikov. A diffeomorphism of simply connected manifolds. *Dokl. Akad. Nauk SSSR*, 143:1046–1049, 1962.

[180] S. P. Novikov. The homotopy and topological invariance of certain rational Pontrjagin classes. *Dokl. Akad. Nauk SSSR*, 162:1248–1251, 1965.

[181] S. P. Novikov. Topological invariance of rational classes of Pontrjagin. *Dokl. Akad. Nauk SSSR*, 163:298–300, 1965.

[182] S. P. Novikov. On manifolds with free abelian fundamental group and their application. *Izv. Akad. Nauk SSSR Ser. Mat.*, 30:207–246, 1966.

[183] S. P. Novikov. Algebraic construction and properties of Hermitian analogs of K-theory over rings with involution from the viewpoint of Hamiltonian formalism. Applications to differential topology and the theory of characteristic classes. I. II. *Izv. Akad. Nauk SSSR Ser. Mat.*, 34:253–288; ibid. **34** (1970), 475–500, 1970.

[184] R. Oliver. *Whitehead groups of finite groups*. Cambridge University Press, Cambridge, 1988.

[185] K. Pearson. Algebraic K-theory of two-dimensional crystallographic groups. *K-Theory*, 14(3):265–280, 1998.

[186] E. Pedersen and C. Weibel. A non-connective delooping of algebraic K-theory. In *Algebraic and Geometric Topology; proc. conf. Rutgers Uni., New Brunswick 1983*, volume 1126 of *Lecture notes in mathematics*, pages 166–181. Springer, 1985.

[187] M. Pimsner and D. Voiculescu. K-groups of reduced crossed products by free groups. *J. Operator Theory*, 8(1):131–156, 1982.

[188] M. V. Pimsner. KK-groups of crossed products by groups acting on trees. *Invent. Math.*, 86(3):603–634, 1986.

[189] D. Quillen. Elementary proofs of some results of cobordism theory using steenrod operations. *Adv. in Math.*, 7(2):29–56, 1971.

[190] D. Quillen. On the cohomology and K-theory of the general linear groups over a finite field. *Ann. of Math. (2)*, 96:552–586, 1972.

[191] D. Quillen. Finite generation of the groups K_i of rings of algebraic integers. In *Algebraic K-theory, I: Higher K-theories (Proc. Conf., Battelle Memorial Inst., Seattle, Wash., 1972)*, pages 179–198. Lecture Notes in Math., Vol. 341. Springer-Verlag, Berlin, 1973.

[192] D. Quillen. Higher algebraic K-theory. I. In *Algebraic K-theory, I: Higher K-theories (Proc. Conf., Battelle Memorial Inst., Seattle, Wash., 1972)*, pages 85–147. Lecture Notes in Math., Vol. 341. Springer-Verlag, Berlin, 1973.

[193] F. Quinn. A geometric formulation of surgery. In *Topology of Manifolds (Proc. Inst., Univ. of Georgia, Athens, Ga., 1969)*, pages 500–511. Markham, Chicago, Ill., 1970.

[194] A. Ranicki. Foundations of algebraic surgery. In T. Farrell, L. Göttsche, and W. Lück, editors, *High dimensional manifold theory*, number 9 in ICTP Lecture Notes, pages 491–514. Abdus Salam International Centre

for Theoretical Physics, Trieste, 2002. Proceedings of the summer school "High dimensional manifold theory" in Trieste May/June 2001, Number 2. http://www.ictp.trieste.it/~pub_off/lectures/vol9.html.

[195] A. A. Ranicki. Algebraic L-theory. II. Laurent extensions. *Proc. London Math. Soc. (3)*, 27:126–158, 1973.

[196] A. A. Ranicki. The algebraic theory of surgery. I. Foundations. *Proc. London Math. Soc. (3)*, 40(1):87–192, 1980.

[197] A. A. Ranicki. The algebraic theory of surgery. II. Applications to topology. *Proc. London Math. Soc. (3)*, 40(2):193–283, 1980.

[198] A. A. Ranicki. *Exact sequences in the algebraic theory of surgery.* Princeton University Press, Princeton, N.J., 1981.

[199] A. A. Ranicki. The algebraic theory of finiteness obstruction. *Math. Scand.*, 57(1):105–126, 1985.

[200] A. A. Ranicki. *Algebraic L-theory and topological manifolds.* Cambridge University Press, Cambridge, 1992.

[201] A. A. Ranicki. *Lower K- and L-theory.* Cambridge University Press, Cambridge, 1992.

[202] A. A. Ranicki. On the Novikov conjecture. In *Novikov conjectures, index theorems and rigidity, Vol. 1 (Oberwolfach, 1993)*, pages 272–337. Cambridge Univ. Press, Cambridge, 1995.

[203] A. A. Ranicki. *Algebraic and geometric surgery.* Oxford Mathematical Monographs. Clarendon Press, Oxford, 2002.

[204] J. Rognes. $K_4(\mathbb{Z})$ is the trivial group. *Topology*, 39(2):267–281, 2000.

[205] J. Rognes and C. Weibel. Two-primary algebraic K-theory of rings of integers in number fields. *J. Amer. Math. Soc.*, 13(1):1–54, 2000. Appendix A by Manfred Kolster.

[206] J. Rosenberg. K-theory and geometric topology. To appear in the handbook of K-theory, 2004.

[207] J. Rosenberg. C^*-algebras, positive scalar curvature and the Novikov conjecture. III. *Topology*, 25:319–336, 1986.

[208] J. Rosenberg. *Algebraic K-theory and its applications.* Springer-Verlag, New York, 1994.

[209] J. Rosenberg. Analytic Novikov for topologists. In *Novikov conjectures, index theorems and rigidity, Vol. 1 (Oberwolfach, 1993)*, pages 338–372. Cambridge Univ. Press, Cambridge, 1995.

[210] J. Rosenberg and S. Stolz. A "stable" version of the Gromov–Lawson conjecture. In *The Čech centennial (Boston, MA, 1993)*, pages 405–418. Amer. Math. Soc., Providence, RI, 1995.

[211] C. P. Rourke and B. J. Sanderson. *Introduction to piecewise-linear topology*. Springer-Verlag, Berlin, 1982. Reprint.

[212] J. Sauer. *K*-theory for proper smooth actions of totally disconnected groups. Ph.D. thesis, 2002.

[213] T. Schick. Finite group extensions and the Baum–Connes conjecture. in preparation.

[214] T. Schick. A counterexample to the (unstable) Gromov–Lawson–Rosenberg conjecture. *Topology*, 37(6):1165–1168, 1998.

[215] T. Schick. Operator algebras and topology. In T. Farrell, L. Göttsche, and W. Lück, editors, *High dimensional manifold theory*, number 9 in ICTP Lecture Notes, pages 571–660. Abdus Salam International Centre for Theoretical Physics, Trieste, 2002. Proceedings of the summer school "High dimensional manifold theory" in Trieste May/June 2001, Number 2. http://www.ictp.trieste.it/~pub_off/lectures/vol9.html.

[216] H. Schröder. *K-theory for real C*-algebras and applications*. Longman Scientific & Technical, Harlow, 1993.

[217] J.-P. Serre. Groupes d'homotopie et classes de groupes abéliens. *Ann. of Math. (2)*, 58:258–294, 1953.

[218] J.-P. Serre. *Linear representations of finite groups*. Springer-Verlag, New York, 1977. Translated from the second French edition by Leonard L. Scott, Graduate Texts in Mathematics, Vol. 42.

[219] J.-P. Serre. Arithmetic groups. In *Homological group theory (Proc. Sympos., Durham, 1977)*, volume 36 of *London Math. Soc. Lecture Note Ser.*, pages 105–136. Cambridge Univ. Press, Cambridge, 1979.

[220] J.-P. Serre. *Trees*. Springer-Verlag, Berlin, 1980. Translated from the French by J. Stillwell.

[221] J. L. Shaneson. Wall's surgery obstruction groups for $G \times Z$. *Ann. of Math. (2)*, 90:296–334, 1969.

[222] W.-X. Shi. Deforming the metric on complete Riemannian manifolds. *J. Differential Geom.*, 30(1):223–301, 1989.

[223] J. R. Silvester. *Introduction to algebraic K-theory*. Chapman & Hall, London, 1981. Chapman and Hall Mathematics Series.

[224] G. Skandalis, J. L. Tu, and G. Yu. The coarse Baum–Connes conjecture and groupoids. *Topology*, 41(4):807–834, 2002.

[225] V. Srinivas. *Algebraic K-theory*. Birkhäuser Boston Inc., Boston, MA, 1991.

[226] J. Stallings. Whitehead torsion of free products. *Ann. of Math. (2)*, 82:354–363, 1965.

[227] C. W. Stark. Topological rigidity theorems. In R. Daverman and R. Sher, editors, *Handbook of Geometric Topology*, Chapter 20. Elsevier, 2002.

[228] C. W. Stark. Surgery theory and infinite fundamental groups. In *Surveys on surgery theory, Vol. 1*, volume 145 of *Ann. of Math. Stud.*, pages 275–305. Princeton Univ. Press, Princeton, NJ, 2000.

[229] N. E. Steenrod. A convenient category of topological spaces. *Michigan Math. J.*, 14:133–152, 1967.

[230] S. Stolz. Manifolds of positive scalar curvature. In T. Farrell, L. Göttsche, and W. Lück, editors, *High dimensional manifold theory*, number 9 in ICTP Lecture Notes, pages 661–708. Abdus Salam International Centre for Theoretical Physics, Trieste, 2002. Proceedings of the summer school "High dimensional manifold theory" in Trieste May/June 2001, Number 2. http://www.ictp.trieste.it/~pub_off/lectures/vol9.html.

[231] R. E. Stong. *Notes on cobordism theory*. Princeton University Press, Princeton, N.J., 1968.

[232] D. P. Sullivan. Triangulating and smoothing homotopy equivalences and homeomorphisms. Geometric Topology Seminar Notes. In *The Hauptvermutung book*, volume 1 of *K-Monogr. Math.*, pages 69–103. Kluwer Acad. Publ., Dordrecht, 1996.

[233] R. G. Swan. Induced representations and projective modules. *Ann. of Math. (2)*, 71:552–578, 1960.

[234] R. G. Swan. Higher algebraic K-theory. In *K-theory and algebraic geometry: connections with quadratic forms and division algebras (Santa Barbara, CA, 1992)*, volume 58 of *Proc. Sympos. Pure Math.*, pages 247–293. Amer. Math. Soc., Providence, RI, 1995.

[235] R. M. Switzer. *Algebraic topology — homotopy and homology*. Springer-Verlag, New York, 1975. Die Grundlehren der mathematischen Wissenschaften, Band 212.

[236] R. W. Thomason. Beware the phony multiplication on Quillen's $\mathcal{A}^{-1}\mathcal{A}$. *Proc. Amer. Math. Soc.*, 80(4):569–573, 1980.

[237] T. tom Dieck. *Transformation groups*. Walter de Gruyter & Co., Berlin, 1987.

[238] T. tom Dieck. *Topologie*. Walter de Gruyter & Co., Berlin, 1991.

[239] J.-L. Tu. The Baum–Connes conjecture and discrete group actions on trees. *K-Theory*, 17(4):303–318, 1999.

[240] A. Valette. *Introduction to the Baum–Connes conjecture.* Birkhäuser Verlag, Basel, 2002. From notes taken by Indira Chatterji, With an appendix by Guido Mislin.

[241] K. Varadarajan. *The finiteness obstruction of C. T. C. Wall.* Canadian Mathematical Society Series of Monographs and Advanced Texts. John Wiley & Sons Inc., New York, 1989. A Wiley-Interscience Publication.

[242] K. Vogtmann. Automorphisms of free groups and outer space. To appear in the special issue of Geometriae Dedicata for the June, 2000 Haifa conference, 2003.

[243] J. B. Wagoner. Delooping classifying spaces in algebraic K-theory. *Topology*, 11:349–370, 1972.

[244] F. Waldhausen. Algebraic K-theory of generalized free products. I, II. *Ann. of Math. (2)*, 108(1):135–204, 1978.

[245] C. T. C. Wall. Determination of the cobordism ring. *Ann. of Math. (2)*, 72:292–311, 1960.

[246] C. T. C. Wall. Finiteness conditions for CW-complexes. *Ann. of Math. (2)*, 81:56–69, 1965.

[247] C. T. C. Wall. Finiteness conditions for CW complexes. II. *Proc. Roy. Soc. Ser. A*, 295:129–139, 1966.

[248] C. T. C. Wall. *Surgery on compact manifolds.* Academic Press, London, 1970. London Mathematical Society Monographs, No. 1.

[249] C. T. C. Wall. *Surgery on compact manifolds.* American Mathematical Society, Providence, RI, second edition, 1999. Edited and with a foreword by A. A. Ranicki.

[250] N. E. Wegge-Olsen. *K-theory and C^*-algebras.* The Clarendon Press Oxford University Press, New York, 1993. A friendly approach.

[251] C. A. Weibel. K-theory and analytic isomorphisms. *Invent. Math.*, 61(2):177–197, 1980.

[252] C. A. Weibel. Mayer–Vietoris sequences and module structures on NK_*. In *Algebraic K-theory, Evanston 1980 (Proc. Conf., Northwestern Univ., Evanston, Ill., 1980)*, volume 854 of *Lecture Notes in Math.*, pages 466–493. Springer, Berlin, 1981.

[253] M. Weiss and B. Williams. Assembly. In *Novikov conjectures, index theorems and rigidity, Vol. 2 (Oberwolfach, 1993)*, pages 332–352. Cambridge Univ. Press, Cambridge, 1995.

[254] T. White. Fixed points of finite groups of free group automorphisms. *Proc. Amer. Math. Soc.*, 118(3):681–688, 1993.

[255] G. W. Whitehead. *Elements of homotopy theory.* Springer-Verlag, New York, 1978.

[256] H. Whitney. The singularities of a smooth n-manifold in $(2n - 1)$-space. *Ann. of Math. (2)*, 45:247–293, 1944.

[257] G. Yu. The Novikov conjecture for groups with finite asymptotic dimension. *Ann. of Math. (2)*, 147(2):325–355, 1998.

[258] G. Yu. The coarse Baum–Connes conjecture for spaces which admit a uniform embedding into Hilbert space. *Invent. Math.*, 139(1):201–240, 2000.

Index

Notation

Schedules

Schedule for Sunday, 25.1.04

20:00 – 20:20   Welcome

20:20 – 21:00   Kreck, M.
"A motivating problem"

Schedule for Monday, 26.1.04

9:00 – 9.40   Lück, W.
"Introduction to the Novikov Conjecture and the Borel Conjecture"

9:55 – 10:35   Kreck, M.
"Normal Bordism Groups"

10:50 – 11:30   Kreck, M.
"The Signature"

11:45 – 12:25   Kreck, M.
"The Signature Theorem, Higher Signatures and the Novikov Conjecture"

16:00 – 16:40   Lück, W.
"The Projective Class Group and the Whitehead Group"

17:00 – 18:30   Kreck, M. and Lück, W.
"Discussion"

Schedule for Tuesday, 27.1.04

8:45 – 9.25   Lück, W.
                   "Whitehead Torsion"

9:40 – 10:20   Lück, W.
                   "The Statement and Consequences of the s-Cobordism
                   Theorem"

10:50 – 11:30   Lück, W.
                   "Sketch of the Proof of the s-Cobordism Theorem"

11:45 – 12:25   Kreck, M.
                   From the Novikov Conjecture to Surgery

12:28   Photo

16:00 – 16:40   Kreck, M.
                   "Surgery Below the Middle Dimension I: An Example"

17:00 – 18:30   Kreck, M. and Lück, W.
                   "Discussion"

Schedule for Wednesday, 28.1.04

8:45 – 9.25   Kreck, M.
                   "Surgery Below the Middle Dimension II: Systematically"

9:40 – 10:20   Kreck, M.
                   "Surgery in the Middle Dimension I: The Surgery
                   Obstruction Groups L_m^h"

10:50 – 11:30   Kreck, M.
                   "Surgery in the Middle Dimension II: The Surgery
                   Obstructions"

11:45 – 12:25   Kreck, M. and Lück, W.
                   "Discussion"

20:00   Party & Discussion

Schedule for Thursday, 29.1.04

8:45 – 9.25   Kreck, M.
                "Surgery in the Middle Dimension III: Results"

9:40 – 10:20   Kreck, M.
                "The Assembly Map and the Surgery Version of the Novikov
                Conjecture"

10:50 – 11:30   Kreck, M.
                "The Novikov Conjecture for Finitely Generated Free Abelian
                Groups and Some Other Groups"

11:45 – 12:25   Varisco, M.
                "Poincaré Duality and Algebraic L-Groups"

16:00 – 16:40   Lück, W.
                "Spectra"

17:00 – 18:30   Kreck, M. and Lück, W.
                "Discussion"

Schedule for Friday, 30.1.04

8:45 – 9.25   Lück, W.
                 "Classifying Spaces of Families"

9:40 – 10:20   Lück, W.
                 "The Assembly Principle"

10:50 – 11:30   Lück, W.
                 "The Farrell–Jones Conjecture"

11:45 – 12:25   Lück, W.
                 "The Baum–Connes Conjecture"

16:00 – 16:40   Lück, W.
                 "Relating the Novikov, the Farrell–Jones and the
                 Baum–Connes Conjectures"

16:55 – 17:35   Lück, W.
                 "Miscellaneous"

17:50 – 18:30   Kreck, M. and Lück, W.
                 "Final Discussion"

20:00 – 22:00   Table soccer tournament

List of Participants

Name	place
Becker, Christian	CMAT Palaiseau
Boccellari, Tommaso	Mailand
Brookman, Jeremy	Edinburgh
Crowley, Diarmuid	Penn State. Univ.
Eppelmann, Thorsten	Heidelberg
Friedl, Stefan	München
Fulea, Daniel	Heidelberg
Grunewald, Joachim	Münster
Ji,Lizhen	Ann Arbor
Khan, Qayum	Bloomington
Korzeniewski, Andrew	Edinburgh
Krylov, Nikolai	Bremen
Kuhr, Hohannes	Bochum
Küssner, Thilo	München
Macko, Tibor	Bonn
Minatta, Augusto	Heidelberg
Mozgova, Alexandra	Avignon
Müger, Michael	Amsterdam
Müllner, Daniel	Heidelberg
Schmidt, Marco	Müinster
Schröder, Ingo	Göttingen
Schütz, Dirk	Münster
Sheiham, Des	Bremen
Sixt, Jörg	Edinburgh
Strohm, Clara	Münster
Su, Yang	Heidelberg
Varisco, Marco	Münster
Waldmüller, Robert	Göttingen
Weber, Julia	Münster
Yudin, Ivan	Göttingen

Oberwolfach Seminars

The workshops organized by the
Mathematisches Forschungsinstitut
Oberwolfach are intended to introduce students and young mathematicians to current fields of research. By means of these well-orga-
nized seminars, also scientists from other fields will be introduced to
new mathematical ideas.
The publication of these workshops in the series Oberwolfach
Seminars (formerly DMV Seminar) makes the material available to
an even larger audience.

Your Specialized Publisher in Mathematics
Birkhäuser

For orders originating from all over the world
except USA/Canada/Latin America:
All countries excluding those listed below:
Birkhäuser Verlag AG
c/o Springer Auslieferungs-Gesellschaft (SAG)
Customer Service
Haberstrasse 7, D-69126 Heidelberg
Tel.: +49 / 6221 / 345 0
Fax: +49 / 6221 / 345 42 29
e-mail: orders@birkhauser.ch

For orders originating in the
USA/Canada/Latin America:

Birkhäuser
333 Meadowland Parkway
USA-Secaucus
NJ 07094-2491
Fax: +1 201 348 4505
e-mail: orders@birkhauser.com

*Your Specialized Publisher
in Mathematics*

Birkhäuser

Falk, M., Universität Würzburg, Germany / **Hüsler, J.**, Universität Bern, Switzerland / **Reiss, R.-D.**, Universität-Gesamthochschule Siegen, Germany

Laws of Small Numbers: Extremes and Rare Events

Second, revised and extended edition

2004. 392 pages. Hardcover
ISBN 3-7643-2416-3

Since the publication of the first edition of this seminar book in 1994, the theory and applications of extremes and rare events have enjoyed an enormous and still increasing interest. The intention of the book is to give a mathematically oriented development of the theory of rare events underlying various applications. This characteristic of the book was strengthened in the second edition by incorporating various new results on about 130 additional pages.

Part II, which has been added in the second edition, discusses recent developments in multivariate extreme value theory. Particularly notable is a new spectral decomposition of multivariate distributions in univariate ones which makes multivariate questions more accessible in theory and practice. One of the most innovative and fruitful topics during the last decades was the introduction of generalized Pareto distributions in the univariate extreme value theory. Such a statistical modelling of extremes is now systematically developed in the multivariate framework.

The theory of rare events of non iid observations, as outlined in Part III; has seen many new approaches during the last ten years. Very often these problems can be seen as boundary crossing probabilities. Some of these new aspects of boundary crossing probabilities are dealt with in this edition. This book is accessible to graduate students and researchers with basic knowledge in probability theory and, partly, in point processes and Gaussian processes. The required statistical prerequisites are minimal.

$$\sum_{j=1}^{n} s_{jj} \leq \sum_{p,q=1}^{n} |a_{pq}| \left(\sum_{j=}^{n} \right.$$

$$\leq \sum_{p,q=1}^{n} |a_{pq}| \left(\sum_{j=}^{n} \right.$$

For orders originating from all over the world except USA and Canada:
All countries excluding those listed below:
Birkhäuser Verlag AG
c/o Springer Auslieferungs-Gesellschaft (SAG)
Customer Service
Haberstrasse 7, D-69126 Heidelberg
Tel.: +49 / 6221 / 345 0
Fax: +49 / 6221 / 345 42 29
e-mail: orders@birkhauser.ch

For orders originating in the USA and Canada:
Birkhäuser
333 Meadowland Parkway
USA-Secaucus
NJ 07094-2491
Fax: +1 201 348 4505
e-mail: orders@birkhauser.com

Monografie Matematyczne

New Series

Managing Editor:
Przemysław Wojtaszczyk, IMPAN and Warsaw University, Poland

Starting in the 1930s with volumes written by such distinguished mathematicians as Banach, Saks, Kuratowski, and Sierpinski, the original series grew to comprise 62 excellent monographs up to the 1980s. In cooperation with the Institute of Mathematics of the Polish Academy of Sciences (IMPAN), Birkhäuser now resumes this tradition to publish high quality research monographs in all areas of pure and applied mathematics.

Your Specialized Publisher in Mathematics
Birkhäuser

For orders originating from all over the world except USA/Canada/Latin America:

Birkhäuser Verlag AG
c/o Springer GmbH & Co
Haberstrasse 7
D-69126 Heidelberg
Fax: +49 / 6221 / 345 4 229
e-mail: birkhauser@springer.de
http://www.birkhauser.ch

For orders originating in the USA/Canada/Latin America:

Birkhäuser
333 Meadowland Parkway
USA-Secaucus
NJ 07094-2491
Fax: +1 201 348 4505
e-mail: orders@birkhauser.com

■ **Vol. 63: Schürmann, J.**, Westfälische Wilhelms-Universität Münster, Germany

Topology of Singular Spaces and Constructible Sheaves

2003. 464 pages. Hardcover.
ISBN 3-7643-2189-X

Assuming that the reader is familiar with sheaf theory, the book gives a self-contained introduction to the theory of constructible sheaves related to many kinds of singular spaces, such as cell complexes, triangulated spaces, semialgebraic and subanalytic sets, complex algebraic or analytic sets, stratified spaces, and quotient spaces. The relation to the underlying geometrical ideas are worked out in detail, together with many applications to the topology of such spaces. All chapters have their own detailed introduction, containing the main results and definitions, illustrated in simple terms by a number of examples. The technical details of the proof are postponed to later sections, since these are not needed for the applications.

■ **Vol. 64: Walczak, P.**, University of Łódź, Poland

Dynamics of Foliations, Groups and Pseudogroups

2004. 240 pages. Hardcover.
ISBN 3-7643-7091-2

Foliations, groups and pseudogroups are objects which are closely related via the notion of holonomy. In the 1980s they became considered as general dynamical systems. This book deals with their dynamics. Since "dynamics" is a very extensive term, we focus on some of its aspects only. Roughly speaking, we concentrate on notions and results related to different ways of measuring complexity of the systems under consideration. More precisely, we deal with different types of growth, entropies and dimensions of limiting objects. Invented in the 1980s (by E. Ghys, R. Langevin and the author) geometric entropy of a foliation is the principal object of interest among all of them. Throughout the book, the reader will find a good number of inspiring problems related to the topics covered.

■ **Vol. 65: Badescu, L.**, Università degli Studi di Genova, Italy

Projective Geometry and Formal Geometry

2004. 228 pages. Hardcover.
ISBN 3-7643-7123-4

The aim of this monograph is to introduce the reader to modern methods of projective geometry involving certain techniques of formal geometry. Some of these methods are illustrated in the first part through the proofs of a number of results of a rather classical flavor, involving in a crucial way the first infinitesimal neighbourhood of a given subvariety in an ambient variety. Motivated by the first part, in the second formal functions on the formal completion X/Y of X along a closed subvariety Y are studied, particularly the extension problem of formal functions to rational functions.

Printed in the United States
By Bookmasters